The Russian Theory of Activity:
Current Applications to
Design and Learning

SERIES IN APPLIED PSYCHOLOGY

Edwin A. Fleishman, George Mason University
  Series Editor

Teamwork and the Bottom Line: Groups Make a Difference
Ned Rosen

Patterns of Life History: The Ecology of Human Individuality
Michael D. Mumford, Garnett Stokes, and William A. Owens

Work Motivaton
Uwe E. Kleinbeck, Hans-Henning Quast, Henk Thierry, and Hartmut Häcker

Psychology in Organizations: Integrating Science and Practice
Kevin R. Murphy and Frank E. Saal

Human Error: Cause, Prediction, and Reduction
John W. Senders and Neville P. Moray

Contemporary Career Development Issues
Robert F. Morrison and Jerome Adams

Justice in the Workplace: Approaching Fairness in Human Resource Management
Russell Cropanzano

Personnel Selection and Assessment: Individual and Organizational Perspectives
Heinz Schuler, James L. Farr, and Mike Smith

Organizational Behavior: The State of the Science
Jerald Greenberg

Police Psychology Into the 21st Century
Martin I. Kurke and Ellen M. Scrivner

Benchmark Tasks for Job Analysis: A Guide for Functional Job Analysis (FJA) Scales
Sidney A. Fine and Maury Getkate

Stress and Human Performance
James E. Driskell and Eduardo Salas

Improving Traning Effectiveness in Work Organizations
J. Kevin Ford, Steve W. J. Kozlowski, Kurt Kraiger, Eduardo Salas, and Mark S. Teachout

Team Performance Assessment and Measurement: Theory, Research, and Applications
Michael T. Brannick, Eduardo Salas, and Carolyn Prince

Job Feedback: Giving, Seeking, and Using Feedback for Performance Improvement
Manuel London

The Russian Theory of Activity: Current Applications to Design and Learning
Gregory Bedny and David Meister

The Russian Theory of Activity: Current Applications to Design and Learning

Gregory Bedny
Essex County College
Newark, New Jersey

David Meister
Human Factors Consultant
San Diego

 **Psychology Press**
Taylor & Francis Group

NEW YORK AND LONDON

First Published 1997 by
Lawrence Erlbaum Associates, Inc.

Published 2014 by Psychology Press
711 Third Avenue, New York, NY 10017

and by Psychology Press
27 Church Road, Hove, East Sussex, BN3 2FA

First issued in paperback 2014

*Psychology Press is an imprint of the Taylor & Francis Group,
an informa business*

Cover design by Kathryn Houghtaling

Library of Congress Cataloging-in-Publication Data

Bednyĭ, G. Z. (Grigoriĭ Zakharovich)
    The Russian theory of activity : current applications to design
and learning / Gregory Bedny, David Meister.
        p.   cm.
    Includes bibliographical references (p.   ) and indexes.
    1. Human engineering. 2. Human-machine systems. I. Meister,
David. II. Title.
    TA166.B35 1997
    620.8'2—dc21   97-12608
                                                                        CIP

ISBN 13: 978-1-138-87651-4 (pbk)
ISBN 13: 978-0-8058-1771-3 (hbk)

Publisher's Note
The publisher has gone to great lengths to ensure the quality of this reprint
but points out that some imperfections in the original may be apparent.

To the memory of my parents

—G. Bedny

Contents

Acknowledgments

The authors would like to thank the many scientists from the former Soviet Union whose contributions to the book were indispensable. Unfortunately, the number of these makes it impossible to list each one separately. Special mention should, however, be given to Dr. Lev Landa for the material he provided for section 8.2. Finally, the senior author (Bedny) owes the most heartfelt thanks to his wife Inna Bedny and daughter Marina Bedny, without whose continuing support this book could not have been written.

Series Foreword

Edwin A. Fleishman
Series Editor

There is a compelling need for innovative approaches to the solution of many pressing problems involving human relationships in today's society. Such approaches are more likely to be successful when they are based on sound research and applications. This *Series in Applied Psychology* offers publications which emphasize state-of-the-art research and its application to important issues of human behavior in a variety of societal settings. The objective is to bridge both academic and applied interests.

It may seem at first glance that the *Series in Applied Psychology* is a strange venue for a book describing Russian human factors ergonomics. A little consideration will show that it is not. The authors indicate in their Foreword that a discipline so long segregated from the West has now emerged to full light. This book is an attempt to describe what that discipline consists of.

The value of examining a discipline from a very different cultural context is that it highlights the special characteristics of one's own discipline. If one assumes that a discipline has a purpose, goals, assumptions, and a direction of effort, it becomes essential to examine all of these to ensure that the progress of that discipline proceeds efficiently. Russian human factors ergonomics allows one to examine Western counterparts more closely by illustrating the differences between them. At the same time, a discipline nurtured in a very different conceptual culture may well have certain features to offer its Western colleagues. One aspect of the new integration of Eastern with Western sciences is the role of Russian human factors ergonomics. Its existence demands that it be integrated with Western work in this area. This cannot be accomplished until Western ergonomists become familiar with its outlines. It may be possible to have a Western and an Eastern philosophy, but the unity of science requires that the different aspects of human factors ergonomics be integrated, regardless of their provenance.

Here is a discipline that has risen from an entirely different cultural context, one that was segregated from the West for many years. Western concepts did cross geographic and ideological barriers, although the interaction was mainly in one direction only. Books and papers passed from the West to the East, but there were only a few glimpses in the West of what colleagues were doing in the East. This book attests to the enduring strength of many Western concepts in this field, which can be recognized, although somewhat transformed by the Russian need to divorce itself from "bourgeois" Western influences. There are, however, real differences and these must also be attended to in the West.

The field of human factors ergonomics prides itself as being a distinctive discipline and indeed it is. In its 50-year history, however, it has been profoundly influenced by its predecessor discipline, psychology. Although human factors is practiced in an engineering context, its fundamental concepts are derived from psychology. Indeed, a recent survey of 5000-plus members of the Human Factors and Ergonomics Society revealed that half of them considered that what they were doing was a species of psychology. It is important for the applied psychologist to learn how basic psychological paradigms have been adapted in this new discipline.

Psychological concepts must be modified to fit a technological framework. The question that should be concerning psychologists is what those changes portend. For example, what is the significance of the "system concept," as used in the field of human factors, to psychology as a whole? Does traditional psychology take sufficient account of the system and the changing technological milieu? The classical psychology of 19th and 20th centuries was preeminently the study of the human as largely an isolated independent entity, sensing and responding to stimuli of various kinds and itself emitting stimuli. Although the human must still sense and respond to stimuli, we are now living in what is literally a technological civilization. In some sense, the system is seen as controlling the human. The role of the human relative to the machine has changed significantly, because of technological changes such as those produced by the computer.

Until recently the operator of equipment controlled it directly, and much of these operations consisted of linear, step by step procedures. Now the computer, although programmed by humans, controls the machine, with the human acting as supervisor to ensure that the machine performs within specified parameters. When it does not, the operator must interpret symptomology, diagnose what has gone wrong, and repair the machine. This requires of the operator what can be termed "non-linear" thinking. This is a much more difficult process, because one must now respond concurrently to complexes of stimuli. The impact on the human abilities that are required is very pronounced; motor ability requirements have been reduced (although by no means eliminated) and cognitive abilities of a highly sophisticated nature are now much more required.

One now must ask whether the technological developments that have changed the role of human beings so markedly have also changed the psychological concepts required to explain their performance in that role. For example, is it necessary to add higher orders "system concepts" describing human interactions with the machine, that are separate from our current descriptors of human and machine capabilities? Some human factors specialists in the United States, including David Meister, have suggested that, indeed, basic behavioral concepts and processes need to be augmented to account for the impact of what is termed the "human-technology relationship." In reading this book, the reader will be interested in seeing the extent to which the 19 "function blocks" of Russian human factors ergonomics depend on molecular psychological processes or on system processes.

New concepts have indeed entered the human factors ergonomics field. The system concept assumes that the human-technology relationship creates a new entity with dimensions and variables distinctly different from those of either the human or the machine alone. How will this new entity interact with and influence psychological concepts that describe the human alone? This presents a new problem for psychologists and ergonomists to ponder. However this problem is solved, Russian human factors ergonomics suggests an enduring relationship of this field with the larger discipline of psychology.

Nevertheless, the new discipline requires us to examine its structure very deliberately. Human factors ergonomics assumes that the interrelationship between the human and technology creates a dynamic tension, a dissonance, which can be disruptive to the human. The machine may be more difficult to operate and this may lead to a rejection of technology's potential. How many people refuse to use a VCR or automated teller machine or personal computers because these are not "user friendly"? Human factors ergonomics seeks to improve the human condition by reducing the demand imposed by technology on the human. It does this by designing the machine into a form more consonant with human capabilities and limitations. This may require the transformation of the behavioral factors and principles with which psychologists are familiar into a new form. It is possible that a psychology that does not pay sufficient attention to how this transformation occurs will become less relevant in the new century.

Thus, human factors ergonomics suggests that psychology is experiencing a new challenge. Technology may eventually develop a consciousness that interacts with human consciousness. Under these circumstances old familiar concepts may have to be re-examined. Russian human factors ergonomics, as described in this volume by Bedny and Meister, offers another perspective on such questions.

Author's Foreword

David Meister

The opening of the former Soviet Union to the West has given us an opportunity to describe Russian human factors/ergonomics and to compare American theories and methods with it. Bedny is a doctor of science and former professor at the University of Odessa. It should be noted that in the former Soviet Union there exist two scientific degrees. First is a candidate's degree, roughly equivalent to an American PhD. The second is the doctor of science, a postdoctoral degree by comparison to the American PhD. The doctor of science is awarded for developing theoretical concepts or outstanding contributions in the field of specialization. In the former Soviet Union there existed about 200 doctors of science in psychology. Dr. Bedny has provided the Russian discussion; Dr. Meister has added a "gloss" on that material that attempts to interpret it in the light of American thinking. Comparisons between Russian and Western approaches, and particularly the comments concluding each chapter, are those of the American coauthor, Meister.

Although this book principally is dedicated to describing the theory of activity as it applies to issues of design and training, this work is also offered to a general audience of psychologists and interested lay readers. This is the first comprehensive, systematic description of the Russian theory of activity in the English language. Existing attempts to translate the theory of activity into English suffer from certain limitations. Among the more important of these limitations is that the theory of activity, considered as one of the more important accomplishments of Soviet psychological science, has an extensive history dating back to the work of Vygotsky and his followers. Subsequent development of the theory took place within various schools of thought, which differ significantly from one another. Second, in the former Soviet Union no psychological theory could be advanced unconnected to Marxist–Leninist ideology. Accordingly, theoretical formulations were subject to their own version of "political correctness." Books published in this field were addressed only to other scientists with backgrounds in the field. Finally, the technical terms in Russian psychology frequently resist translation in the absence of the context of the debates in which they were being used. Thus, because they were written in a specialized and politicized environment for Russian audiences, the simple translation of books in this field is really not a particularly sensible or worthwhile undertaking.

The publishing of this book is auspiciously important at this time because attempts are being made in Western psychology to overcome the traditional separation between cognition and motivation using an approach called the "psychology of action." Psychologists who are now working in this area are not familiar with the Russian theory of activity, which integrates cognitive, motivational, and behavioral components into a holistic system. Both approaches are based on some general principles; however, they are not identical. The conceptual apparatus in these approaches differs significantly. For example, the understanding of what constitutes action is not the same in the theory of activity as it is in the psychology of action. Russian scientists disagree with the concept in the psychology of action that activity and action are indistinguishable from one another. From activity point of view, regulative functions of cognitive processes and their connection to behavior are not examined thoroughly enough in the psychology of action. Russian psychologists also differ from Western psychologists in their understanding of self-regulation.

This book is addressed in the first instance to Western psychologists. It compares, among other things, analyses of work from the former Soviet Union with the work from the West. Applications of activity theory to design and learning were paramount in the Soviet Union. In this book we also have comparative analysis of the various schools working in activity theory as seen from our own theoretical perspective. The authors hope that this book may facilitate the exchange of ideas between Russian psychological scientists and Western psychologists working in ergonomics, human factors, education, and related areas where the theory of activity may find general application.

A little history is in order to provide a backdrop for the ensuing discussion, as the progress of engineering psychology and ergonomics in the former Soviet Union is not well known to American and Western European specialists.

Following the 1917 revolution, between the 1920s and the 1930s, there were many specialists in this field. Unfortunately, during this time there was pressure to introduce Marxist philosophy into psychology. In spite of that, psychologists in the USSR were on a par with those in Britain and the United

States. However, continued criticism of the Soviet psychologists by Marxist ideologists regarding their purported "bourgeois" ideology led to, in 1936, the closing of all working psychology and psychophysiology laboratories and the imprisonment and even killing of many of these psychologists.

The need for industrial recovery following the Second World War, as well as the militarization of the USSR, led to the rehabilitation of work psychology. From 1950 to 1958 at the Department of Psychology in Moscow University, K. Platonov taught this subject for the first time after the long hiatus. In 1962 the first national conference of philosophers, physiologists, and psychologists was convened, at which issues of work psychology were extensively discussed. From this time on there was a step-by-step recovery of work psychology.

At the end of the 1960s, a new direction emerged in engineering psychology and ergonomics. In 1959 at Leningrad University the first engineering psychological laboratory was created. In this university, as well as at Moscow University, the teaching of engineering psychology in its own right began.

The first Conference on Engineering Psychology in 1967 can be considered the official beginning of Soviet ergonomics. Because of the extreme militarization of science and technology, ergonomics and engineering psychology initially evolved for military applications, utilizing what would be considered, compared to Western practice, a small group of experts.

Efforts were made to substantiate that ergonomics is necessarily based on science. These efforts were critical because of the precarious political position of the discipline. Specialists justified their utility on the grounds of technical military failures attributable to human errors, a justification not unknown in the West. They also gathered examples of the poor design of controls and displays, prepared normative data to provide standards for military equipment, and conducted experimental studies of components of military personnel work activity.

All of these efforts were profoundly influenced by Western European and American literature in these fields. For example, American standards such as MIL-STD 1472 were adopted. At the same time, Soviet ergonomists analyzed American experience as described in research reports. Many basic works from the United States and Western Europe were translated into Russian during the period 1970–1980, including books by the following: Chapanis, Woodson and Conover, Meister, Siegel and Wolf, and Sheridan and Ferrell (these are only a representative sample).

Nevertheless, the small (compared to America) professional group of ergonomists in the former Soviet Union had their own original theoretical studies, critical analysis, and refinements of Western work. Because the development of mathematics and physics was quite sophisticated in the USSR, these had extensive influence on the development of ergonomics. Soviet ergonomists devoted much effort to trying to deploy mathematical methods to formalize ergonomic theory. However, the practical outcome of these efforts was quite limited. The creation of concepts and theory that were at once mathematically rigorous and relevant to practical concerns turned out to be extremely difficult.

Certain characteristics of Russian ergonomics, which are exemplified in the text, need to be explained. There are, for example, very distinct differences between Russian and American terminology. We have retained the former (although it may seem odd to Western readers) and have attempted to show that, in translation, the concepts described by Russian terminology are not as dissimilar to American concepts as would appear on the surface.

The reader will find numerous references to the work of American and British researchers. That is because this work has been accepted by Russians and forms part of the corpus of Russian thinking. In other cases a Western source is mentioned critically.

For reasons that are not quite clear (possibly Russian history and Marxist ideology), Russian ergonomists have been much influenced by cognitive psychology, philosophy, physiology, and cybernetics. The impact of cybernetics is shown by the manner in which they describe functions ("function blocks") in computer software scenarios.

In spite of the fact that Russian scientists use different approaches to solve ergonomic and engineering psychological problems, including those that are used by American scientists, the most important approach for them is the concept of activity. During the analysis of activity, the structure of the activity, defined as the logical organization of actions and operations in space and time, which are directed to achieve a conscious goal, is of major importance. Analysis of activity can be performed from different viewpoints. From a functional position, major attention is paid to motives, goals of activity, and process of self-regulation. From a morphological viewpoint, major attention is paid to actions and operations performed by an individual. Western theorists postulate that goals have both cognitive and motivational features, whereas Russian psychologists state that goals can be only cognitive in nature. This enables activity to be described as a goal-directed, self-regulative system, which is principally distinguished from a homeostatical self-regulative system. The process of self-regulation can be described as different stages of processing information, which involve different psychological mechanisms. Each stage is called a *function block*, because it performs a particular function. Between different function blocks exist forward and backward interconnections.

During morphological analysis of activity, when major units of analysis are actions and operations (smaller parts of actions), it is important that their classification is made according to particular criteria. When operating with materials, or real objects, actions are called practical (physical or motor). Transformations of images, concepts, or propositions in our mind are called mental actions. Often mental actions and mental operations are considered synonymous. Classification of mental actions can be performed according to subgoals that are achieved during execution of action. Classification of mental actions also is made according to psychological processes that dominate during performance of action. This requires studying separate psychological processes. Some similarities between the approaches to studying psychological processes in the Russian theory of activity and in American cognitive psychology exist. However, there are also differences. In chapter 3, psychological processes are described from a position of the theory of activity. This information should prove interesting for Western psychologists who are unfamiliar with these data.

Because the structure of activity is first of all a logical organization of actions, the algorithmical description of activity becomes important during analyses of tasks performed by operators. In the theory of activity two notions are distinguished: human and nonhuman algorithms. A human algorithm is described as a logical sequence of actions and operations that are performed by individuals. This method should be distinguished from other algorithmical methods that often look like human algorithms. For example, human algorithms are often confused with a flow chart's description of task performance. However, not every flow chart can be considered a human algorithm (e.g., if a flow chart does not describe a human action, it cannot be considered a human algorithm). For example, when learning how to draw a particular object, one might encounter the following algorithmical instruction (flow chart): "Check to see whether the draft is satisfactory: If 'yes,' ok. If 'no,' try to revise it." This fragment of instruction (from Dicherson, 1965) does not describe what particular actions should be performed by the individual; therefore, this cannot be considered a human algorithm.

Algorithms of activity (human algorithms) can be divided into algorithms of identification and algorithms of transformation. Algorithms of identification provide categorization, formation of concepts, analysis of situations, and so forth. Algorithms of transformation provide manipulation of information and external objects.

In the system of self-regulation of activity, motivation includes emotional evaluation and inducing components. This is described with the help of two function blocks in a model of self-regulation. The first block is called "assessment of sense of task," whereas the second is called "formation of the level of motivation." Sense of task is connected with the evaluation of "subjective significance of goal and task performance." Sense of task, from which the personal significance derives, should be distinguished from "assessment of meaning of input information." Changing the significance of task or its components influences the next function block, which is called "subjectively relevant task conditions." This function block presents dynamic representation of different components of tasks (operative image). Of particular importance is the interaction between different function blocks, changeable subjective criteria of evaluation of performance, and so on. From this it follows that activity can be considered as a dynamic, self-regulated, and organized system. This theory permits scientists to efficiently solve different practical problems, which are demonstrated in multiple examples throughout this book.

One important recommendation is that, during experiments, psychologists should reproduce basic characteristics of real tasks with their goals, motives, significance, and so forth. As shown in chapter 5, the significance of the task (which is derived from the self-regulation mechanism sense of task) influences design decisions. In one experiment, instrument configurations are also evaluated. It was discovered that sometimes subjects developed their own scale of significance of task while using different instrument configurations. As a result they demonstrated the best performance with the worst design configuration. Apparently, the pilots considered their ability to work with the worst version a sign of their professional qualifications.

In the theory of activity as it pertains to developing analytical procedures of design, algorithmical methods of description of activity are used that, if required, can be supplemented by description of time structure of activity and its quantitative evaluation of complexity. Graph theories and some other methods are also used. The major units of analysis are actions (mental and motor). Different practical examples of using these approaches in solving ergonomic design problems are demonstrated. Learning how the functions between the pilot and the aircraft are allocated is important to the solution of different problems of design in aviation. Regarding this, an interesting principle, called *joint flight*, was developed that follows from the theory of activity. Joint flight helps the pilot sustain a more active physical and mental state during flight and, as a result, increases the reliability of the flight. Also from the point of view of activity, this book suggests a possible interpretation of the operator errors.

The theory of activity, its algorithmical description, and the model of self-regulation of activity are the theoretical bases for presenting theories of learning in this book. From the behavioristic point of view, the major aspect of learning is a *stimulus-response* relationship. In cognitive psychology, learning is considered from the point of view of human information processing. In concepts of learning suggested in this book, the major units of analysis are actions (mental and behavioral). According to the presented material, learning can be performed based on conscious or unconscious levels of self-regulation and the complex relationship between them. The unconscious level of self-regulation is connected with blind trials and errors and corrections of behavior according to them. Conscious level learning is connected with the formation of voluntary actions based on the conscious level of self-regulation and acquiring complicated rules or algorithms of behavior.

In this introduction, only a few examples are described to give a more general overview of theoretical concepts that are discussed throughout this book. The following chapters demonstrate how this theory of activity can be applied to design, safety, learning, and so forth. The major theoretical data of the theory of activity are described in the first three chapters.

The purpose of this introduction is to facilitate a more meaningful interpretation of the material presented in this book. According to Ausubel (1968), introductory data can be considered as advance organizers that make it easier for readers to understand a book containing unknown material. This overview should make it easier for readers unfamiliar with the Russian theory of activity to understand these concepts.

Russian ergonomists received so much from American and Western sources that it is now only right to give something back. Despite the considerable similarity between Russian and American theories and methods, the special "spin" the former put on their work may stimulate new thinking on the part of their American colleagues.

Although the two are often confused, it is necessary to distinguish between engineering psychology and human factors/ergonomics. The first is primarily psychology; the second is primarily system development, operation, and testing. The former deals with concepts and research that describe human performance in the technological context, and presumably supports the practice of human factors. This book describes engineering psychology primarily, although there are attempts to relate the psychology to system design. Engineering psychology approaches ergonomics from the standpoint of fundamental human processes such as sensation, perception, motoric activity, and cognition. Russian ergonomists in general utilize the techniques of behavioral system development that were initiated in the West; these are not described in any depth, because their description would be repetitious to the Western reader. Moreover, what is distinctive about Russian engineering psychology is the point of view, which in some ways deviates markedly from that of the West.

Quite apart from that, there is historical value in recording concepts and practices of colleagues working in a very different environment. The sum total of Western ergonomics experience can only be enriched by the experience of others. At least, we certainly hope so.

The Concept of Work Activity

1.1 OVERVIEW AND INTRODUCTION

The concept of activity (*deyatelnost*) plays a key role in Russian psychology and ergonomics. Although it translates to the American concept of "behavior," there are significant differences. *Deyatelnost* is a coherent system of internal mental processes and external behavior and motivation that are combined and directed to achieve conscious goals.

Another important concept is that of the goal, which is the image of the desired result in the future. Needs that induce human activity directed to attain goals are called motives.

In the concept of activity awareness of the goal is basic. All other aspects of activity may be conscious or unconscious. The kind of goal utilized by the operator determines other activity aspects.

The parameters of work activity are the result (output) of the work, the method used, the individual's style of work, the work method prescribed by instructions, and the attributes permitting organization of the activity to achieve a goal. Goals and outcomes differ.

Another major concept in activity is self-regulation, which permits the operator during task performance to compare the desired goal with actual outcomes, and to adjust behavior when there is a discrepancy. Self-regulation involves updating information during the performance of a task, deciding whether or not to attempt an action (part of the task), deciding on a program of action, and then performing it.

The task is the basic component of activity. It is viewed in part as a problem-solving process in which the task is a situation requiring achievement of a goal in specific conditions. Task performance starts with an initial situation, which is then transformed by the task performance, and concludes with a final situation. The task has many attributes, including complexity, difficulty, degree of automation, and organization.

Activity can be performed by two types of actions: motor and mental. Motor actions are those that are of, involve, or relate to muscular movement. Mental actions involve direct connection action (e.g., sensory perception) and information-transforming actions. The latter involve diagnosis of a situation, hypothesis formulation, development of decision alternatives, selection of a decision, and evaluation of the effects of implementing the decision.

Here we should emphasize an important distinction between the theory of activity and the area of cognitive psychology called "human information processing." A basic principle in the theory of activity is the unity of cognition and external practical behavior. Psychological processes are considered to be a system of mental actions or operations with ideal objects. Basic to the development of these actions are external material actions. Accordingly, internal mental actions cannot be correctly understood in isolation from external practical actions.

There are two types of motives: sense formative and situational. The former are relatively stable, persistent, related to personality, and determine the individual's general motivational level. Situational motives are related to ongoing task solution. Sense formative motives are intimately connected with needs and emotions.

The question of goal formation is a fundamental problem of motivation. A distinction is made between meaning and sense; the former is cognitive and general, the latter more personal and emotional. The goal of activity must be differentiated from the American concept of goal. Goals can be *overall* or *partial.* An element of activity that is directed to achieve a partial goal is an *action.* The image of a future outcome becomes a goal only when linked to motivation. The relationship of motive to goal provides direction to self-regulation.

1.2 NEEDS, MOTIVES, AND GOALS

The purpose of this and several subsequent chapters is to illuminate the psychological foundations that support Russian ergonomics. Although most American texts on ergonomics (e.g., Proctor & Van Zandt, 1994; Salvendy, 1987; Sanders & McCormick, 1987) spend relatively little space on general psychological issues, Russian ergonomics is so closely tied to fundamental concepts of psychology that it is necessary to describe these in some detail. Otherwise, the reader would not know where Russian ergonomics is "coming from."

The concept of activity (*deyatelnost*) plays a key role in Russian psychology and ergonomics. It is roughly comparable to the English term *behavior*, but it is not the same and the differences are instructive.

Russians consider that the construct of behavior emphasizes the similarities between human and nonhuman activity, and in consequence overlooks some of what they consider the fundamental features of human functioning. To them behavior is associated with the external (stimulus-response) manifestations of activity, as in Skinner's (1974) concepts of learning.

Consequently, the behavior construct is virtually unused in the former Soviet Union. The concept of activity, which is far more comprehensive and, as is demonstrated later, emphasizes the central organization and integration of all aspects of behavior, was initiated after the 1917 October revolution and is currently the central concept of Russian psychology. *Activity* in its most general sense can be defined as a coherent system of internal mental processes, external behavior, and motivational processes that are combined and directed to achieve conscious goals. To the Russians, then, activity is both internal and external.

To understand better the practical and theoretical significance of the activity concept, we must examine its basic characteristics. One begins with human needs and motives that energize the activity. However, needs and motives must be differentiated from activity. The former do not determine the character of the latter. For instance, a motive compelling a man to work may come from the need to earn a living for basic sustenance. The daily work does not directly satisfy his human needs, but does produce goods and services within society that can indirectly satisfy those needs.

Needs must also be differentiated from goals. In the previous example, the content of the work activity is not determined by needs but by the goal to make a certain product or perform a certain service. The worker performs particular operations not for satisfaction of his hunger, but toward achievement as a specific goal. The needs and motives that energize activity in most cases diverge from the concrete task goals that direct our activity.

Needs that induce human activity directed to attain certain goals are called *motives* (Petrovsky, 1986, Ed.). The presence of needs and motives creates a system of inducements to stimulate the individual's activity. These can be conceptualized as hierarchically organized.

Needs and motives can be in different relationships to each other in this system, either synergistic or antagonistic. The importance of a motive and consequentially its hierarchical position depends on many factors, including the individual situation. Motives are not always understood by the person who has them.

The distinction among goals, needs, and motives is very important in the theory of activity. Needs may be transformed into motives only in those cases in which they acquire the capacity to induce an activity of a person to achieve a particular goal. Motives may derive not only from needs, but from desire, intention, aspiration, strivings, when these induce our behavior toward particular goal achievement. The more important the goal, the more motivated the worker will be to attain it, and the more she will expend physical and psychological energy toward achieving the goal. As can be seen, one goal may occasion different motives. Thus, motivation can be disitnguished not only on the basis of intensity, but on the basis of the content of the motivation informed by an identical goal. For example, an individual might strive toward a goal during task performance to help her friend perform a particular task, or she may attempt the same goal out of intrinsic interest or to gain some advantage. In one case, the desire to help is the dominant motivation. In the other case, more egotistical motives are at play. Thus, the reasons for a person's actions are distinct from the objects toward which he or she strives. For example, the activity of an air traffic controller is directed toward safe takeoffs, landings, and so on. However, this kind of activity is also possible using totally different actions that involve evaluation of position of aircraft, the relationships to other aircrafts, weather conditions, and so forth. The same goal may serve different motives.

The goal is both sensori-perceptual and cognitive. In the first instance, the goal is the image of a desired result in the future. (It must be pointed out that, as we see in greater detail later, to Russian theorists imagery plays an important role in conceptual processes.) Goals also include a conceptual form of future action such as that encoded verbally, which is associated with logical reasoning concerning attainment of desired results. The goal, therefore, includes both sensory and conceptual components of future results and in fact represents to the operator a form of information.

Goals are complex and dynamic and may have many different personal interpretations. At various stages of performing an activity an individual may be consciously aware of different aspects of his or her goal. The goal has both objective and subjective aspects, which interact to influence the manner in which the activity is performed. As is illustrated later, the goal in Russian psychology is differentiated from the process of its formation, which is influenced by human needs, motives, emotions, and other idiosyncratic variables. Goal formation is primarily important in analyzing those work activities that are self-initiated, but even when goals are presented to individuals by means of instructions or requirements, the interpretation of the meaning of the goal can vary depending on idiosyncratic factors and situational factors such as context. The more complex the task, the more likely this is to occur.

For example, in experimental conditions where subjects should precisely and quickly react to different signals, psychologists use instructions like, "React to different signals as rapidly and precisely as possible." This goal may be interpreted by subjects in diverse ways. Some subjects may attend more to precision, whereas others may attend more to speed. How the subject idiosyncratically formulates the goal affects the "program of performance" and ultimately the achieved result. As can be seen, instructions do not strictly determine goal formation process. The meaning of a goal varies according to its context.

In the Russian concept of activity, awareness of the goal is critically important. The other aspects of the activity (motives or the method used to perform the activity) can be conscious or unconscious. Thus, a person can be aware or unaware of the motive or method of activity, but must always be aware of the activity's goal. For example, possessing a highly automatized skill allows one to carry out an activity without consciously attending to its method of implementation. However, because the goal directs the activity, it is necessary for the individual to be aware of it, except in a few pathological conditions such as somnambulism.

Goals can be self-induced or imposed in instructions. Even where the goal is imposed as a requirement, the clarity and detail of the goal will determine the selection and interpretation of information. Russian theorists view the goal as a species of information, about which more or less can be known. Thus, if the goal is obscure, there may be attempts to elucidate it by gathering more information and/or interpreting the goal in various ways and changing the activity method in accordance, to see whether the results of the changed activity tend (or do not) to produce better results in accomplishing the goal.

Several competing goals may be present in one situation. For example, when a nuclear power facility malfunctions, the goals of stabilizing the system, maintaining production output, and locating the source of the malfunction may challenge each other (Meister, in press). In such a situation, the choice of the goal to be attempted is part of the decision-making process. Indeed, depending on circumstances the reactor team may shift from one goal to another. We see than that the goal formation process (which includes decision making) may be even more complex than implementation of the activity after the goal is selected.

We see later another goal aspect: acceptance of a goal earlier selected provisionally. The sequence, then, in self-initiated activity is goal formation→goal selection→goal acceptance. In "instructed" activity (the task is prescribed) the sequence is goal recognition→goal interpretation→goal acceptance.

The goal of an operator's activity may emerge as an image of a future result that must be achieved during the process of activity. The goal determines the specific information selected and the specifics of regulation of activity during task performance. For example, what kind of signals will be extracted from a general flow of information by an operator depends to a large extent on the goal. The goal determines the mode in which the information is coded and transformed by the operator, and what kind of strategies for each attainment will be used. Inadequate understanding or formulation of goals can cause accidents.

Among pilots controlling a flight according to different display information, for instance, cases of goal deviation were discovered. Sometimes pilots change the goal of activity involuntarily. Instead of controlling the aircraft, they start to control pointers on the different display apparatus. In other words, instead of using the display patterns to control the aircraft, they focus their attention on controlling the patterns. This can result in the pilot's losing orientation of aircraft position during flight (Beregovoy, Zavalova, Lomov, & Ponomarenko, 1978). Instances of goal switching frequently occur during automatic flight when the pilot, heretofore a long-term passive observer performing a backup function, finds himself in a stressful situation when he urgently must mobilize for activity, and he loses sight of the goal of the activity. In such cases, his behavior exhibits chaotic features.

Once a goal has been established, interpreted, and accepted, the operator must formulate a plan to accomplish this goal and perform different actions. Each action has a separate subgoal, which must be reached to attain the overall goal of the activity. For example, as mentioned earlier, the activity of an air traffic controller is directed toward safe takeoffs, landings, and so on, yet this kind of activity is possible using totally different actions that involve evaluation of position of aircraft, the relationships to other aircraft, weather conditions, and the like. The same activity can be performed by a variety of different actions, among these object (or objective) actions and mental actions (Leont'ev, 1977).

Mental actions are exclusively psychic processes, such as those associated with sensation, memorization, and thinking. All mental actions are concerned with achieving intermediate goals.

Operations are the methods by means of which actions are carried out. The special characteristic of operations is that they are not determined by a motive or a goal, but by the conditions in which the activity must be performed. Such conditions include both external factors that influence the specific manner in which the activity is implemented and internal factors that determine the capacities of a person to perform the activity.

The task, according to Leont'ev (1977), is a goal pursued in special conditions. For example, a driver can slow down the speed of an automobile by various operations, either by braking or by releasing the accelerator pedal. The choice of operations depends on the conditions of the road, the existence of certain road factors, and the driver's skill level. As the driver becomes more skilled, the process of releasing pressure on the accelerator becomes a way to perform higher level actions, and assumes the nature of a skill that can be used as an operation to achieve more complex actions. In this case, an action was transformed into an operation.

Internal mental actions are originally formed on the basis of external (objective) actions. Internalization, which the Russians trace back to very early child development, is the process of transforming objective external actions into idealized internal actions. During child development external behavior and speech are converted into intrapsychological processes. American psychologists became acquainted with this concept through the work of Piaget (1952). Psychologists in the former Soviet Union emphasize to a somewhat greater degree social interaction and verbalization (Gal'perin, 1957).

Transition of one type of unit of activity into others permits one to conclude that a psychic phenomenon is considered a process, such that it is important to study the development of these processes. This idea determines developmental or genetic explanations of work activity. From this also follows historical understanding of psychic development. In the theory of activity, historical analysis played a particularly important role in studying higher mental functions. Founders of this approach to the theory of activity were Vygotsky (1960), Leont'ev (1977), Rubinshtein (1946), Anokhin (1955), and Bernshtein (1966). From this follows the important methodological principle of the study of work activity. For purposes of understanding how operators perform particular tasks the specialist should know how the subject performs tasks at different levels of skill acquisition.

The Concept of Actions is becoming more and more important in Western psychology. We now turn our attention to a scientific approach that looks at motivation and volitional process. This approach attempts to integrate the motivational and cognitive components of activity for the purpose of discovering their effect on the behavior of a person. This approach has been named "the psychology of action." We should note that the understanding of action in the theory of activity and in the "theory of action" differ fundamentally. For example, Gollwitzer (1996) developed a linear model of action and presented it as a sequence of phases including: predisional phase (motivational), preactional phase (volitional), actional phase (volitional), postactional phase (motivational). When we examine the model from the point of view of the theory of activity several contradictions arise. The first is that action is analyzed to a large extent from the point of view of motivation and volitional process. Cognitive and behavioral aspects of action are not fully considered. Second, in this model, action is presented as a linear model which contradicts the principles of the self-regulation of activity. As we shall see later, the theory of activity action and activity as a whole is described using functional models having loop structures. Further, the theory of action lacks the understanding of mental action. Cognition is considered only as a process which precedes behavioral actions. In the theory of activity cognition is considered as real activity which consists of inner mental actions. Each mental action includes intermediate goals that involve the achievement of the general goal. In this way, people perform not only external, practical actions but also internal, mental actions which are closely inter-related and affect one another. Cognition is considered not as an isolated system of psychological processes but as a system of mental actions which just like the external activity has a certain structural organization. A person can perform the same task, and solve the same problem in different ways, while also performing related mental and behavioral actions. This is why the study of the structure of activity and the method of system-structural analysis connected with this obtain central meaning in the theory of activity. These questions will be further discussed later.

The preceding discussion emphasizes again the close links Russian ergonomics has with fundamental psychological principles.

Behaviorists often use the concepts of result or outcome and goal as if they were synonymous, but in terms of the theory of activity they are not the same. An outcome is the result of an activity, whereas the goal is the image or model of future desired outcomes. The result of an activity may achieve its goal, but it may not. The achieved result and the original goal of the activity are not always matched. If the result of an activity does not coincide with the operator's goal, then she or he needs to reformulate the strategy used to achieve this goal or the goal itself.

The parameters of work activity have been enumerated by Shadrikov (1982). These are both quantitative and qualitative aspects by which one can evaluate the effectiveness of the work activity in terms of its quality, quantity, and reliability. The parameters are (a) result—index of work accomplished, used for comparison with goal, (b) method—manner in which the goal is achieved, (c) individual style—method of work dependent on individual preference (the operator can select how she or he will perform the work), (d) standard style—method of work activity prescribed by instructions, and (e) informational basis of activity—the sum of all attributes permitting the organization of the activity to achieve a goal.

Of the five parameters, the ones of greatest interest are (c) and (e). It is noteworthy that even when the method of work activity is specified by instructions, the worker still can select a strategy for goal achievement that appears most suitable for him or her. Even when procedures are very rigid (one step must be followed by a second specific step), the manner in which each step is performed can be varied slightly (e.g., performed faster or slower, with greater or lesser deliberation). Most tasks permit at least slight variations in the way in which the activity must be performed. Individual styles of work activity allow individuals with different capacities and personalities to create individual strategies to achieve the same goal. This also enables people to compensate for personal weaknesses by using their own strengths.

Parameter (e) suggests that the activity, like the goal, is viewed as a form of information and hence the attributes of quantity and clarity are important. Remember that the goal of the activity as it is idealized provides information to the operator. Information is also provided as feedback from the performance of subprocesses or subactions (intermediate actions); this information enables the operator to reorganize the structure of his or her task.

Let us consider examples with pilots. A pilot may control an aircraft according to goals that are extrinsically supplied but may be idiosycratically interpreted. In other cases, the pilot may formulate the goal himself depending on the situation. Based on the evaluation of the significance of the goal, or the different components of the task, the pilot may develop different strategies of gathering information from different displays. Pilots do not simply receive information, but also evaluate it from the point of view of expectations or of the desired goal. Similarly, a pilot may evaluate the result of his own actions according to goals and his motivational intentions. Due to feed-forward and feedback influences in self-regulation processes, he may continually correct his activity and can even transform the goal of his activity. For example, depending on the goal created by the pilot during the performance of his tasks, the same display apparatus may perform different functions. Some display patterns in particular conditions can be used for evaluation of flight parameters. Others might be used to check or examine the functions of various display apparatus or to correct different actions. Still other display patterns might serve a control function. Depending on how the pilot uses different patterns of display, information can be divided into checking and correcting functions (Dobrolensky, Zavalova, Ponomarenko, & Tuvaev, 1975). The function of a single display apparatus can change as a function of the goal of the activity.

In all these situations, interactions with information from different display patterns may be different. The duration and frequency of visual attention to different instruments might be significantly altered. The pilot's attitude to the information received depends on the significance of the information. Indeed, it was discovered that we cannot identify the direction of visual fixation with the process of receiving information. A pilot may look at a particular instrument but fail to recieve information from this instrument. We know that pilots manipulate through images of light. As has already been established, the strategy for receiving information depends on the goals generally and specifically on self-regulation of activity. A pilot attends to different patterns of the display apparatus not as totally new manifestations, but to compare with an image of flight at a particular moment. In other words, a pilot does not search for totally novel information from the instruments, but tries to fulfill her forecasts. So a pilot does not react to each different signal from the display apparatus but rather, voluntarily receives information from different instruments. She then acts voluntarily in accordance with this information. The major units of analysis are not signal and reaction interconnection, but voluntary goal-directed actions. Depending on the goal and conditions, a pilot consciously regulates her own actions and develops different strategies for attaining the goal. She shapes her dynamic image of the situation, which influences the specificity of processing information. From this it follows that the study of "input–output" characterisitics and psychological processes are required but are not in and of themselves sufficient to analyze a human-machine system. These approaches need to be supplemented by the activity approach, the basic purpose of which is the study of the structure of an activity. As part of the structure of activity we should understand the content of actions and their interrelatoinship during task performance. These questions are considered in more detail later.

1.3 THE UNITS OF ANALYSIS OF WORK ACTIVITY

The selection of units of analysis has a fundamental meaning in the theory of activity. Pavlov (1927) considered the conditioned reflex to be a basic behavioral unit. It was also a psychic phenomenon because he identified the conditioned reflex with the association. Unlike Pavlov, Skinner (1974) deduced the basic units of behavior from behavior itself without resorting to other sciences, physiology in particular. He used the stimulus and response relation (S-R) as a basic unit of analyzing behavior. In psychology, other units of analysis were also used. For example, Piaget (1952) used mental operations. Vygotsky (1962) was the first scientist to pay attention to units of analysis from an activity point of view. According to him, units of analysis should not be allowed to lose the specificity of the analyzed phenomenon.

Gordeeva and V. P. Zinchenko (1982) systematized the principles underlying the selection of analytic units:

1. The units should be organized into a structure that functions by means of feed-forward and feedback mechanisms.
2. Individual units should possess the properties of the larger units in which they are embedded.
3. The units should be capable of evolving; that is, they must possess the capability of transforming their properties and mutual conversion into each other.
4. The taxonomic approach should be used in order to develop the units of analysis.
5. The units of analysis that have been used as separate components should allow the specialist to assemble a holistic activity.

The reader has already encountered four major types of analytic units: activity, task, action, and operation (Leont'ev, 1977). Another approach to the development of analytic units is based on the concept of self-regulation, which plays an extremely important role in Russian engineering psychology.

The first to develop the principle of self-regulation were Anokhin (1935) and Bernshtein (1935). The basis of their principle is the concept of self-regulating units of activity (Anokhin, 1955).

According to Anokhin, the neurophysiological basis of an activity is a functional system that can be considered a dynamic organization. This organization selectively integrates different central and peripheral neural mechanisms; their interaction permits the achievement of the desired result. A functional system is considered a closed loop having continual feedback information about the success of performance.

There are two aspects to this conceptual system: (a) a drive toward a continuing effort to adapt or modify the course of the activity to the requirements imposed by the goal, and (b) continuing information reception, processing, and decision making. The latter is, of course, equivalent to the Western information orientation toward behavior. The self-regulation mechanism is a continuing comparison of the desired activity result or outcome with the actual result or outcome, a comparison that produces feedback information.

As early as 1935 Anokhin analyzed animal behavior to show that any reaction or reflex must be evaluated on the basis of feedback. The stimuli resulting from performance of an action was labeled *backward afferentation* (feedback). This notion was understood long before the concept of feedback was introduced into cybernetics by Wiener (1958). Any consequence of activity can be evaluated by an organism, but only those outcomes that satisfy the organism's needs are retained. Those that do not satisfy are or must be modified to bring them closer to the desired goal. Any behavioral action is an adaptive self-regulative act that has a loop structure with feedback (Anokhin, 1962). The basis of the self-regulation system is therefore evaluation (comparison of actual with desired results), feedback, and the ability to modify behavior.

The Anokhin model is illustrated in Fig. 1.1. It consists of five components: A, B, C, D, and E.

A. *Afferent Synthesis.* The stimuli received by the organism are of two types, those relevant to the activity goal (major initiative stimulus) and those present in the activity environment but that are irrelevant to the activity (situational stimuli but still have some influence on major stimuli). The information provided by the relevant stimuli impinge on the organism, under conditions of the organism's current motivation and its memory (learned habits).

B. *Decision Making.* A decision is made whether or not to attempt the action.

C. *Action Acceptance.* There is an anticipation of whether the action to be performed is appropriate to the desired goal of the action. The concept

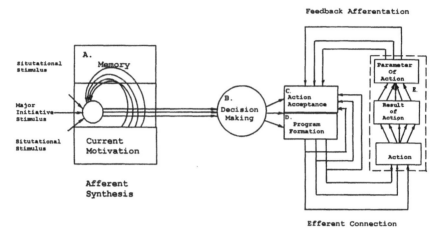

FIG. 1.1. Model of self-regulation of condition reflex. From Anokhin (1962).

of the goal is inherent in the concept of action acceptance, because if there were no goal, there would be nothing to be accepted. However, here a goal is considered in physiological aspects and not always connected with consciousness. A goal is considered an end state toward which is directed behavior. Implicit in this component is a comparison and an evaluation of the result of the action with action acceptance.

D. *Program Formation.* A program is formed for the performance of the reflex.

E. *Reflex Implementation.* This consists of three subprocesses: (a) The action is performed and (b) the results of the action are noted in terms of (c) certain desired parameters of the action, which in effect represent the accomplishment (or nonaccomplishment) of the goal. These parameters are then compared (by means of feedback) with action acceptance (c) parameters; if the two sets of parameters match in terms of neural impulses, the reflex action is completed and stops. If there is a discrepancy between the two sets of neural impulses, corrective actions directed at achieving the desired match are activated.

In this model psychic processes are correlated with physiological (neural) processes but are viewed as systemic entities. The Russians have a tendency to conceptualize their models in terms of neural processes, but they also recognize that there can be no one-to-one relationship between neural and psychological functions. This tendency may make it more difficult for Western ergonomists to understand Russian models, because they have long discarded the use of neural analogies.

The concept of feedback from the result of an action and its use to compare the desired result (goal-oriented) with the actual result of the action is entirely coordinate with Western psychological concepts. The Russians consider that Anokhin's concepts represent a functional system rather than a reflex arc, even if it is described in neural terms. At the same time they consider that Western thinking about behavior (e.g., Wade & Tavris, 1990) is fixated on linear stimulus-response arclike concepts. The use of neural models to explain task behavior is almost completely foreign to modern American thinking, although one finds it in neural models used in the design of computers that attempt to mimic human processes (see Meister, 1971). In this connection, we have already referred to the strong influence of cybernetics (computerized) thinking on Russian psychological concepts.

Bernshtein (1966) considered the organism as an active system that is continually attempting to solve tasks. The task in part can be considered as a goal or model of the desired future. The individual creates a plan or intention of action in the process of synthesizing stimuli and evaluating his or her situation. Once the task is initiated, he or she determines whether it is necessary to change the ongoing system of activity, when this should be done, and how his or her activity will modify the situation into the desired and anticipated one. The individual plays the critical role in controlling and correcting the components of activity, evaluating intermediate and terminal outcomes, and providing adaptive flexibility to the conduct of the activity.

The extent to which the individual controls his or her activity depends, of course, on how rigidly structured the task is by external instructions. Major self-regulation occurs only in self-initiated activity, although some self-regulation occurs even in activities that are highly programmed by instructions.

Units of behavior as self-regulative systems were also suggested by Linhart (1970). He conceptualized three major types of afferentation: informational-orientational, motivational, and feedback (proprioceptive and psychological), of which the last plays an especially important role.

It appears then that any unit of behavior can be considered as a self-regulative subsystem. The basic unit of activity is an action directed to achieve a conscious intermediate goal. All actions have a loop structure. The starting point of any action is the moment when the goal is formulated or accepted. The end of the action occurs when the results of the action are evaluated. This permits a continuing flow of activity divided into individual units delimited by the goals of the activity (intermediate and terminal) and the evaluation of the outcomes of the action.

The size of the activity unit can be influenced by its overall context of activity. For example, Newtson (1976) presented subjects with a film depicting the performance of various tasks. They pressed a button to mark their perception of the beginning and the end of each action. It was found that in more complex tasks subjects separated smaller action units.

Units of analysis possess attributes such as hierarchical structure, integrity, and modifiability. Structure is inherent in the notion of intermediate goals and activity subprocesses. Intermediate actions may have more molecular actions within them, and all subactions are nested in the overall activity. These intermediate, more molecular actions do not possess their own conscious goal and were called psychic operations (Leont'ev, 1977.) Integrity follows from the notion of structure: Activity components are interconnected physiologically, motivationally, and rationally. These two features permit the activity units to develop and modify themselves through learning. Because of this, an individual is capable of acquiring new forms of behavior.

The preceding concepts have also been suggested by the seminal work of G. A. Miller, Galanter, and Pribram (1960), who also characterized behavior as a hierarchial structure. The differences between their approach and that of Russian psychologists is that Miller et al. did not give the same emphasis to goals, motives, and methods of implementing activities. In order to conduct microstructural analysis of activity, one should use even smaller units of analysis. Such units are called function blocks.

In accordance with principles of cognitive psychology, the Russian activity viewpoint considers that mental activity has the following characteristics:

1. Any psychological phenomena (and behavioral actions including perceptual reactions to stimuli) that appear to the individual as instantaneous actually occur in short but measurable time periods and can be represented as a series of subprocesses.

2. These psychological phenomena are implemented through a series of function blocks (as explained in the Foreword the term function block is a reflection of computer software thinking, which has greatly influenced Russian thinking), which are actually information-processing phenomena. This concept is discussed in greater detail later.

3. The function block must not be thought of as something physical or as an observable process. It is a construct that is inferred as a result of certain chronological (time measurement) experiments and the qualitative analysis that forms a significant part of Russian analysis of work activity. Precisely what goes on within the block is unclear to Russian theorists. Nevertheless, although the function block is a theoretical construct, Russians view it as an actual functional system, with mechanisms that form particular functions.

4. All the function blocks are interconnected by feed-forward and feedback loops; in toto, they serve as self-regulative mechanisms.

5. It is important to point out that context is extremely important in influencing the function blocks, their degree of development, and how they perform. The content of a function block is not always the same, and how it is utilized will vary according to the individual situation.

Although Russian theorists may not know precisely what is going on inside the block, they feel it is very important to know how much time is spent processing information within the block. This accounts for the extraordinary emphasis placed on time measurement in Russian psychology. Presumably, duration of functional activity provides clues as to what is being done within the block. Although they talk about the function block as a construct, they tend to think of it as having physical characteristics, such as an entrance and an exit. Knowing what has happened at the entrance and exit and the duration of what has gone on within the block, they can hypothesize about the supposed mechanisms within the function block. This methodology they characterize as a "black box" approach, which is acceptable to them because it is used in other sciences such as physics. The duration and character of the interconnections among function blocks is felt to have both theoretical and practical value.

A function block from position of cognitive psychology can be considered a particular stage of information processing when specific objectives of mental activity are attained. Each stage of information processing can be depicted as boxes or blocks. The sequence of these blocks presents stages of information processing. The first person that introduced the idea that cognition can be divided into sequences of different stages was Sternberg (1969). For example, Lindsay and Norman (1992) described perceptual process as a sequence of the following stages:

- Stage 1 is connected with recording sensory images.
- Stage 2 is tied to the process of extracting and analyzing the sensory features.
- Stage 3 characterizes the process of creating and revising expectations during interpretation of information. This stage also includes the process of decision making about the final image.
- Stage 4 involves operations with information from memory.

Each stage can be considered as a function block. From the viewpoint of the theory of activity, it is not simply a sequence of stages but rather a functional system that has not only forward, but also feedback interconnections. Accordingly, the model of recognition can be presented as in Fig. 1.2 (Bedny & Zelenin, 1980). This is a functional system because it organizes different stages for performing particular functions and can be presented as functional blocks.

When we study activity as a whole, not only as a separate psychological function, function blocks can be a much more complicated phenomenon. This is discussed in chapter 2. When we use function blocks as units of analysis, and try to discover the relationship between them, this is a functional description of activity. Activity can also be described on the cognitive level when we study psychological processes. However, activity can also

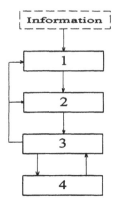

FIG. 1.2. The model of recognition.

be described as a system of external and internal actions and operations. This is morphological analysis of activity. Hence, activity can be described from different viewpoints and with different levels of decomposition.

Units of activity are not cast in concrete. They may change over time and transform into other units, for example, the transformation of an action into an operation and vice versa. In the process of mastering a skill, the importance of some operations may be altered. During the automatization of a skill, the importance of feedback will decrease because the operator needs to play less conscious attention to the regulative effect of feedback. From that standpoint regulation becomes less of a conscious process.

McCormick and Ilgen (1985) suggested the following units of analysis for work behavior as shown in Fig. 1.3. As can been in this case, in contrasted with the stimulus-response (S-R) model, these authors included variables involved in human information processing.

In the Russian concept of activity, the basic unit of analysis is an action. This action can be decomposed into more detailed units of analysis, as seen in Fig. 1.4. It is a mere convenience to present this model as a linear sequence of steps followed by the feedback influence. In reality it has a recursive loop structure with multiple forward and backward interconnections. Russian theorists emphasize that the formation of goals implies the integration of informational and motivational processes. They distinguish activity stages in terms of their relationship to the entire activity. For example, at the stage of "information input" the direction of the action is not yet determined by

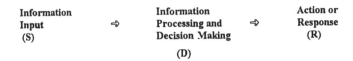

FIG. 1.3. Unit of analysis of work behavior. From McCormick and Ilgen (1985). *Industrial and Organizational Psychology* (p. 45). Englewood Cliffs, NJ: Prentice-Hall.

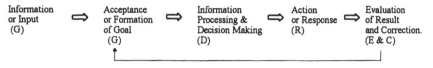

FIG. 1.4. Unit of analysis of work activity.

the stimulus. The critical point at which the action is initiated takes place at the stage of "goal formation" or "goal acceptance." The action terminates at the stage of "evaluation of results," which iterates into the next cycle by means of corrective processes.

Within the total activity the intermediate action is a self-regulative subsystem, with conscious subgoals and concomitant motivational components. Nevertheless, motive and goal on the certain stage of analysis may be excluded from the analysis after the goal is accepted, after which one can concentrate on the mechanisms of activity implementation. (This is called the operational stage of analysis.)

The stimulus input is considered by Russian theorists as presenting information only, and thus, lacks a motivational component. For them the external stimulus cannot cause one to perform a task until a goal is formed and one has a motive to achieve this goal. Russian ergonomists would say, as Western ergonomists would, that the motive to perform a task must exist even before the initiating stimulus is received, because without that prior motivation, the initiating stimulus would be meaningless (in an activity sense) to the individual.

The notion of "set," which is a familiar one to Western psychologists, was introduced into the Soviet Union in 1930 by Uznadze (1961). This concept is as important to Russian activity theory as it is in the West. A stimulus situation interacts with prestimulus motivation and, as a result, a set emerges of which the individual may or may not be aware. This is a tendency to react in a special way to a particular situation, and means that behavior cannot be inferred directly from a stimulus situation.

The problem, however, is to determine in advance of performance, or even following performance, what the individual's set was, how important it was in influencing that performance, and in fact how the set functioned. All of this is not easy to do, even in an explanatory mode.

Russian psychologists feel that the stimulus-response paradigm can be employed only for analysis of involuntary reactions, that it cannot be used to explicate conscious goals and directed actions. At the same time, they understand that Western ergonomists have not ignored internal mental processes. As McCormick and Ilgen (1985) illustrated in Fig. 1.3, information processing and decision making do enter into the analysis of work behavior. This concept goes back as far as Tolman (1932), who was the first to introduce intervening cognitive and inducing variables between the stimulus and the reaction.

The principal distinction between the Russian activity approach and the behavioral approach is the existence of the conscious goal, which determines the specificity of the selection of information and influences the strategy of its attainment. A person does not react to the stimulus or simply process information, but actively performs in a given situation based on the goal and existing motives.

Presented material permits one to draw a conclusion that activity is not an aggregation of different elements. Rather, it is a system with its own structure. This is why the systemic-structural analysis of work activity becomes very important. When one studies work activity, it is important to study not only separate actions but also their relationship and how they influence each other. Interaction between separate elements of activity can be determined by the specificity of these elements, the goal of the activity, and strategies used for their attainment. From this follows the necessity to study the dynamic features of the structure of activity. Including or excluding certain components of activity can change the entire structure of the activity. Because of that, activity cannot be considered as a sum of responses to different stimuli.

In summary, if one wishes to abstract units of work activity as actions, one must define the appropriate units of information that lead to the formulation of conscious subgoals of action, then abstract the cognitive, response, and evaluative components of that action.

1.4 THE TASK AS A BASIC COMPONENT OF ACTIVITY

The task is the basic component of activity. Our lives can be conceptualized as a continuing attempt to solve various tasks. The development of human-machine systems often begins with task analysis and task design. For example, the interpretation of input information depends on the task the operator is attempting to perform. Changing any equipment characteristic influences the method of task performance. One can evaluate the behavioral adequacy of equipment design only by assessing it in the context to task performance. By analyzing the task that is being performed by the worker, we can estimate the efficiency of his or her work performance.

In the theory of activity, the definition of the task was suggested by Leont'ev (1977), who stated that a task is a situation requiring achievement of a goal in specific conditions. A more narrow definition of the task is that it is a situation that requires the discovery of the unknown based on what we already know (Kostuk, 1962).

Brushlinsky (1979) offered a definition similar to that of Kostuk. He claimed that the task is a situation in which we must find an unknown when we are not given the method by which to solve the problem. In American ergonomics, the task and task analysis were defined by R. B. Miller (1953), although since that time there have been many definitions. There is general agreement that a task is a set of human actions that can contribute to a specific functional objective and ultimately to the output goal of a system.

Note that the American concept lacks the problem-solving aspect, or, if it does include it, considers it only in the context of problem-solving situations. The Russian notion that the task is inherently a problem-solving endeavor reflects the greater philosophical emphasis of Russian thinking.

The elements of the task are the initiating stimulus, the response required, and the goal that organizes all task elements as a whole. In any task we can distinguish and extract requirements and conditions. Anything that is presented to the operator or known by him or her is a condition of the task, the requirements of which include the finding of a solution (or what we need to prove).

When the task requirements are accepted by the individual (a necessary step in task performance), those requirements become part of the individual's personal goal. This transformation is one of the major means by which the goal is formed. Great differences may sometimes exist between specified task requirements and the goal as finally accepted by the individual.

The goal of the activity is embodied in the search for the required outcome. For example, when one solves mathematical problems, the goal is to find the solution, which is presently unknown. Suppose that in an experimental task two indicators, red and green, are part of the task. Two numbers are presented to the subject and she or he is told that if the red light is illuminated, the numbers must be added; if the green light is illuminated, the numbers must be subtracted. The goal is to derive the correct number. The green and red lights are a condition of the task. Although it may be difficult in some cases to distinguish between task goal and task conditions, it is necessary for an adequate task analysis to do so.

Conditions may include the following (Tikhomirov, 1984): (a) habitual and nonhabitual situations that determine the operator's orientation in the situation and the possibility or impossibility of reaching a desired result by an existing method, (b) the specificity of the representation of the situation, which can be either a verbal description of the situation or an observation of the situation (both may provide information to the operator), and (c) the degree to which key elements for solving the problem can be distinguished from less significant elements.

Parameter conditions can include the following: The elements of a situation are relatively independent of each other but nonetheless interact, there are rules for the transformation of a situation, and a list of alternatives for that transformation.

For example, during a chess game we can point out the number of squares on the chessboard that are full and empty and their configuration. The elements of conditions can exist in specific space and functional relations.

Functional relations determine the rules of a transformation situation. For example, a chess player tries to "move to the opponents's square," or "create an obstacle for his opponent to make a specific move." Strategic alternatives consist of a number of correct transformations that can bring a situation to a required result. Theoretically, the number of alternatives can be equal to zero, which results in an unsolved problem, or can have unlimited possibilities.

Task performance involves an initial situation (the problem presented before task performance begins), a transformed situation (the actions taken to solve the problem), and a final situation (the problem has been solved and the initial situation has been changed). Elements of a situation possess certain meanings that change as the situation changes. For example, at any one stage of chess playing, the meaning of the board configuration at this time will vary.

Any transformation has a value ("good" or "bad" situation) relative to the final outcome of task performance. Another means of evaluating a situation is by its relationship to general rules of operation (e.g., chess-playing rules). However, specific actions associated with changing an existing situation can be "weak" in relation to general rules of operation for performing an activity but be "correct" in relation to the final result. For example, the operator's strategy for task performance can be inefficient but can guarantee achieving the required result.

Task structure includes requirements that help specify the goal. For example, instructions can precisely prescribe what the final situation should be. However, in creative task performance or when the operator alone determines what the final situation should be, task requirements will be imposed by the worker alone. The task is also characterized by the relationship between its initial conditions and requirements for solution of the problem. For example, when all elements of the condition are presented to the operator, some may be extraneous and some necessary elements may be absent. The relationship between the initial task conditions and requirements for solution may also vary according to the mode of presentation. The task condition and the task requirement can even contradict each other. For example, the requirement may demand very precise execution of the task, which is all but impossible in distracting conditions.

The motivational process is crucial for solving the task. Motivation is associated with task characteristics such as its complexity and significance (importance to the performer). We consider these two attributes in greater detail later. During task performance, motivation can change. If, for example, the task is very difficult and the operator has not realized this initially, his or her motivation may initially be high. When he or she experiences difficulty, his or her motivation can decrease.

In understanding the process from external to internal, motivation is very important. Here two notions must be considered, the meaning of the task and the personal sense of the task. The meaning of the task is the operator's understanding of what the task is and how to perform it. The sense of the task is its significance to the performer. The latter adds to or diminishes the individual's motivation in performing the task.

Another aspect of task performance is the subjective image of the task formed by the operator. Bartlett (1932) demonstrated that while performing a task people create an image of the task and the task situation, an image that he termed *schema*. This is not a precise copy of task reality but a model in the performer's mind of the present and the future task situation; the schema guides the performer's action, affecting in particular the way in which she or he receives, interprets, and processes sensory input information. For example, the same mathematical problem can be solved either geometrically or algebraically.

The task schema is particularly important in indeterminate tasks in which the goal is unclear and the method of implementation is unknown or ambiguous. Under such conditions the performer will be especially subject to the effect of schemata. This does not mean that people do not develop schemata for more proceduralized tasks, because they do, and this affects the resultant performance; but, in highly proceduralized tasks, the schemata must work harder to controvert instructions, which themselves create counterschemata.

Kozeleski (1979) suggested that the subjective or mental representation of the task has the following features:

1. The mental representation depends, of course, on the structure of the task.
2. It is a dynamic phenomenon that can change during task performance.
3. Representation determines task performance (it is possible to see the same task from varying aspects).
4. Success in solving the task depends on the performer's personnel representation (how the task is interpreted).

Although most tasks are presented to people in written or verbal form, in special cases (tasks performed by composers, artists, writers, scientists) the representation may be in the form of visual or auditory images.

A task is embedded in its environment. Howard (1968) suggested that the task and task environment can be considered within three-dimensional space. According to Kozeleski (1979) that space has attributes of unpredictability or uncertainty, dynamism (changeability), and complexity, as shown in Fig. 1.5. When events occurring in the environment cannot be predicted with reasonable precision, it lacks certainty. Obviously, the task and task environment can be modified and change with time. Tasks and task environments can be more or less complex in terms of their interactions among static and dynamic components and phenomena. Any task can be graphically described as a point

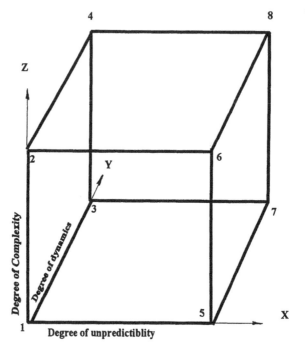

FIG. 1.5. Task environment as a three dimensional space. From Howard (1968). The foundations of decision analysis. *IEEE Transactions of Systems Science and Cybernetics, 3,* 211–219.

in space. In Fig. 1.5 coordinates x, y, and z, and points 1, 2, 3, and 4 represent deterministic tasks such as those that are simple and static (point 1), complex and static (point 2), simple and dynamic (point 3), and complex and dynamic (point 4). In reality, completely deterministic situations are quite rare (exemplified by death and taxes). Risky tasks (highly indeterminate and uncertain) can be simple and static (point 5), complex and static (point 6), simple and dynamic (point 7), and complex and dynamic (point 8).

The nature of the task is, of course, largely determined by the instructions provided to the operator. When task instructions do not contain a sufficiently detailed description of how to achieve the goal and consequently the task requires conscious deliberation about how to accomplish that goal, the task becomes one of problem solving. Landa (1982) defined two major groups of problem-solving tasks: algorithmic and nonalgorithmic. The latter class was divided into three subgroups, semialgorithmic, semiheuristic, and heuristic. The basis for this categorization was the relationship between the instructions and the mental operations required to solve the problem. The distinctions among the types of tasks are relative, not absolute and are defined by: (a) how and to what extent the instructions specify the actions to be performed by the subject, (b) whether or not the actions are relatively elementary for the subject (if an action can be performed virtually without error, it can be considered as relatively elementary), (c) how well the instructions specify the "field" of objects of search for which the goal object is to be selected, and (d) how the instructions delineate or specify the criterion to be used to identify the goal object.

If the instructions completely define the physical and mental actions to be performed by the subject, do not require any independent actions to be performed by one to whom the instruction is addressed, and guarantee successful solving of the problem, such instructions can be called algorithmic.

If the instructions contain some uncertainty resulting from vagueness of criteria that determine a logical sequence of activity and, therefore, require the subject to create his or her own methodology, the instruction is semialgorithmic.

If the uncertainty is even greater, the set of instructions is termed semiheuristic. Semiheuristic problems may include algorithmic and semialgorithmic subproblems.

The major criterion for heuristic instructions is an undefined field of objects from which a goal object must be selected. Some examples are:

Suppose there is a bookshelf with books among which there is only one book in a red binder. A person says, "Give me the red book, please." This is an algorithmic problem because the subject to whom the instructions were addressed knows exactly what to do. If there is the same bookshelf, but there exist two books in red binders and one uses the same instructions, in this case, some degree of uncertainty emerges. The instructions did not specify which of the two red books should be selected. This is a semialgorithmic problem (task).

In the next case, one might ask for an "interesting book." This request does not define specific criteria for selection. What does "interesting" precisely mean? Here, the degree of uncertainty is greater; therefore, we encounter a semiheuristic problem (task).

One of the most important goals of ergonomic design is to reduce the degree of uncertainty in solving problems or in performing specific tasks. The examples are given in terms of selecting something. However, Landa's (1982) categorization can be applied to all types of tasks.

Certain attributes are particularly important in analyzing the task.

Complexity is defined in terms of the number of static and dynamic components of the task and task situation, and the number of interactions among these components. Other task characteristics influence complexity: specificity of instructions, inadequate information, excessive work requirements, and indeterminism.

An aspect of complexity derived from reliability engineering theory is the degree of dependency of one task and action on another (Meister, 1976). If tasks and actions are independent of each other, meaning that one does not influence the performance of the other, complexity is less. The greater the amount of dependency, the more complex the task, because dependency involves interaction, which leads to complexity. The dependency need not be physical, but rather logical. For example, the performance of puddlers in a steel mill depends on a supply of coal, which could break down; this means in analyzing jobs in a steel mill one must consider this relationship.

If one could measure task/job complexity and relate that complexity to human performance (producing a prediction), one could evaluate the efficiency and reliability of activities and of equipment designs and forecast the learning of tasks. The overarching concern of ergonomics is to reduce work stress (hence human-engineered design) and this can be done only by knowing how complex the task (and the equipment required by the task) is.

Although in most cases complexity and *difficulty* are considered synonymous, they must be differentiated. If complexity is an objective characteristic of the task situation, then difficulty is the performer's subjective evaluation of the effects of that complexity. We hypothesize that the more complex the task situation, the greater the probability will be that it will be difficult for the operator. However, depending on the skills the operator brings to the task, complex tasks will be evaluated by him or her as relatively more or less difficult. Complexity does not have a subjective component; therefore, the operator cannot experience complexity itself per se, but rather as subjective difficulty. Cognitive effort depends on the task difficulty.

The *structure* of the task is the spatiotemporal organization of the actions performed by the operator. Obviously, the individual actions are organized logically in relation to others, because certain actions require others to be performed either before or after those actions. Instructions usually specify these relationships, but in self-initiated tasks the operator has more freedom to organize his or her work as he or she feels is most appropriate.

Automation is important because it allocates the functions formerly performed manually to the machine. This has the effect of changing the role played by the human. Whereas in manually operated equipment the operator was responsible for making the machine perform, with automation the operator's tasks change. Once the machine is energized, it is programmed by computer to perform certain functions on its own. Now the operator monitors the equipment to see that its performance stays within specified parameters. Should the machine exceed these parameters, the operator detects the malfunction and must diagnose its cause. Those human functions that were once psychomotor are now perceptual and highly cognitive. The importance of automation is that, increasingly in the future, systems will become highly computerized (Meister, in press).

The kinds of questions that must be asked in introducing automation are:

1. What functions can be automated?
2. What improvement or decrement in human performance will result from automation and how significant are these?

3. How does the automation influence the operator's tasks?

In some situations a high level of automation can be unfavorable because it reduces the reliability of the system. As a result, the principle of the "active" operator (not excessively passive, not completely automated) was introduced by Lomov, Zavalova, and Ponomarenko (1977).

The term *hardening* is probably unfamiliar to Western readers. It refers to the physical effort that the operator must exert during the work activity. Just as mental effort is required and should be considered in analysis of the task, physical effort is required and, if that effort is great, should also be considered. Hardening is related to complexity and difficulty that may affect the physical effort required. Although one can measure physical effort, it also has a subjective component (a feeling of stress) that is affected by the individual's physical condition. For example, women can exert on average only 80% of the effort exerted by men.

1.5 PSYCHOLOGICAL COMPONENTS OF WORK ACTIVITY

Activity theory pays much more attention to selecting appropriate units of analysis of activity during task performance, so there is need to spend some time discussing them.

1.5.1 Motor Components

Leont'ev (1977) divided activity into the following hierarchical units: activity, task, action, and operation. The motor component of activity can be divided into motor actions, which in turn can be subdivided into yet smaller units, motions or operations. Zinchinko et al. (1978) in addition proposed dividing motions into function blocks. Function blocks as units of motor analysis of motion are discussed later in this section.

Automatization causes the transformation of actions into operations. During automatization, awareness of the conscious goal of performing individual motor operations ceases to be conscious and is submerged in an awareness of a more complex action goal. For example, when a paraplegic learns to walk again, he is at first conscious of the individual goals of standing, raising the leg, moving the leg forward, and so forth. As he learns to walk, he is no longer aware of these individual motion goals and is aware only of the goal of walking.

On the other hand, the motor actions involved in a higher order task can disintegrate. Walking down a staircase is in almost all cases a highly auto-mated process. If, for any reason, the walker stops to consider consciously a motion involved in that process, she or he becomes aware of the individual goals associated with each walking motion and the process slows down and regresses to an earlier learning state.

The flexible transition from operation to action suggested by Leont'ev (1977) brings us to the possibility of polysemantic descriptions of different units of analysis, which cannot be permitted as it can result in an ambiguous task description. To choose correctly adequate units of activity analysis, it is important to consider them as self-regulation subsystems. Because every self-regulative subsystem has initial and terminal points (goal formulation and output evaluation), these can be used as one criterion for selection of task units. At the same time, according to motion and time study in the former USSR, motions cannot be considered as actions (Gal'scev, 1973) because actions must include no less than two or three motions (Bedny & Zelenin, 1988). This means that a motion, even in cases when its goal is distinctly realized (the individual is aware of that goal), should be considered only as a motion, not as an independent action. For example, during task performance an operator can carefully control his or her hand motions and be aware of the motion goal (to control the hand), but nevertheless the motion is only a motion, not an action. Similarly, a motor action consisting of several motions, as a consequence of automatization, may become unconscious to the operator.

Motions can be simple or combined. For instance, a bending of the body could be accompanied by a simultaneous turn, while the hand holds an object.

At the same time, when such combinations of body movements have conscious goals—for example, to avoid a dangerous situation in equipment operation—than it is said to be body action. However, most real-world body phenomena are motions that, once habituated, lack a conscious goal.

Gordeeva et al. (1975) and Gordeeva and V. P. Zinchenko (1982) studied the microstructure of motion. The experimental tasks included shifting a white dot of light from one position to another on a computer screen by manipulating a handle in a three-coordinate system (x, y, and z). They discovered three relatively separate stages or subsystems of movement to which they gave the following names: the latent stage or the programming block (motor instructions), the executive stage or implementation of the motion; and the control or corrective block. The first and third blocks were cognitive in nature, whereas the second block was motoric in nature. In the first stage of skill acquisition, cognitive components of movement played a very important role. In the final stage of skill acquisition, these components were significantly reduced. The second block (executive components) was characterized by many microshiftings of the subject's arm. During skill acquisition this shifting was significantly reduced. The authors concluded that these microshiftings are not executive implementation components of motion but play a cognitive role in exploration of the movement space. Micromotions that are essentially unconscious (the individual is rarely aware of them) are conceived by Russian psychologists as having exploratory cognitive functions. This presents the Western scientist with an interesting concept: cognition at a physical, unconscious level. The analogue of this is unconsciousness at a mental or cognitive level, as was postulated by Freud. What makes these micromotions cognitive is their exploratory character.

Cognitive functions for motions are simplified and materialized as concrete performances during skill acquisition. Gordeeva et al. (1975) and Gordeeva and V. P. Zinchenko (1982) also discovered that during skill acquisition the importance of each motor function block and its relative significance changes. In the final stage of skill acquisition, the more significant element is the executive function block.

For the purposes of standardizing the description of motor actions, Bedny and Zelenin (1988) introduced the following definition: Motor actions are complex standard motions performed by the body, unified by a single goal and a constant set of objects and work tools. This definition of action requires a definition of standard motion.

Various "predetermined time systems" contain different lists of motor elements and motions with the times for their performance. A comparison of these systems reveals that the motoric taxonomies in these systems vary widely. However, this does not mean that they are not of any practical importance.

The introduction of strictly defined rules and criteria allows users of such systems to extract clear and well-defined motions in the work process. Russian ergonomists prefer MTM-1 (methods-time measurement system).

This system, which was developed from cinematic studies of industrial operations, analyzes any manual operation or method into the basic motions required to perform it and assigns to each motion a predetermined time standard that is determined by the nature of the motion and conditions under which it is made (Karger & Bayha, 1977; Maynard, Stegemerten, & Schwab, 1948). Some Russian ergonomists believe that the application of this system to engineering-psychological and ergonomics research can be helpful.

Because motor activity occurs over time, in describing this activity it is useful to employ a taxonomy of standard activity elements and their duration. The establishment of a nomenclature of standardized motions and the previous definition of action allows one to extract individual motor actions. Recall that we regard a motor action as a set of two or more standardized motions unified by one conscious goal. For example, the motion "reach and grasp an object" may be considered as a motor action involving two standard motions "reach" (R) and "grasp" (G).

Actions that form logically related items in the work process but are being performed by different parts of a human body cannot be considered as one action. They should be considered as different motor actions performed simultaneously. For instance, simultaneous movement of two arms cannot be considered as one action even if they have one goal.

The shifting of the body or head in most cases could be considered motions that are combined with the actions of the hands or legs. This is because, in most cases, such motions do not have conscious goal destinations; they only play an auxiliary role.

From an activity perspective, it is important to distinguish involuntary (automatic) reaction (such as blinking and the posture reflex) from voluntary actions associated with conscious goals. These involuntary reactions occur within a short period of time (sometimes .1 seconds). Such reactions must be taken into account when designing safe work methods. For instance, an aircraft pilot, while bending to turn on a switch, could accidentally by an involuntary reaction lean on the plane's control handle, resulting in the plane's beginning to lose altitude (Zarakovsky & Pavlov, 1987). Automatic reactions can also have a positive effect. For instance, a pain stimulus can be used as a signal of danger.

Vocal actions and motions are a special type of motor activity. Vocal actions are usually associated with cognitive functions. So-called internal speech is well known and reveals itself through changes in electrophysiological characteristics of the throat.

Russian ergonomists prefer a level of task description that is much more detailed than that of Western ergonomists. They would prefer a more detailed step-by-step description such as one finds in an operational sequence diagram (OSD) (see Meister, 1985), although even an analytic technique like the OSD does not adequately describe mental operations or motions. However, it is more important to Russian scientists that the units of analysis used in different symbolical models very often do not allow them to describe the structure of activity.

When examining an operator's activity in a manual or semiautomatic system, Russian ergonomists can describe that activity at various levels of detail or decomposition. They start the analysis at a macrolevel by analyzing the goals of the activity and the motives of the operator. At a more detailed level, they switch to the analysis of the content of actions made during task performance. Following this, they analyze individual actions from the perspective of the operations or function blocks (i.e., the constructs) that are included in the actions. This often results in a highly detailed analysis.

The requirement for very detailed level of description stems, one can speculate, from the Russian interest in describing the structure of activity. So, for example, depending on the operator's position and type of control she or he uses, the operator will employ a different method of motor action implementation; that is, a given action may contain different motions. The same method of decomposition is applied to the mental (indeed to all) components of activity.

1.5.2 Mental Components

The description of the mental components of work activity is much more complex than that of motor components, because the former are unobservable. Nevertheless, the requirements for their description are much the same as those for motor actions. Usually mental components are described verbally, which ensures some ambiguity, because the mental activity does not possess explicit external referents. Each verbal term denoting a specific mental operation should possess distinctly individual attributes, but of course, there is overlap among them.

Usually, in cognitive psychology, mental actions are not extracted. Scientists extract only mental processes (short-term memory, long-term memory, thinking, making decisions, etc.). However, these units of analysis are not sufficient for analyzing the structure of solving problems by operators. While solving particular problems, mental processes can be presented as a set of mental actions (Leont'ev, 1977). In this case, mental actions are organized in certain systems that vary according to the task: Psychic processes can be the same for different tasks or problems. However, the content of mental actions or operations can be different. This is why the notion of mental actions or mental operations is useful in describing the structure of an operator's mental activity during task performance.

Mental actions are classified by the type of dominant psychic process at any given moment (Bedny & Zelenin, 1988; Dguravlev, 1981; Zarakovsky & Pavlov, 1987). It is possible to distinguish sensory, perceptual, thinking, and mnemonic (memory) actions. Psychic operations at the microlevel (similar to motor motions or operations) are characterized by the relative homogeneity of their structure and by an absence of a conscious goal, although, of course, mental actions that incorporate several mental operations are associated with conscious goals.

Mental actions can be represented as a sequence of stages of processing information that function over a very short time frame. Such stages in the transformation of information have been termed function blocks. Graphically each function block is illustrated in the shape of a box; the mental action as a whole is shown as a sequence of such boxes. For example, in the study of mnemonic and perceptual actions, we can deduce one function block as "iconic memory," another as "scanning block," and so on. The iconic block in the visual system serves the function of short-term storing of sensory information. This gives time for the scanning mechanism to select the necessary information from the point of view of the goal.

The function block is the smallest unit of analysis to be identified by cognitive psychological research. The analysis of functions of information transformation enables one to describe a detailed differentiated structure of mental activity and not simply a single homogeneous system of mental processes.

Zarakovsky and Pavlov (1987) distinguished between direct connection actions and transformational actions. Direct connection mental actions proceed without there being distinct differentiated steps in the process and, for example, immediate recognition and identification of signals, images, symbols, and so forth. Transformational mental actions involve more deliberate examination and analysis of stimuli.

Transformational mental actions also have two subclasses, reproductive and productive: The first involves actions associated with the formation or transformation of images and logical thinking; productive mental actions involve creative processes.

The two types of mental actions require different levels of attention. Highly automatized actions of direct connection demand minimal attention, whereas transformational actions require a higher level of attention. The transition from one action to another also requires a shift of attention, specifically that of goal awareness and conscious evaluation of action.

In the learning process some actions become automated and lose their conscious goal. As a result, these turn into psychic operations, relatively homogeneous, and become direct connection actions.

Because mental actions have a short duration they often are called mental operations. In the future, terms such as *mental actions* and *mental operations* are used as synonymously.

In the context of activity, a mental action can be defined in the following ways:

1. The first group of actions are those of direct connections. They involve (a) sensory actions, which permit detection of a signal from noise, discrimination of similar stimuli, or require a decision about the signals at the threshold level; (b) simultaneous perceptual actions, which are involved in perceiving clearly distinguished stimuli, are well known to the operator, and require only immediate recognition; and (c) involuntary memorization actions, which are involved in the process of involuntary memorization without significant mental effort.

2. The next group of actions consists of transforming actions. They involve: (a) successive perceptual actions, which are involved in the interpretation of information from unfamiliar stimuli, require the creation of a perceptual image of a new object, and are involved in more deliberate examination and analysis of stimuli (during skill acquisition, one can observe the transition from successive to simultaneous perceptual actions); (b) if, after receiving information, the operator must determine which actions are needed based on a logical analysis of the situation, such acts can be related to decision-making acts on a verbal thinking level; (c) recording actions, which occur when the operator transforms one kind of information into another; (d) categorization actions, which include processing information

that results in the division of some signals into a series of separate subsets, categories, or classes (these are also thinking actions); (e) deductive actions, which can be considered thinking actions when a person applies a general rule for making a conclusion about newly acquired knowledge from existing information (mathematical operations can also be related to thinking actions); and (f) imaginative actions, which involve manipulation of different images and are actions based on a perceptual process and memory and those involved in more creative thought processes.

3. Yet another kind of mental actions are mnemonic ones. Their major purpose in the work process is to extract information from long-term memory, store necessary information, and keep information in the working memory.

4. The last class of mental actions are creative actions. They include psychic acts or operations that allow the acquisition of new knowledge from old knowledge in a logical way. Creative actions are fundamentally different from reproductive actions. Reproductive actions are connected with convergent thinking processes. Creative actions are productive actions connected with divergent thinking.

Any actions have their duration in time. This is why chronometric analysis is very important to the study of mental and motor actions. The Russian emphasis on chronometric analysis is peculiar to them, although others working in cognitive psychology (e.g., Donders, 1862; Sterenberg, 1969) have utilized this type of measurement in their experiments. To Russian ergonomists, one of the drawbacks of cognitive psychology is the assumption that a person reacts to any situation with maximal speed. Therefore, when conducting chronometric measurements, it is necessary to study not only cases when a human is asked to respond as fast as possible, but also to determine the duration of individual mental elements carried out with a pace corresponding to that in real-world activity. This is another illustration of the Russian emphasis on very detailed measurement of molecular activity elements.

Another method of determining the duration of mental action involves electrophysiological methods to measure the duration of nervous impulses. When used as chronometric standards, these units of analysis can be employed for determining the duration of larger psychological units.

All actions can be classified based on a preliminary cognitive analysis of the task. Russian theorists recognize the commonly accepted division of psychological processes into perception, memory, thinking, and so forth. However, this cognitive approach is considered insufficient. As was discussed earlier, the same psychic processes can be involved in different tasks. This requires studying the structure of an activity, which is a system of actions involved in task performance that are logically organized in time and space. From this it follows that the major criteria of classifying mental actions are psychic processes dominant in a particular moment and intermittent goals achieved by particular actions.

The purpose of classifying actions is to present an activity as a structure with a systemic organization. All actions are organized as a system because of the existence of a general goal of activity. Any changes in individual action can immediately influence other actions. A person continually evaluates his or her own activity from the viewpoint of the goal and conditions under which he or she performs the activity.

Let us consider how we can decompose activity into hierarchically ordered units of analysis. Subtask as units of analysis can be used during the decomposition of activity. In the algorithmic description of task, a subtask is called a "member of an algorithm." A member of an algorithm is composed of one to three homogenous, tightly interconnected actions. This is considered in more detail later. For now, though, an example of a subtask or member of an algorithm might include the following: "Move right arm, take a lever, and place in a particular position." Here we have two motor actions—first, move arm and grasp lever, and second, move lever into particular position. Together these constitute a subtask or a member of an algorithm, called an *operator*.

Now let us consider an example of how we can decompose a subtask into discrete units. Suppose a worker performs the following task: If a red bulb and white bulb are both lit, a worker should take some part from a right-hand bin and install it into the air-operated clamping device. If both a green bulb and white bulb are lit, the worker should take the part from the left-hand bin, and install it into the same clamping device. The method of decomposition of the subtask is presented in Table 1.1.

1.6 MOTIVATION AND GOAL FORMATION

1.6.1 Motives

Regulation of behavior, by which is meant maintaining behavior within parameters directed to goal achievement, can be performed in two ways. In one the regulation can be carried out by external stimulation, as when pretask instructions have been given or during task performance stimuli are presented that guide performance. For example, an aural alarm is sounded in a nuclear control station when an equipment component is exceeding limits; this is a signal to the reactor operator to take a specified highly automated action. This type of regulation is essentially reactive.

Regulation of behavior can also be accomplished by goal direction. In the well-trained operator, the need to achieve a specified goal requires certain actions; the goal sets limits that the individual tries not to exceed.

TABLE 1.1
Decomposition of Activity During Task Performance

Subtask (Operator)	Actions	Operations (or Motions)
Look at one bulb, then another	1. Look at bulb #1 2. Move eyes 3. Look at bulb #2	1. The same 2. The same 3. The same
	1. Decide to choose left bin or right bin	1. The same
Grasp part and install in air-operated clamping device	1. Grasp part	1. Move right arm to part 2. Grasp part
	2. Install part in clamping device	1. Move part to clamping device 2. Align part with clamps 3. Insert part into fixed secure position
	3. Clamp part into clamping device	1. Move left arm to the handle of air-operated clamping device 2. Grasp handle 3. Turn handle

Obviously, in both cases, but even more so in the latter, motivation is essential. Motivation is considered a source of energy to drive the activity. The theory of "internal energy" as a driving force for activity was suggested by Freud (1920/1960), but probably far antedates him. The concept that it is possible for an individual to insert energy into his or her thinking is critically important to Russian ergonomists.

One must distinguish between an individual motive and motivation in general. Motivation has a broader meaning and includes a hierarchy of individual motives, any of which may be conscious, semiconscious, or even unconscious. A person's motives can be divided into two groups: sense formative and situational (Leont'ev, 1971). Sense formative motives are relatively stable and determine a person's general motivational direction. Situational motives are connected with immediate ongoing activity and the solving of specific tasks. As a result, situational motives are more flexible. The sense formative motive is connected with personality and may be significant in the selection of people for different jobs. The situational motive is more involved with task performance and is, therefore, more significant for ergonomics. The content of situational motives, their hierarchical organization, and their relative weight can be changed, depending on the character of the tasks to be solved, the temporal stages of task performance, and informational feedback about task solution. Motives can be needs, attractions, sets, desires, or the like. Motivation is always connected with emotions (Merlin, 1973). There is no clear-cut distinction between these notions, and it is an unresolved issue in psychology (Hilgard, R. L. Atkinson, & R. C. Atkinson, 1979). However, in the theory of activity, this notion is clearly distinguished.

Reykovski (1979) wrote that the motivation is emotions plus the directness of action to a specific goal. Bayton and Conley (1957) demonstrated the point by having subjects perform motor tasks. The feedback that was provided was manipulated by the experimenter who provided false information to the subjects. They found that if the subjects received positive feedback over a long time period, performance progressively degraded. When the positive feedback was changed to negative (failure) feedback, after a short time performance improved. Further negative feedback reduced performance. It is apparent from this study that the order in which success and failure are experienced is also important. Obviously, emotional-motivational components of activity are important for task performance. The theory of self-regulation specifies that feedback effects are both cognitive and motivational.

Reykovski (1979) found similar results. In his experiments, subjects performed perceptual-motor and cognitive tasks. They were informed that if they performed incorrectly, they would be punished with electric shock, although in reality shock was not used. Consequently, subjects perceived their performance as successful. Again, after repetitive successful performances, degradation occurred, followed by improved performance. Subject debriefings revealed that after a series of successful performances, they felt positive emotions were relaxed, had a lower motivation to perform, and had a feeling of tiredness. However, this emotional state was of short duration and performance again improved.

Reykovski (1979) came to the following conclusions. Although usually positive emotions are linked to success and negative emotions are associated with performance degradation, nevertheless, in some situations positive emotions can degrade performance. This may occur when success feedback is conveyed with very strong emotions. The same thing can be observed when success requires a great effort, after which relaxation is necessary.

In some cases, negative emotions can improve performance. This may occur when failure follows a sequence of successful performances. However, this improvement has a short duration.

As a whole, emotions, from an activity point of view, cannot be considered as only disorganizational components. Emotions perform an important role in the regulation of activity. They are important components of regulation activity. Leeper (1943) also wrote that emotions perform an important adaptive function in human behavior, and cannot be considered only negative factors.

Motives are associate with various needs (Maslow, 1954), which can be categorized in a hierarchy. The most basic needs are associated with the self: self-preservation, food and shelter, perhaps love of family. At a somewhat less personal level is the need for self-recognition and feelings of worth, and at a more abstract level, concern for the society in which one functions.

Needs, however, do not directly generate activity; they merely create dispositions directed toward their satisfaction. Activity is derived only from those cases in which the person consciously imagines specific objects that can satisfy his or her needs. The image then becomes the goal.

The factors that determine a person's effort to reach the goal are called motives. When activity is discussed, one must presuppose a motive directed toward achieving a specific goal. Simonov (1982) stated that emotion is a reflection by the brain of a specific needs quality and achievement probability. Based on that, he developed a need informational theory of emotions. In this theory, emotions (E) are determined by the following formula:

$$E = f[P(Vn - Vc)] \qquad (1.1)$$

where P is the force and quality of actual need; Vn is the information about means, resources, and time that is required for the satisfaction of needs; Vc is the information about means, resources, and time that the individual possesses at the present moment; and Vn − Vc is the estimate of the capability of satisfying needs. The greater the difference between Vn and Vc, the less an individual can satisfy his or her needs and the more his or her reaction will be emotional. The purpose of this equation is a demonstration of relationship between different variables.

Zarakovsky and Pavlov (1987) theorized about inducing and regulatory components of emotions. Inducing components of emotions have only one function: to direct the person to achieve a specific goal. The regulatory components of emotions have four functions: switching, reinforcing, compensation, and organization. If an individual has several concurrent motives, the switching function enables him or her to concentrate on the behavior most closely related to the goal of the activity that has more subjective value for that individual. The reinforcing function provides rewards and, thus, reinforces desired behaviors. For example, reward increases response, punishment decreases response. The compensation function enables emotion to be transformed into an increasing level of motivation. For instance, time constraints increase the emotion intensity of the operator and motivates him or her to mobilize effort to achieve the desired goal. It increases the speed of task performance. The organizing function of emotion promotes recognition of any conscious discrepancy between existing and required methods of achieving a goal, and, thus, tends to a more correct organization of activity. This function is connected with the selection of correct strategies of activity depending on the emotional state of the person and the requirements of task.

When Russians study the emotional-motivational components of activity, the concept of will becomes very important. The will contains mechanisms that sustain activity under conditions in which obstacles to goal achievement appear. The willing process can appear when a conflict exists between motives in the operator. In such cases any change in the activity can be mediated by a conscious act of will.

Heckhausen (1986) specified three orientations toward studying motivation. The first considers the dissimilarity of people's behavior based on their individual differences. In this case, the major ergonomics problem is the selection of personnel for specialized work. The second orientation considers differences in people's behavior produced by changing situational factors, such as experience. The third orientation melds the first two, and explains variations in behavior by complicated relationships between idiosyncratic and situational features that influence each other.

Magnusson and Endler (1977) formulated four basic principles:

1. Actual behavior is a function of a continuous multidirected process of interaction or backward interconnection between an individual and a situation in which she or he is involved.
2. In this situation, an individual appears as an active player who pursues his or her own goals.
3. The significant cause of behavior is the interaction of cognitive and motivational factors.
4. From the situational side of the interaction, the most crucial cause of behavior is the psychological meaning that the situation has for the individual.

The task situation cannot, therefore, be considered as a completely objective stimulus, because the individual always interacts with the situation, putting his or her own interpretation on the stimuli it provides. From those multiple stimuli, the individual will select only those informational items that are relevant to the task, the goal, and his or her interpretation of that task and goal (Heckhausen, 1986).

The more indeterminate the task situation, the more options are open to the individual and the less precisely specified the method of performing the task, the more the individual interacts with the situation, and the more active she or he becomes in the interpretation of its stimuli.

1.6.2 Goal Formation

How goals develop is one of the fundamental questions in the theory of activity. The relationship between goals and motives in human activity is quite complex. Russian psychology considers this problem from the standpoint of self-regulation, which is one of the major components of activity theory. Figure 1.6 presents the model of formation and acceptance of a goal developed by Bedny. This model is considered as a dynamic subsystem consisting of the following function blocks: (a) mechanism of orienting reflex, (b) assessment of the meaning of input information, (c) assessment of the sense of input information, (d) experience, (e) motive, and (f) goal.

The orientation reflex is largely physiological; it includes external movements such as the turning of eyes or head to the stimulus, altering the sensitivity of different sense organs, a change in blood pressure or heart rate, and changing the rate of breathing. At the same time, there appears to be some electrophysiological change in brain activity, perhaps in the cortex. Orientating reflexes appear to play an important role in the functioning mechanisms of involuntary attention. Specific kinds of neuron detectors of novelty were discovered in the brain's cortex (Sokolov, 1963). It was determined that in healthy people electrical activity of the frontal part of the brain significantly increased when they concentrated. However, patients with damaged frontal lobes exhibited a very short attention span.

The regularity of orientational reflexes as a mechanism of involuntary attention should be taken into consideration, if at all possible, during the

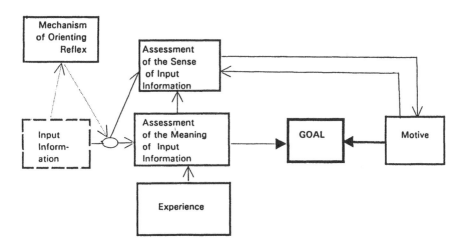

FIG. 1.6. Model of goal formation process.

design of displays. Sometimes displays must attract our involuntary attention process. In some cases, involuntary attention can distract an operator, which can cause errors to be made. A repeated stimulus weakens and loses its influence over the involuntary attention process, causing the operator to ignore it. If, after attracting our attention, information is evaluated as important, involuntary attention can be transformed into voluntary attention. It follows that the meaning of input information can be elicited directly when we talk about voluntary attention, or indirectly when mechanisms of orientational reflexes are involved. Input information can come not only from external sources but internally from memory.

The problem of meaning and sense is basic to the Russian theory of activity. This problem was first analyzed by Vygotsky (1956) but the authoritative formulation was provided by Leont'ev (1977).

Leont'ev (1977) differentiated objective "meaning" from subjective personal "sense." Through meaning, which is an essential component of human consciousness and is embodied in language, we apprehend different phenomena in the environment. Meaning emerges as a reality independent of words denoting objects. At the same time, it is reflected in individual consciousness as an image or concept, and the relationship between them. Meaning is a form of the presentation of reality in consciousness.

Meaning functions in social relations, and at the same time in individual consciousness in the form of perceived objects and images which, however, do not lose their objectivity. For example, the driver of an automobile is aware of the importance of his brakes, or the pilot of her landing gear. These notions are formed on the basis of experience gathered by many other drivers and pilots in the past, as well as from their own experience. However, together with the objective assessment of the role of a given object, there is also a subjective assessment derived from the influence of their needs and motives. For example, if the driver or pilot has ever been involved in an accident, this experience will undoubtedly affect the way in which they examine their instruments.

Obviously, commonly accepted objective meanings are translated into idiosyncratic impressions for each individual. Individual interpretation is fundamental to human consciousness (Leont'ev, 1957). This requires a translation of the objective meaning of a perception into an internal subjective sense; this is a necessity for explaining the goal-directed activity of individuals. Russian psychologists believe that classical conditioning cannot explain the voluntary goal-directed behavior of humans or animals. Conditioning effects can explain reactivity, but not activity. "Significance" and "nonsignificance" performs for human consciousness and human activity in a way that is analogous to "essential" and "nonessential" survival characteristics of animals (Bassin, 1973). Behavior and activity are shaped by criterion of significance, in accordance with the "sense" or the feeling of the individual.

Anything that is significant for one has sense. Specifically, the transition from meaning to sense, along with needs and motives, gives behavior a goal-directed character.

At the same time, sense is a theoretical construct. From a psychological point of view, sense unifies cognitive and emotionally evaluative components. In sense block they represent a complex unity of these elements, because emotional behavior has expressive features undirected toward any goal.

On the other hand, motivation is distinctive for its essential link with goal directedness. The defining feature of motivation consists in its function of providing and maintaining goal directedness. During self-regulation, the goal-directedness of behavior calls for two distinct function blocks, one that subsumes sense and another that subsumes the energy and goal-directed quality of the self-regulative system. In the model (see Fig. 1.6) this quality is represented as a vector "motive-goal" and designated by a broad arrow. To summarize, the function block "sense" refers to cognitive, emotionally evaluative components of activity, whereas the function block "motivation" determines directedness and energetics for achieving a specific goal. These two function blocks are intimately interconnected through loop structures and together comprise the motivational subsystem of self-regulation (see Fig. 1.6).

Thus, sense is in some way connected with intentions—any desire, plan, aspiration, purpose, or belief that is oriented toward some goal. An intention is less concrete (more diffuse) than a motive, less conscious to the person. Although an intention to perform anticipates the task situation, that intention, which persists during task performance, becomes part of the task conditions.

A motive is more action oriented than an intention, because it is more concretely directed behavior than is the intention. Russian psychologists attribute the emotions accompanying the execution of a particular action not to the action per se, but to the individual's interpretation of the action, which produces a sense that informs the action. For example, the driver after an accident caused by malfunctioning brakes develops a very different sense of the notion "brakes."

Sense is the tension and emotional coloring that the action has for the subject. Sense reflects the relationship of people to ongoing situations. Intentional components have special meaning in the beginning of the formation of goal and increase a person's sensitivity to anything that is connected with the goal. This intentional process, as was indicated earlier, anticipates the stimulus situation, but at the same time motives develop during the activity and are directed to achieve a conscious goal. Intentional components are determined by personal traits, as well as temporary psychological states. Some motives have a more temporary character because they developed through ongoing activity and become apparent only through that activity.

Meaning is a more lexical construct. Sense is associated with subjective evaluation of significance of activity; that is, sense determines how events, people, ideas, and so on, assume relevance for one. This point of view is congruent with Kahneman's (1973) theory of attention. Kahneman wrote that errors appearing during the solution of complex problems are associated not only with the limitations of attentional resources but also with underestimation of the significance of the task. As stated previously, personal motives according to Leont'ev (1971) can be divided into sense formative, which determines the general direction of motivation in a relatively stable manner, and situational, which are connected with an immediate situation, an ongoing stimulus presentation, and/or an ongoing problem-solving task. Thus, a sense of activity is more evaluative. At the same time, the motivation that follows from the sense is more action oriented. The sense evaluation of information determines its significance or relevance for us. Motivation determines the intensity of our induction to the goal. Information can be understandable for people, but their reactions to that information may not follow if the information lacks significance. I can, for example, read and understand that the temperature in Moscow is 60°F but because I live in San Francisco and am not planning to visit Moscow, the information lacks significance and, therefore, means nothing to me.

On the other hand, significant information may fail to motivate goal formation and goal achievement, if there is lack of will in the individual; one might consider this a pathological condition. In principle, all these processes that have been discussed are interconnected in a coherent motivational system describing human activity. The sense aspect of motivation and its inducing aspects may be in conflict (Leont'ev, 1977). Personal significance, value, valence, or utility of the goal bears on the affective (emotional) evaluation of the task or goal. One may consciously be aware of the importance of a task for another, can understand that the task is complex and solving this task is very prestigious for others and for oneself, and that one has enough ability to solve this task, but for the individual (I personally) this task may not be significant or valuable. In this situation one will not be motivated to solve this task.

From the self-regulative point of view, these two aspects of motivation are treated as distinct functions: "assessment of the sense of the task" and "formation of the level of motivation" (see Fig. 1.6).

The notion of meaning and sense can also be characterized in the following way. Meaning has two important characteristics: level of complexity and informativeness. To what extent is the meaning of the task or situation informative to a person? The level of complexity reflects the other aspects of meaning connected with a situation; how comprehensible is the meaning of the task to a person? Sense also has two characteristics: the values solving a task has for a person, and the positive emotions accompanying the achievement of a goal. This type of situation has positive significance. Goal achievement and task solution threatened by obstacles, such as danger, are accompanied by a negative emotional state. This situation has negative significance. The foregoing is described graphically in Fig. 1.7. This short description of this concept underscores the number of functions important to the formulation of a goal.

In conclusion, the function block called "experience" (see Fig. 1.6) must be discussed. Our past experiences, attitudes, skills, and knowledge about the world influence what we perceive, how we interpret input information. According to the concept of activity, and also cognitive psychology, uncertainty can be resolved only by discovering the interaction with something that is already known. During the thinking process, new objects or phenomena evolve into new connections and relations with something already known. As a result, from two objectively similar stimulus situations, the operator can extract alternative meanings and understanding of these situations.

This process is called "analysis by synthesis" in Russian psychology because the extraction (analysis) of new features in objects can be performed by relating (synthesizing) a new object with another, well-known one. Something unfamiliar can be comprehended and interpreted by being associated with what is already known (Rubinshtein, 1958). This process is presented in Fig. 1.7 as the interaction of "input information," which is actively selected by the subject, with "assessment of the meaning of input information," and "experience."

As can be seen from Fig. 1.6, the interaction of the past experience and new input information results in the formulation of the meaning of the new input information. Simultaneously, the assessment of the sense of information and possible action is being formed.

Information about the external situation is connected with the present need that dominates present motivation, as well as being associated with

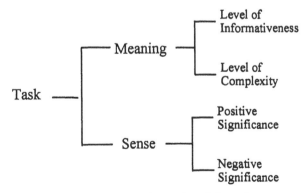

FIG. 1.7. Meaning and sense in task formation.

set and instructions (momentary intentions). The individual's motivation that precedes the reception of task information belongs to the dominant (primary) motivation. This primary motivation is connected with the function "assessment of the sense of task." This function influences the function "formation of the level of motivation" (secondary, action-oriented motivation). Because of feedback, the activity is tuned to goal formation and goal directedness.

The analyses of the model presented in Fig. 1.6 shows that an environmental situation can trigger the goal formation process. The specifics of this process depend on the inner state of the subject and the features of the situation. This means that the goal formation process can be voluntary and conscious or unconscious. Goal in this model emerges as a conscious determinate that predicts specific selection of information and method of execution. A goal also governs the other functions.

"Goal of activity" is a conscious image or logical representation of desired future results. At any given moment one can be conscious of some components of the goal.

We also need to distinguish the "overall, terminal, goal" from "partial or intermediate goals." Any element of activity that is directed to achieve a partial goal is categorized as "action." An image of a future outcome does not necessarily constitute a goal. Rather, a future result emerges as a goal only when it is joined with motivation. For example, a student may know that an excellent grade will result from 4 or 5 hours of preparation for an exam. Such knowledge does not create a goal unless she or he has some motivation to achieve good grades. Only when an image of a desired future joins with motivation, as when the student actively prepares for the exam, does this future result transfer into "goal of activity."

It is also possible to have different relationships between the same goal and different motives. In these cases goal-directed activity acquires an ambivalent character. If the goal appears to a person to be undesirable, motivation will emerge as avoidance of the goal. Consciousness by the individual of negative and socially unacceptable motivation may cause the person to formulate a new goal, thereby suppressing the inappropriate motive.

The relation "motive-goal" also provides direction to the "self-regulative process." When we try to understand the self-regulative process, we need to determine precisely what the goal of the activity is. Studies show that different individuals may have an entirely different understanding of a goal, even if objectively identical instructions are given (Bedny, 1987; Konopkin, 1980). As a result of that we distinguish "subjective" and "objective" understanding of the goal.

Although in the theory of activity, informational (including goal) and energetical components (motives) are distinguishable (informational components being cognitive in nature, whereas energetical components are emotional motivated), these components are tightly interconnected.

There are several types of informational-energetics interconnections in activity (Vekker & Paley, 1971). The first type was shown in the psycho-physiological study of Vekker and Paley where it was discovered that increasing the intensity of external stimuli results in increasing the sense experience. Another type of interconnection is related to the reticular activating system of the brain, which plays an important role in controlling the state of arousal and awareness. A third group of interconnections is linked to the emotional-motivational components of activity. This fact is often overlooked in cognitive psychology, particularly in the area of human information processing. For example, the model of attention, developed by Lindsay and Norman (1992), ignores the energetics components of an activity.

According to Russian activity theory, the goal is connected with motives and creates the vector "motive-goal" that lends activity a goal-directed character.

The more significant a goal, the more value it has for the person, the more the person is motivated to reach this goal. If a situation has negative meaning for a person, they will be motivated to avoid the situation. Russian and Western psychologists have somewhat different conceptions of the attributes the goal possesses. Western theorists (Locke & Lathman, 1984; Pervin, 1989) postulate that the goal has both cognitive and affective features. Pervin stated that the goal can be weak or intense and Locke and Lathman (1990) considered the goal as a motivational component of behavior. Presumably, the more intense the goal, the more one strives to reach it. Hence, the goal "pulls" the activity.

Differences in the understanding of the goal's position in the structure of activity are seen in the work of Kleinback and Schmidt (1990). They described the volitional process and considered the goal as a source of inducing behavior (see Fig. 1.8). The direction of the volitional process goes from the goal to behavior. In this work, the goal and motive are also not distinguished.

Russians consider it a mistake not to differentiate motive and goal. According to activity theory, the goal is an information component only, and does not have the attribute of intensity. The goal can be precise and detailed or unclear and general, but not intensive. From an activity point of view, the more intense the motive is, the more a person will expend his or her efforts to reach this goal (see Fig. 1.9). Energetics and informational components of activity are interconnected, but they are not similar. The difference between informational and energetics components of activity has for Russian psychologists both theoretical and practical meanings.

An analysis of the data gathered in psychology finds four approaches to understanding the goal (Tikhomirov, 1984):

1. The goal is not a scientific notion. For example, Skinner (1974) described a person's behavior with the following terms: *stimulus, reac-*

tion, and *reinforcement.* Here the goal is not considered as a psychological concept.

2. The goal is the end state toward which the motivated behavior is directed and by which it is completed. This corresponds to the direction in psychology that studies purposeful behavior (Tolman, 1932). This also corresponds to the self-regulation concept of reflex suggested by Anokhin (1962).

3. The goal is considered the physical location of an object or as a formal description of the final situation, which can be achieved during the functioning of technical or biological systems. This corresponds to the cybernetic understanding of the goal.

4. The goal is considered a conscious mental representation of a future result connected with a motive. Only this last approach is connected with understanding the goal in the theory of activity.

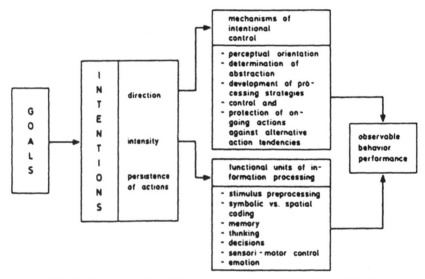

FIG. 1.8. Scheme of the volitional process. From Kleinback and Schmidt (1990). The translation of work motivation into performance (Fig. 2.2, p. 28). In V. Kleinback, H.-H. Quast, H. Thierry, & H. Hacker (Eds.), *Work Motivation* (pp. 27–44). Hillsdale, NJ: Lawrence Erlbaum Associates. Reproduced with permission of Lawrence Erlbaum Associates.

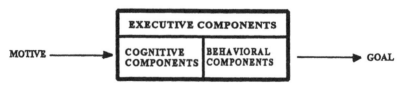

FIG. 1.9. Relationship between goal and motive according to activity theory.

Goal, meaning, and sense of task have important implications for the evaluation of task performance. That significance and sense are related, thereby influencing work performance, is well established empirically both in Russian and American research. For example, Blackwell (1952) showed that a money reward enhanced visual discrimination and lowered the visual threshold. Kotik (1974, 1987) showed that significance affects the reliability of work activity.

The studies by Krinchik and Risakova (1965) and Leont'ev and Krinchik (1964) demonstrated that significance influences reaction time. The information value of the stimuli varied from 1 to 3 bits of information. Stimuli were given significance through monetary rewards, specific instruction, and electric shock. The results are shown in Fig. 1.10. The dashed line shows the dependence of the time reaction on the signals that carry significant information (J), the more information the slower the reaction time. As can be seen, the significance factor considerably lowers the reaction time. Although the linear relationship in Hicks' law is preserved, the reaction time to significant signals increases insignificantly with the increase in the number of bits of information. This is apparent when we compare the reaction time to the

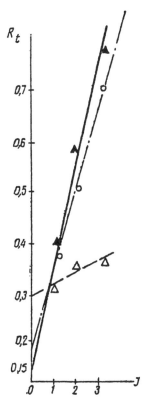

FIG. 1.10. The dependence of the reactive time on the amount and significance of information processed. From Leont'ev and Krinchik (1964). J—amount of information processed in bits; R_t—reaction time; Δ—reaction time for significant information; ▲—reaction time for nonsignificant information; O—reaction time after disregarding significance.

significant signals with the reaction time to nonsignificant signals (solid thick line). In a second experiment, subjects were told to ignore the significance of the signal. The results are shown by the dash point line. Reaction time increased but insignificantly less than that of the first experiment.

Similarly, Neumann and Timpe (1975) provided data about reaction time in relation to significant and nonsignificant displays, where the total number of displays is not more than 24 (see Fig. 1.11). Obviously, there is a significant difference in reaction time between significant and insignificant displays at all ratios.

The strength of a motive obviously influences skill acquisition. When task performance is highly skilled, it increased the motivation to perform. The negative side of this is that under strong motivation, undesirable events seem less probable, which can increase risky behavior.

J. Atkinson (1957) concluded that people select tasks at different levels of difficulty based on motivation: aspiration to success and/or avoidance of failure. The relationship between these two components has an important meaning for work safety. For example, if workers neglect devices necessary for work safety, it may be that they have high aspirations to reach a goal and,

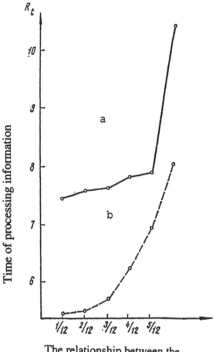

The relationship between the significant and nonsignificant indicators

FIG. 1.11. The dependency of the time of processing information on the significant and insignificant indicators. From Neuman and Timpe (1975).

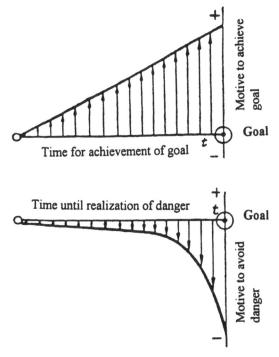

FIG. 1.12. Changes in the strength of motivation in the process of approaching the goal conflicts between the motives. From Kotik (1987). a—changes in the strength of the motivation in the process of achieving the goal; b—changes in the strength of the motivation as the danger level increases.

therefore, neglect the possibility of failure. Rules for safety sometimes restrict the worker's behavior. As a result, they face an ambiguous situation. The achievement of high productivity can oppose the requirements of safety. In this case, the workers' strategy depends on which function is dominant. Very often people ignore high-risk, low-probability possibilities for more probable and valuable ones. The process of reaching a goal can change the relationship between the motives of achieving the goal and the avoidance of danger. As shown in Fig. 1.12, if a goal is valuable for a person and connected with danger, the strength of profit motives during the stages of achieving the goal are significantly increased.

1.7 CONCLUDING COMMENTS

Almost all the fundamental theoretical concepts of Russian psychology and ergonomics can be assimilated to American theory. Although differing from the latter in terms of emphasis, primarily, the Russian conceptual structure is quite complex and sophisticated.

Russian theory is highly syncretistic, meaning that it has been heavily influenced by Western, particularly American, concepts. There is, however, some difficulty in translating these concepts into American equivalents because their terms are very different.

As in American ergonomics, the task is a fundamental component. A significant aspect of the Russian treatment of the task is that the latter is viewed in part as a problem-solving process in which the solution is not necessarily foreseen at the start. This concept has significant implications for understanding the task performance of personnel in the increasingly automated systems of the future.

Russian ergonomists emphasize the importance in task performance of goals and motives. Russian specialists reject theories that are based on behavioristic principles. They pay more attention to American cognitive psychology and particularly those directions that are called human information processing. They try to adapt these data in the theory of activity. For example, when they study human information processing, they introduce the notion of cognitive actions. When they study cognitive processes, they pay attention to how people accept or promote different goals of activity. They emphasize regulative functions of cognitive processes.

The concept of goal has also become more important to American psychologists in recent years. Many scientists pay attention to the goal's importance in regulating behavior. However, the notion of goal in Russian psychology is significantly different from that in American psychology. In American psychology, the goal is considered a motivational phenomenon. In Russian psychology, the goal is also only a cognitive component of activity. The motive, on the other hand, is considered an energetics component.

Because of the existence of goal, behavior is no longer considered a system of reaction to different stimuli. Behavior includes flexible and adaptive elements, which are called actions. Actions are not only external or motor, but also internal or mental. There is a complicated relationship between these two types of actions. Internal cognitive processes are considered to be an activity that consists of different actions directed toward achieving particular goals. There is an opportunity for mutual transition from external activity to internal activity, and vice versa, as a result of internalization and externalization. Because of different actions, an activity acquires a complicated dynamic structure. All elements of an activity are tightly interconnected and influence each other. From this, the requirement of using the systemic structural method of study follows. As can be seen, an activity is a process. However, this process consists of a sequence of a hierarchically organized structure of elements (actions and operations).

Some points need to be made about the emphasis in Russian ergonomics thinking about self-regulation. The self-regulation theory blends all aspects of the individual experience: cognitive, emotional, motivational, and behavioral are considered as a conscious, striving process. Self-regulation cannot, therefore, be considered simply as a homeostatic mechanism (i.e., a physiological process translated into psychological terms) but involves much more. Psychological self-regulation does not seek to maintain a constant internal state but creates a goal (or even more than one goal) and develops a strategy to achieve this goal.

Self-Regulation of Activity

2.1 OVERVIEW AND INTRODUCTION

The previous chapter dealt with general principles of the theory of activity and introduced the concept of self-regulation. This chapter applies the principles of self-regulation to behavior and motivation.

The concept of self-regulation is widely used in psychology, but often incorrectly. It has no real meaning unless it is tied to the goal of activity.

Motor skills are performed in three consecutive stages—an initiation phase, an intermediate phase, and a conclusion phase—which correspond to stages of learning. In the first stage, vision is most important, whereas in subsequent stages proprioception is primary. Motor actions can be decomposed into small micromovements, which are cognitive and exploratory in nature. Movement has three functions: program formation, realization, and control and correction. These lead to the development of action "images." The self-regulatory process involves control and correction of movement based on feedback. The same motor act has two fundamentally different types of regulation: programmed and afferentational. The relation between the two types of regulation depends on the amplitude, precision, speed, stability, and feedback of the movement.

Various regulation models of activity, in which cognitive elements exist, are described and discussed. An important mechanism of self-regulation is something called the *subjective model of activity conditions*. Studies indicate that humans are capable of self-programming or self-regulation of activity independent of instructions. They can do this by developing various strategies in accomplishing the goal. The subjective model of activity conditions referred to previously involves prediction of what and when stimuli will appear. When the prediction is difficult to develop, productivity is reduced. When performance criteria of success are not provided to the individual, they will develop these criteria in the course of the task performance. Even when objective criteria exist, people will develop their own, and objective and subjective criteria may not agree.

Sense of task plays an important role in goal formation. Task significance produces emotional involvement with the task. The precision with which signals are detected depends in part on their significance to the operator.

Studies show that people develop various strategies during task performance. An example is the speed-accuracy trade-off when one can maximize either speed or accuracy in performing the task.

The notion of feedback outside the overall context of self-regulation is not very fruitful. Cognitive psychology has failed to develop an effective concept of self-regulation. Self-regulation is based in part on past experience, evaluation of the present situation, and the anticipated future. Strategy is fundamental to self-regulation and a major problem is the requirement for continuing reconsideration of strategies when internal and external conditions change.

A model of self-regulation consisting of 18 parameters is described. Each of these parameters is interconnected with all others, from which it appears that the self-regulation process is much more complex than alternative motivational models.

The self-regulation model is what was called by Hoyos (1987) a process model. All the elements of the self-regulation model can be found in alternative models, but the interpretation of the interactive mechanisms differs.

2.2 THE REGULATION OF MOVEMENTS
AND MOTOR ACTIONS

The concept of self-regulation is widely used in contemporary psychology. However, very often this notion is not used correctly. For example, Bandura (1977) described the process of self-regulation as four consecutive steps: observing, judging, rewarding, and regulating oneself. This linear scheme of the process of self-regulation is used in current psychological textbooks (e.g., Gibson & Chandler, 1988). However, this process is better described as a recursive, loop structure with definable function blocks, that have both forward and backward interconnections.

Self-regulation as currently used in psychology has been made synonymous with such notions as willpower, ego strength, and volition (Kuhl, 1992). Kanfer (1996) considered self-regulation as a motivational mechanism which sustains attention and effort over time in the face of difficulty and failures. This understanding of self-regulation is erroneous. Self-regulation is a process of not only living beings but nonliving things. Nonliving things lack consciousness, will, and motivation. Examples of this are the most complex cybernetic and technical devices such as missals and robots, etc. Self-regulation is not a psychological notion, rather it's a cybernetic notion. Informational processes hold foremost importance in self-regulation. One fails to see what is gained by trying to translate traditional psychological notions into modern cybernetic idioms under the rubric of "self-regulation." All psychological functioning and behavior can be construed as a self-regulation process. The concept of self-regulation becomes meaningful only when the self-regulation model is developed. One defines it in terms of a function block model. In the absence of the specification of function blocks and feedback mechanisms, the use of the concept of self-regulation is little more than a metaphor. An example of a model that uses the concept correctly is G. A. Miller et al.'s (1960) TOTE model.

Any self-regulative system is defined in the first instance by its goal orientation (Ackoff, 1980). According to Wiener and Rosenblueth (1950), a system is goal oriented if it continues to pursue the same goals through variable behavior, as environmental conditions change. This system can also change its goals and methods of their accomplishment as a result of processing other system functions.

This permits a more precise definition of self-regulation. Self-regulation is an influence on a system that derives from the system in order to correct its behavior or activity. It is necessary to differentiate between self-regulation and regulation. Regulation is an external influence on the system; self-regulation is internal. Self-regulation is an intrinsic self-organizing tendency of the system itself.

Bernshtein (1947) considered self-regulation with regard to voluntary movements and actions. According to Bernshtein, movement of different segments of the body can be considered the result of the collective influence of both internal and external forces. The external force field acts outside of the human body. The internal force field operates within the body; it depends on the interaction between segments, muscles,and other internal organs.

Human movement is executed by various limbs, which can be thought of as multiple kinematical chains. These movements are executed within complicated force fields that are never static. Therefore, it is very difficult to coordinate a specific movement within these dynamic force fields. No individual's neural impulses can supply precise directions to the body segments by themselves. Moreover, the same neural impulses can result in different effects because of the changeability of external and internal force fields. Thus, one-to-one interconnection between central neural impulses and movements cannot exist. In this situation, coordination of movements is provided by the coordination of central neural impulses and dynamic phenomena with peripheral body segments. This occurs because of feed-

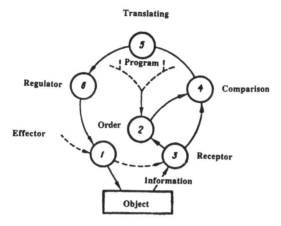

FIG. 2.1. Self-regulative model of human movement. From Bernshtein (1966).

back; the ongoing performance of the action and the state of the muscle enables one to introduce specific corrective impulses during the movement execution. Figure 2.1 presents a model governing human movement, which is derived from studies of the physiology of activity (Bernshtein, 1966). This model also has a loop structure as did the reflex model presented in chapter 1. Bernshtein's model has six basic components: (a) effector (motor) apparatus, the work of which is governed according to specific parameters, (b) order apparatus, which is based on a specific program to introduce required meaning control parameters, (c) receptor apparatus, which perceives the ongoing meaning of parameters and their stimuli, (d) comparison apparatus, which evaluates any discrepancy between the desired and actual parameters, (e) translating apparatus, which translates data from comparison apparatus into corrective impulses that are used by the regulator, and (f) regulator apparatus, which governs the effectors.

According to this model, the essence of coordination is overcoming superfluous degrees of freedom in joint movement. The secret of coordination is not expending neural impulses to depress reactive forces, but using them to perform actions. Acquiring motor skills can be described as consisting of three consecutive stages:

Stage 1: Initiation phase of movement in kinematical chains by using muscles to restrict superfluous degrees of freedom, with the intent of excluding reactive forces that interfere with desired movements. At this state of motor action, movement is very jerky.

Stage 2: Intermediate phase of movement with greater degree of freedom of movement than in the first stage. This stage minimizes the obstruction of reactive forces to the desired movement: Some obstructive forces are overcome by using rapid muscle impulses. At this stage, movement is executed more efficiently and effortlessly.

Stage 3: The conclusion phase of movement consists of the final acquisition of motor skill when the individual can use not only muscle forces but also reactive forces from different body segments to perform gracefully.

In this self-regulation process of movement, different senses are in operation. At the first stage, vision is the most important, whereas during the later stages, proprioception increases in importance and to some extent is substituted for vision. Proprioception is essential for an internal cycle of movement self-regulation, whereas external senses are essential for an external cycle of self-regulation. These two cycles interact, their relationship depending on the stage of motor skill acquisition. As skill increases, the internal cycle becomes paramount. At times the two cycles can be in opposition and their relationship influences the association between conscious and unconscious processes during skill acquisition. The internal cycle reflects unconscious processes, whereas the external one reflects conscious ones. Sometimes a person who consciously tries to control an acquired and hence, unconscious motor skill will deter its performance, or even lose it entirely if it is a complicated skill. This is commonly observed among gymnasts, for example, and also among nongymnasts. Walking down a flight of steps is ordinarily a highly automatized motor activity, but if, during the process, one stops to consider how one is doing it, the activity becomes disorganized, awkward, and slowed.

Kinesthetic stimuli within the body have a shorter response time (.1–.2 seconds) than response times to external stimuli. That is why, if someone is asked to operate a device as fast as they can, they should be allowed to use kinesthetic and touch stimuli as well as external (visual) stimuli.

The findings of Anokhin referred to in chapter 1 and those of Bernshtein have profound implications for the development of physiology and psychology, not only in the former Soviet Union, but also in other countries. During the development of the closed-loop theory of motor learning (J. A. Adams, 1968, 1987), Adams used Anokhin's and Bernshtein's ideas about acceptor of actions, ordered apparatus, and comparison apparatus, in addition to the neural model of the stimulus (Sokolov, 1960). Based on these ideas, Adams introduced the concepts of memory trace and perceptual trace, in which execution of movement generates a perceptual trace. Control of movement is produced by comparison between ongoing feedback of movement and the perceptual trace of that movement.

J. A. Adams (1968, 1987) hypothesized the existence of a special mechanism called a memory trace. Because no feedback is available at the beginning of a movement, a second construct, memory trace, is needed to explain the selection and initiation of the movement and its accompanying perceptual trace.

According to J. A. Adams (1968, 1987), some motions are carried out based only on a memory trace, particularly ballistic motions. Correction of such motions is usually carried out on the basis of an estimation of preceding action results, involving an open-loop cycle. This concept suggests that motor actions and movements can be regulated based on opened and closed-loop cycles. Unfortunately, Adams' theory suffers from lack of parsimony, because it has to postulate that for every movement there exists its own perceptual or memory trace. Such a concept could easily overload the memory.

Schmidt (1975) tried to overcome this problem by utilizing the concept of schema, originally introduced by Bartlett (1932). Schemata (there are a number of these) are constructs describing hypothesized structures, presumably within the brain, that perform necessary physiological functions involving memory, perception, cognition, and in the present case, motoric actions. A construct is an inference; something must be there that causes behavior (which can be observed) but the causal structure cannot be seen. Many of the concepts of Russian engineering psychology are of this nature, as are those of Western ergonomics theorizing. Schemata guide our actions and organize our experiences.

Schmidt (1975) distinguished between recall and recognition schema. In order to plan a movement, two inputs are required: desired outcome and initial conditions. The specification of how to perform a movement is determined by the relationship between desired outcome and past movement specifications. When the response specifications have been determined, the movement is then carried out by means of "motor programs." The recognition schema generates the expected sensory consequences. These consequences can be compared against the actual sensory consequences during or after movement performance. The sensory consequences may involve proprioceptive, visual, and auditory feedback. The recall schema is concerned with response specifications whereas the recognition schema is needed to recognize feedback from the movement.

Schmidt (1975) suggested that individuals develop a motor response schema. This schema includes the following elements: (a) initial conditions: information about body and limb position and the state of the environment in which the movement is to be made, (b) response specifications: requirements for force, direction, speed, and other dimensions (it is assumed that a general motor program for generating movement exists), (c) sensory consequences: feedback generated by movement, and (d) response outcome. Schmidt's concept is, to some extent, similar to Anokhin's and Bernshtein's theories.

Gordeeva, Devishvily, and V. P. Zinchenko (1975) and Gordeeva and V. P. Zinchenko (1982) discovered that actions split into a series of directed micromovements. These micromovements are not errors. They appear to have an explorative function and to be cognitive in nature. Based on these movements, an "image" of the space parameters of the movement is constructed that is partly perceptual, partly kinesthetic. The implementation of the movement, in several stages, involves at first cognitive components. Initially, a total action does not exist but consists of separate movements in different space coordinates. As skill develops, the micromovements are gradually reduced in number and integrated into a harmonious whole.

As was mentioned in chapter 1, the aforementioned authors discovered three function blocks of motor movement: program formation, realization (implementation), and control and correction. Based on these function blocks, they developed several criteria, two of which are:

1. Cognitive loading represents the relationship between the time in which the program formation block and the control and correction block are initiated to the time in which the realization block is completed. The first two blocks have a cognitive function, and the last implements motions. Cognitive loading relates the cognitive (planning and control) and motor components of motions. Because in different stages of acquisition of a motor skill the duration of these components can be reduced, this ration or coefficient has less sensitivity for evaluating motions at different stages of learning. As a result, Zinchenko introduced a new coefficient, which he designated as the coefficient of "operationability."

2. Coefficient of "operationability" represents the relationship between cognitive loading time and the total time occupied by the action (which included the program formation, realization, and control and correction blocks of motions). This coefficient demonstrates the relationship between the various function blocks as compared to the change in the duration of motion.

The motor image of space integrates with the perceptual image and, based on this integration, an image of the motor action as a whole is constructed. This process is associated with the development of *situational images* and *action images*, which are considered later in more detail. Translated into Western terms, there is a kinesthetic image and a perceptual image, and over time the two images meld into each other.

American scientists also pay attention to the image of motor actions. For example, Welford (1974) wrote that the instantaneous image of a situation was constructed by analysis and integration of data from different points of space. This image was then transformed into an image of motor programs, which in turn became an action.

From the Russian point of view, it is very important to pay attention to the integration of information from different actions. For example, Gordeeva and V. P. Zinchenko (1982) recorded the interaction of eye and arm movement. They discovered that images of motor actions are constructed based on perceptual and motor elements.

Any self-regulation process involves control and correction of movement based on feedback. A problem arises with ballistic actions because there are no tracking corrections with these movements. However, the results of any movement can be compared with movement criteria after their completion. If there is a discrepancy, the next movement can correct the previous mistake. Schmidt and Russell (1972) discovered that very quick movements that have 160 ms execution time can be regulated without feedback. However, a movement that has more duration depends on feedback. This suggests that the time component of feedback is critical to motor actions.

The human is more adaptable to delay or distortion of visual feedback when they have more training or experience. For example, a skilled typist types without looking (blind method), but a novice typist cannot (Long, 1976).

K. U. Smith (1962) showed that visual feedback delay influenced the precision of actions, an influence that increased with the increasing speed of action. Similar data were produced by A. I. Novikov, Sidorova, and Federov (1980) when they studied discrete tracking. The introduction of visual feedback delay splits the movement into two separate stages. The first stage is a ballistic motion, the second, a series of corrective motions.

V. P. Zinchenko (1978) pointed out that the motor components of action (excluding the program formation parts) can be divided into a ballistic stage (acceleration stage) and a slowing stage. The ballistic stage is not sensitive to visual feedback but the slowing stage of the movement is highly sensitive. This means that the same motor act has two fundamentally different types of regulation: programmed (ballistic) and afferentational (slowing). The relationship between these two kinds of regulation depends on the amplitude, precision, speed, stability, and feedback.

It is possible to postulate two situations. The first is connected with movements performed as part of an open-loop cycle. The second is based on a closed-loop cycle. For example, in the former, when an operator works with a mechanical manipulator, feedback delay causes changes in the strategy of movement performance. The operator switches from a strategy of continuous movement to a strategy of interrupted movement. The second strategy is characterized by alternating stops and movements. The last strategy is considered by Vertut and Coiffet (1985) to be a regulation based on an open-loop structure.

Based on data gathered by Gordeeva and V. P. Zinchenko (1982), the pause phase is not passive but involves an evaluation of the early part of the movement and the formation of the subprogram for the next part of the movement. Cognitive function evaluating and programming components of actions are performed during this phase. Dividing regulation of action into opened and closed loop cycles is not precise. A movement can be performed initially without evaluation of the execution process. However, evaluation

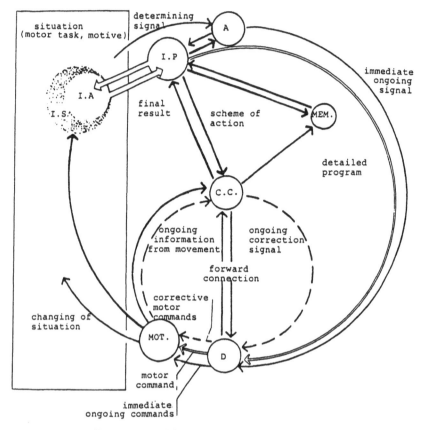

FIG. 2.2. Self-regulative model of human movement. From Gordeeva and Zinchenko (1982). A—poly model afferentator; D—differential programs; I.P.—integral program/plan of action; C.C.—control and correction; Mot.—motor component; I.A.—image of action; I.S.—image of situation; MEM—memory.

of the movement when completed can result in the creation of a program for a new action. This can be considered closed-loop structure regulation.

Gordeeva and V. P. Zinchenko (1982) proposed a model of self-regulative motor actions (see Fig. 2.2), which is composed of the following function blocks:

Image of the situation (IS): This is the subjective picture of the situation in which motor actions are performed. The picture is constructed of sensory impressions of the environment and past experiences with such environments. The image of the situation is not a simple objective reflection of the surroundings of a performed action, but is constructed from the point of view of the individual.

Image of the action (IA): This is the subjective picture of the action itself. It is based on sensory feedback from the motor action.

The preceding two images are integrated and form the basis for developing the integral program.

Integral program (IP): This function block contains a general program of action that helps to shape and interact with subsequent function blocks. This block analyzes the meaning of the action and the sequence of several actions, as well as makes decisions about action modification on the basis of feedback.

It also develops programs connected with shaping the image of the situation and image of the action.

Differential program (D): This block includes both elementary programs that do not require learning and other more complex ones that do. For example, the general program for performing motor actions can be decomposed into more specific programs including time performance, amplitude, and required effort. A major function of this block is the detailing of the general plan of action to develop the required commands for performance.

Control and correction (CC): This block has various levels of receiving and processing information from the differential and integral program blocks. One level of control and correction deals with general programs, another with detailed programs. This block contains general data about action results and the commencement of other actions, as well as making decisions about corrections and action termination. Information is also derived from the motor components block (MOT).

Memory (MEM): This block stores information about the manner in which previous actions have been performed. Because detailed descriptions about these actions do not exist, the block depends on a schema for such actions. Consequently, the action is not reproduced, but is rebuilt over and over again.

Polymodel afferentator (A): This function block selects relevant information from the environment, although a small amount of irrelevant information may also be introduced.

Motor component (MOT): This block is responsible for implementing the motor action, as well as anticipating and selecting impending actions.

According to activity approach, it is functionally structured analysis. This analysis suggests dividing the analyzed process into separate functions with the following description relationship between them. The Gordeeva and

Zinchenko model of motor functioning has much in common with cognitive theories used in computer software design, thus illustrating two major influences on Russian engineering psychology.

There are in the Gordeeva and Zinchenko model and other models described later similarities to the Barnard, Wilson, and McLean (1988) interacting cognitive subsystems model, in which subsystems operate in specific processing domains, that is, subsystems for sensory (acoustic, visual) and effector (limb, articulation) processes, as well as subsystems that deal with higher level processes like memory and decision making. Interestingly, each of Barnard's subsystems has an internal organization involving three kinds of resources, including an image record that preserves all input to the subsystem, a copy process, and a process that recodes one type of information to another. The point of this digression is to indicate to the reader that, although the Russian models may initially appear strange, they contain parallels to Western theorizing about advanced computer architecture.

The description of the Zinchenko model can be expanded by various examples. The first example, which is the simplest, considers a reactive action such as a jerking back of a hand that touches a hot surface. The stimulus from an external situation (the hot surface) is introduced into the polymodel afferentator (A), then the differential program block (D) circles halfway around the loop structure. This stimulus activates learned or inherent programs that are transformed into motor components through the motor function block (M). This is open-loop structure regulation.

When a new voluntary motor action is contemplated, the goal, the task, and the general image of the result of past motor activation are used as preliminary input data to the individual, but cannot by themselves elicit an adequate action. An action must be constructed based on the preceding input data, which enter into the integral program block (IP), and then to the memory block (M). This block starts the search process for a similar method of action implementation. If this information is not present in the block, the individual begins to employ his or her explorative schema, and attempts a trial action. This causes a change in the situation and a formulation of the content of the function block that is responsible for image of action (IA), image of situation (IS), and integral program (IP). This information goes through the polymodel afferentator (A) and then through the external circle, whereas other information about the executed action goes through the internal circle. This information passes through to the control and correction (CC) block. At this stage, this block does not possess the results of actions. This information can be input to this block only after the formulation of integral programs.

The information about movement that is put into the integral program block has two functions. It can be used as a source of image shaping and of program modification for the following actions. As a result of repeated motions, the image of situation, the image of action, and the integral program of a new action are constructed. The integral program contains general information about what, how, and in what sequences an action must follow. Image of situation and image of action are not the result of perception but must be actively constructed. The information introduced into the integral program is enriched because the memory stores information about other movements.

After the image and integral programs are created, movement realization proceeds in the following way: A signal goes through the polymodel afferentator (A) and is transformed in the integral program into an action image. Based on this, a decision can be made about whether the action should be performed. The integral program depends on the image and presents a plan of action that enters the control and correction block and that of the differentiational program (D). In the integral program, general information is needed to evaluate the result of the action. In the differential program, specific information such as amplitude and effort requirements are entered.

Information then passes into the control and correction block and afterward into the motor block. The former possesses information about the projected action before the motor program does (the control correction system is actively built during the execution of the action). A decision about correction develops based on a comparison of information in the control and correction block and ongoing information about motor block actions. Based on this, a correction signal can be developed in the differential program that can then be transformed into correctional commands. The correctional commands can again be introduced into the control and correction block where the process of comparison is repeated.

As can be seen, in the Zinchenko model of self-regulation of motor actions, cognitive components predominate. Motor actions as units of activity have a loop structure and are guided by action and situational images. The action and situational images depend on the goal of the action, although the model does not clearly describe this process.

Parallel to this theory, Anan'ev, Vekker, Lomov, and Yarmolenko (1959) used a high-powered movie camera to study the sense of touch process and arm orientation in space movement. It was discovered that space movement orientation is better for the right hand's kinesthetic features. However, sense of touch is better for the left hand. In this study, micromotions integrated in the whole structure were found. Roze (1963) also discovered these cognitive micromotions. The more complicated the task, the more micromotions occurred.

The following classification of motions was suggested (Anan'ev, 1960):

1. The first category includes major work movements in which an object can be held and manipulated.

2. The second category includes cognitive micromotions and movements whose purpose is the exploration of an object and situation by trial and error.

3. The third category includes accommodating movements associated with adjusting arm posture, error and movement corrections, balance and compensation movements.

All of these movements should be distinguished from random, superfluous movements that can be considered as equivalent to noise in a system. Such "noise" movements are more frequent in less skilled operators.

The importance of self-control follows from understanding movements as self-regulation units. Self-control is an evaluation function. It evaluates the expediency, efficiency, planning, and regulation of motor actions. The success of self-control depends on instructions and mental representations of situations that allow one to extract criteria for successful motor actions. Ongoing self-control is performed during an action; verification control occurs after the action.

In an accident situation, self-control can be increased. Khachatur'yanz, Grimak, and Khrunov (1976) discovered that in such situations, responses slowed down because one tends to repeat actions mentally and only after that implements them.

2.3 PSYCHOLOGICAL MECHANISMS
OF SELF-REGULATION

The previous section considered the problem of self-regulation of motor movements and actions. In chapter 1, we demonstrated the importance of task subjective representation on which success in solving a task depends in part. In this section we analyze the individual mechanisms of activity self-regulation. Konopkin (1980) described the mechanisms of self-regulation in which a "subjective model of activity conditions" (p. 50) is important. Depending on how this model is formed, the process of self-regulation can be realized in different ways. Different existing subjective models of activity produce different programs of performing tasks. In other words, the stimulus does not predetermine an individual's interpretation of it. This means that one can organize behavior for the same stimulus in different ways.

None of this should surprise the reader. Western ergonomists recognize that, particularly in ambiguous situations, the presumably objective stimulus is not completely objective and can be interpreted in various ways leading to different activities. An example of a supremely ambiguous stimulus (although taken from another discipline) is the Rorschach inkblot.

An important aspect of an operator's activity is the speed of processing information. If an operator has preliminary information about the pace of ongoing stimuli, he or she can, in advance, program different time components of activity. Very often, however, the operator lacks precise information about activity conditions or fails to see a change in these conditions. In this case there is an inconsistency between subjectively perceived and actual work conditions. Konopkin (1980) conducted a series of experiments in which work pace was unknown or was imperceptibly changed until subjects discovered the change and adapted their work methods accordingly.

In the first experiment, eight subjects performed a complex choice reaction task. A square screen with dim glass (320 mm × 320 mm) divided into 16 equal subsquares was placed in front of each subject. The stimuli were lights that were individually illuminated in a programmed manner. The subject had to respond at maximum speed by pressing 1 of 10 buttons positioned radially. The experimenter calculated average reaction time (RT) for any specific conditions and also calculated the speed of processing information (V):

$$V = H/T \qquad (2.1)$$

where H is information conveyed by a stimuli according to $H = \log_2 N$; N is the number of equally likely presented stimuli; and T is the average duration of correct reaction determined as the total time of all reaction divided by number of correct reactions.

In the first set of experiments, two conditions were used: 3- and 1-second intervals between the visual stimulus. For any pace, two through eight signals of equal probability and independent of each other were presented. Prior to any set of experiments, the subjects were informed of how many signals would be presented and what their pace would be. The resultant data revealed that in all cases, when a subject worked with a 1-second interval, their reaction time was shorter than when they worked with a 3-second interval. The differences were statistically significant.

In another set of experiments, the subjects were given false instructions about the transition from pace to another. The real pace was 1.5 seconds, but the two false paces were 1 or 2 seconds. Four signals were used in this experiment. When the false 1-second instructions were introduced, the speed of processing information was higher than that of the false 2-second interval instructions. These results were also statistically significant.

In these experiments, subjects received verbal instructions about work pace. In the following experiment, researchers used nonverbal instructions. In a preliminary briefing, sound stimuli were presented to the subjects at a pace equal to the pace of the visual stimuli to be presented later. This required the subjects themselves to determine the pace. The study revealed results similar to those of the previous experiments. The significance of the study was that it demonstrated that humans are capable of self-programming or self-regulation of activity independent of instructions.

Performance was also studied under changing activity conditions. The experiment included five conditions: 3-, 2-, 1.5-, 1-, and .75-second time intervals between stimuli. The transition from one interval to another was conducted slowly. Subjects increased their reaction speed, but only when the signal pace was equal to 1 second or .75 seconds. The increased speed of reaction occurred only as the subject's performance decreased. Then subjects began to miss signals. The absence of precise information about work pace impaired adequate self-regulation of activity. Figure 2.3 shows reaction time during the transition from one activity pace to another. These studies revealed that subjects formed different strategies directed at achieving the goal. The subjective understanding of activity conditions can be considered a specific mechanism of self-regulation.

However, the subjective model of activity conditions cannot determine the programming components of activity directly. Subjects with the same understanding about work pace or goal may use different programs for performing the activity.

The previous experiments utilized pacing of stimuli as the major component of the subjective model of activity conditions. The following experiments involve the unpredictability of events as a major variable of that model. The investigators (Strukov, 1972; Strukov, Gritsevsky, & Konopkin, 1971) studied the activity of operators of a chemical plant who monitored the occurrence of low-frequency signals with different probabilities of occurrence. In one group of operators, the probability was 50%, whereas in another group it was 70% to 90%. The stimuli were unpredictable, occurring randomly, which made it difficult for an operator to create a subjective model of activity conditions.

As a result, the operators displaced stress reactions, as shown by psychophysiological data and later interviews. It was discovered that during different

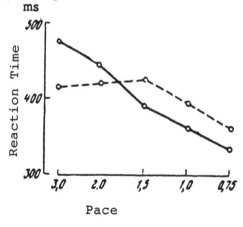

FIG. 2.3. The dependence of RT on the pace of stimulus presentation. From Konopkin (1980). Preliminary information about the pace shown by the solid line. No preliminary information about the pace shown by the dashed line.

shifts events occurred with varying probabilities. This information was introduced during the operator's training and helped workers construct a more adequate subjective model. This reduced their tension and decreased fatigue.

Konopkin and Luchkov (1973) attempted to discover how information about task duration influenced subject performance. The task involved canceling various numbers that were placed in random order in tables of numbers. The numbers could change from 1 to 8, each of which required a specific method of canceling.

Ten tables were used. After 3 minutes into the task, a bell rang and subjects marked how much they had accomplished. This enabled the experimenter to calculate the amount of work performed during successive 3-minute intervals. Instructions required the subjects to work with maximum speed and a minimum of errors. When the productivity of work reached a stable level (usually 12–15 45-minute trials were required) the critical experiment was introduced. This experiment was completed only after 117 minutes (39 3-minute intervals).

Productivity in the longer, unpredictable time duration was lower, indicating that work duration had not become part of the subjects' model of activity conditions.

Nojivin (1974) studied the speed of simple sensory-motor reactions from the self-regulative aspect. The point of this experiment was to introduce into instructions new requirements that could change the subjects' system of self-regulation. There were seven experiments.

In the first experiment, subjects pressed a button after receiving a visual signal. They were asked to respond with any suitable speed; requirements for the self-regulation process were, therefore, minimal. In the second experiment, a new requirement employed a warning signal 2 seconds before the visual stimulus requiring the reaction. Subjects in the third experiment were asked to react with maximum speed. In the fourth experiment, subjects were given information about the speed of their reactions. Each subject in the fifth experiment was required to match their fastest reaction time in the previous experiment. The sixth and seventh experiments repeated the requirements of the first and second experiments, to see how the previous experiments (3–5) had influenced reaction time when all speed requirements were removed.

Figure 2.4 shows that reaction speed increased from Experiment 1 to Experiment 5. Reaction time in Experiments 6 and 7 was less than that of Experiments 1 and 2. These differences were statistically significant. The high-speed performance in the preliminary trials (3–5) had a positive effect on the speed of performance of the subsequent trials (6–7). As a result, the "suitable" pace had changed and the subjects did not even realize it.

It is interesting to describe how the subjects learned their suitable pace in the first series of experiments. At the beginning, subjects decreased re-

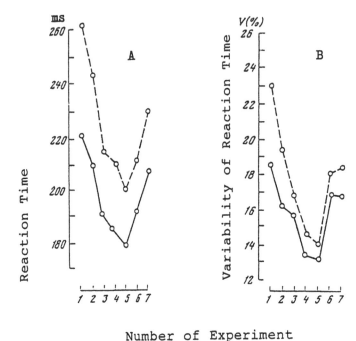

FIG. 2.4. The effects of self-regulation process on speed (a) and variability (b) of reaction time. From Nojivin (1974). Dashed line—women; solid line—men.

action time to a certain level during the first 10–15 trials, and then increased it. The strategy of the subjects' activity was as follows: They started working slowly, then increased the speed, reached the suitable pace and continued to speed up, and then went back to working at the suitable pace. That is because a criterion such as suitable is not sufficiently precise for self-regulation of activity, which is why subjects tested various speeds of performance to try to find a standard for suitable speed.

The results of these experiments reveal that even in strictly predetermined stimulus-response situations, subjects developed different strategies that influenced their response time. This suggests that complicated self-regulative processes take place even in the simplest reactions.

Self-regulation as a goal-directed process requires criteria to permit comparison of obtained results with those that must still be achieved. These criteria permit subjects to confirm that the output of an activity corresponds to the accepted goal. These criteria are part of Anokhin's (1955) model of self-regulation in which they derive from a mechanism called "action acceptance." Bernshtein (1966), in his model of self-regulation of motor actions, called this an "ordered apparatus." However, these criteria may not always be clear to subjects (Morosanova & Stepansky, 1982). Ammons (1956) demonstrated that even if the subject does not have information about response outcomes, he or she can hypothesize about the criteria needed for performance evaluation. This also reflects the model of self-regulation of motor actions developed by Gordeeva and Zinchenko (1982) that was considered earlier in this chapter.

Stepansky (1976) and Sipachev (1976) studied how reaction time changed in subsequent trials when visual stimuli are presented. In each experiment the subject performed 60 trials. Success of performance is evaluated by calculation of average time performance and comparison of individual trials. In the first series of the experiment, the subjects did not receive information about their performance results. In the second series of the experiment, subjects received information about their time performance. In the third series of the experiment, in addition to receiving information about time performance, they were presented the result that was to be achieved (standard criterion for success was presented). This study showed that the subjects in the last series received the best results. However, during questioning of the subjects, it was discovered that even in those cases when the subject did not receive information about reaction time and required standards, they developed their own criterion for success by comparing and evaluating trials over time through subjective experience. These subjects evaluated a response by using either a 2-point scale, *on time* or *late,* or a 3-point scale, *good, normal,* and *bad.* The subjects made their evaluations based in part on the muscle sense during performance of different reactions. Over repeated responses, this self-evaluation will change because the criteria were rough and imprecise; speed of reaction in subsequent trials increased when subjects received more precise information about their performance.

Significance of different reactions in subsequent trials can be changed by presenting different instructions to the subjects. When the subjects perform sequences of reactions and do not receive information about time performance, they develop different subjective standards of success for significant and insignificant signals (Stepansky, 1976).

Seashore and Bavelas' study (1941), based on Thorndike's (1932) experiment where a person drew a line without seeing what they were doing, also demonstrated that a psychological mechanism for evaluating the achieved result on subjective criteria exists. Nevertheless, subjects developed their own subjective standards, which they used in their performance. Although objective criteria influence subjective ones, subjects tend to use the subjective criteria more. American psychologist also use the notion of subjective parameters of success. According to Bandura (1977, 1982) subjective parameters of success can be established through the process of social comparison.

Self-regulation can also be applied to tracking tasks. This involves the relationship between input signals and the operator output reaction (the so-called "transfer function"). From the Russian standpoint, this is a behaviorist approach based on S-R relationships. From the activity approach the transfer function ignores several important features, for example, the significance of the input signal, the operator's motivation, and his or her strategy.

Zabrodin and Cherneshov (1981) conducted a study that may explain the low efficiency of tracking task models. Visual harmonic stimuli were tracked, with the experiment starting with .05 hertz frequency and gradually moving to higher frequency levels, until the operator could no longer perform. The investigators noticed that operator responses contained micromotions with additional harmonics not anticipated in the activity goal. It was evident that the subject did not perform a linear transformation of input signals.

The additional micromotions that are not related to the tracking goal gives the subject information that is useful in improving the accuracy of their tracking performance. Decreasing and/or distorting this information reduces the quality of the tracking.

Zabrodin and Cherneshov (1981) hypothesized two possibilities in relation to these additional micromotions. On the one hand, they can be considered tracking errors; on the other hand, they can be considered useful information. However, from the standpoint of existing mathematical models of motor performance, these motions are unnecessary; hence they are not included in these models. From the activity point of view, these micromotions are additional evidence that an operator works as a self-regulative system.

These additional movements, which are considered elsewhere as errors, can be viewed as explorative motions that represent cognitive functions. Russian ergonomists conclude that when studying the tracking task, attention should emphasize such variables as goal, strategy, feedback, and significance of information.

This short research overview dealt with the process of self-regulation as it relates to informational processes. It is also necessary to consider self-regulation as it relates to energetics activity processes. As previously discussed in chapter 1 (section 1.6), sense of task, which includes emotionally evaluative and cognitive components of activity, plays an important role during goal formation. When we talk about a person's sense of task, we mean the value and significance the task has for him or her. This means that information, goal, and intermediate and final task outcomes affect the personal significance of the task. This emotional reaction depends not only on the specific task situation but on past experience, skill level, amount of effort required, and other idiosyncratic factors.

Ol'shannikova (1962) showed that the significance of a stimulus can affect the physiological law of force, which was discovered by Pavlov (1927). This law states that, with the increasing intensity of the stimulus, reaction time decreases. Olshannikova discovered that if a weak stimulus was very important for an operator, the stimulus reaction time could be decreased in comparison with a stronger, less significant signal. El'kin (1962) found that in significant situations, a person more precisely evaluates the duration of time.

Studies show (Kotik, 1987) that increasing the significance of the situation increases the organism's arousal and the emotional-motivational level of the person. However, increasing the level of motivation is not the sole factor connected with the significance of the situation. Significance also influences the operator's sensitivity to various stimuli and selective features of attention. In Kaidro and Kotik's study (Kotik,1987), the experimenter presented a red light (significant) signal and a white light (insignificant) signal to the subjects. Instructions emphasized the significance of the signals and alerted the subjects that an electrical shock would be administered when their responses were delayed.

In addition, a 100-decibel sound was presented to the subjects, with either a red or white signal. Psychophysical reactions to the noise under these two conditions were measured (pulse rate, breathing rate, and galvanic skin response). When the red light was presented together with the noise, there was no psychophysiological reactions. During the white light/noise combination, the psychophysiological reaction significantly increased. When the signal was significant, sensitivity to the noise was reduced.

Rogovin and Gromov (1979) also studied the significance of information on the selectivity of the perceptual process. The task was to determine which parts of an X ray were the brightest or darkest. For this purpose, they selected two groups of subjects, one group of experienced pathologists and the other without such experience. The purpose of the experiment was to see whether experience could influence a simple psychophysical function.

In one condition the X rays had no pathology; in a second, films presented pathological changes. In the first condition, in which there was no pathology, the task of discriminating brightness was performed more effectively by the nonprofessionals because the radiologists, in view of their training, spent their time trying to find nonexistent pathology. In the second condition, where there was pathology, the task was performed more effectively by the professionals because the radiologists looked for pathology and the pathology was identified by more intense light or dark patches.

The concept of "significance" demonstrates that cognitive, and emotionally evaluative components of activity are tightly interconnected. The American psychologist has drawn more and more of his/her attention to this problem. An example of this is the introduction of the notion of "current concern" by Klinger (1996). In one of his studies the subjects listened to two simultaneously presented narratives, one presented to each ear. It was discovered that the subjects spent more time listening to passages associated with their concerns. The concepts of "significance" and "current concerns" do not always suggest conscious processing of information. They can also be connected with an unconscious level of processing information and emotional reaction to it. Significance, therefore, influenced not only emotional-motivational level but also the selectivity of attention, the sensitivity of the operator to stimuli, thought content, recall, etc.

2.4 SELF-REGULATION OF MORE COMPLEX ACTIVITIES

Previous studies have shown that self-regulation mechanisms function even during the performance of simple sensorimotor reactions. When the actions become more complex, the complexity of the self-regulation process increases also. The studies following demonstrate this point.

Most studies of the self-regulation process have involved relatively simple sensorimotor experiments. One must ask how that process functions with more complex behaviors. Research analyzing the interaction among multiple actions to find out how they integrate into a total activity is, therefore, important. Here the studies of Welford (1960) and Borger (1963) can be useful. The research described next is directed at finding out how people regulate their activity during performance of sequence of actions.

Krilov and Pakhomov (1965) found that the latent period of simple sensorimotor reaction to a second signal depends on the interval between the first and second signals (see Fig. 2.5).

When there are very short intervals between stimuli, the program for the second reaction is initiated during the performance of the first reaction. That is why reaction time for the second signal in Fig. 2.5 is so much faster than for the first signal when the interval between the two is only .1 second. As the intervals between the two signals increase, reaction time for the second signal becomes progressively slower. When the interval between the signals in-

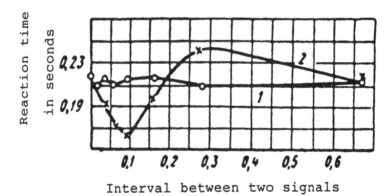

FIG. 2.5. The dependence of reaction time on the interval between two signals. From Krilov and Pakhomov (1965). 1—time reaction for the first signal; 2—time reaction for the second signal.

creases to .6 seconds, subjects react to the two signals independently, which is why the time for the second reaction becomes exactly equal to that of the first. Again, it can be seen that even in simple psychophysical functions, the individual can develop strategies that can vary over time. These strategies occur whether or not the human is a single- or multiple-channel system.

Another way of examining the regulation of the total activity is to compare how an individual performs positioning actions when these are isolated and when they are in sequence with other actions. Real tasks usually involve a sequence of actions, hence studies performed by Bedny (1979, 1987) and Bedny and Podnevich (1975) can be interesting. They studied self-regulation when subjects performed a sequence of actions. One begins with Fitts' (1954) and Fitts and Petterson's (1964) well-known research on positioning actions from a self-regulation point of view. In Fitts' (1954) study subjects moved their arm with maximum speed from one target to another. The width of the targets and the distance between each other were varied, and the time and precision of those motor actions were measured. Fitts' law suggests that time has a linear relationship with the difficulty of an action:

$$t = a + b \ ID \qquad\qquad (2.2)$$

Where a and b are constants, the difficulty of the action is defined as $ID = \log_2 2A/W$; A is the distance of the motor action and W is the width of the targets.

Some have attempted to use this law for the evaluation of hand-controlled computer devices (Epps, 1986; Radwin, Vanderheiden, & Lin, 1990); Walker, Meyer, & Smelcer, 1993). However, Fitts' (1954) study dealt with performance at maximum speed. Leplat (1963) showed that the performance time of a motor action does not depend significantly on its amplitude. A further drawback of Fitts' (1954) study is that he considered only repetitions of the same actions.

Bedny (1987), on the other hand, required positioning actions not only with two targets but with four. In the first group of experiments, the subjects had to hit two targets with a stick, whereas in the second group four targets had to be hit. To conduct this study, a special apparatus was designed that contained brass strips (the targets) mounted on a base. The number of targets, their width, and the distance between the targets could all be changed. One meter counted the number of hand actions from one target to another; another counted the number of errors. A red light warned the subject that a trial would begin in 2 seconds. A green light signaled the beginning of the task. There was also a stopwatch on the panel, which automatically turned on when the task began. After 10 seconds, the stopwatch automatically was turned off and a buzzer informed the subjects that the task was over. The five subjects were required to hit the targets at their maximum speed. Using the number of times the subjects hit a target and the entire task performance time, one can calculate the average time for performing the action.

Three groups of experiments were conducted. In the first group of experiments, only wide targets were used (W = 50 mm). In the second group of experiments, only narrow targets were used (W = 7 mm). In the third group of experiments, both the narrow and wide targets were used. In each experiment the number of targets was varied (two and four); and the distance between targets was either 60 mm or 120 mm.

The purpose of the study was to discover how different strategies of activity were formed while positioning motor actions were conducted in isolation (two targets), as well as when they are conducted in sequence (four targets). Although some of Fitts' (1954) study was reproduced, the conditions of the number of targets was increased to four; in another set of trials, the subjects repeated a single action and a sequence of different actions. After the experiment was over, the subjects were questioned to determine the work strategy used.

Performance time with wide targets was designated as T_1, whereas the performance with narrow targets was designated with T_2; performance of the three actions hitting the four targets was designated T_3. If actions performed in a sequence have no influence on each other, the performance time of three actions is $T_3 = 2T_1 + T_2$, provided T_1 and T_2 are identical. If $T_3 > 2T_1 + T_2$, the task with four targets is more difficult and actions are not independent of each other.

In the first experiment, $T_3 > 2T_1 + T_2$, the difference being significant ($p < .01$). This means that the pace at which the sequential actions are performed is slower than the pace of the actions performed in isolation.

It was discovered from comparison of the experimental data with observation and interviews with subjects that their strategy is not simply predetermined by instructions and task conditions, but was selected on the basis of subjects' personal characteristics and experience. They formed a subjective standard of success that does not always match the objective standard of success. When subjects switched to four wide targets, the target zone used was expanded, but still remained narrower than the entire width of the band. The choice of a subjective standard was determined by the width of a target, distance between the targets, subjective evaluation of what constitutes success, and the subject's ability to achieve the goal. With respect to the last, Hoppe (1930) showed that especially difficult or especially easy goals fail to become criteria for success; instead, other goals are selected by the subject.

Some of the subjects preferred precision, whereas others chose speed (speed-accuracy trade-off). These differences may be due to personal characteristics or the inherent ambiguity of the instructions, which recommended the subject perform the task with maximum speed and precision. When the distance between the targets was increased, especially in the case of a subject switching from two to four targets, one can see a shift in the subjective criterion of success. Some of the subjects tried to keep the high speed of activity at the expense of using a wider part of the target, even though the risk of moving out of the target's range increased (a risk strategy). Others continued to hold their subjective criterion and reduced their speed significantly. When a subject tapped two targets, the average error rate was 1 per task/subject, whereas in the case of four targets, the error rate increased to 2 per task/subject, although the difference was not statistically significant.

With the narrow targets ($W = 7$ mm), as in previous group of experiments, $T_3 > 2T_1 + T_2$ and the difference is significant ($p < .01$).

When the task was changed from two to four targets, one might hypothesize that there would be a greater increase in time performance for the narrower targets, where the action is more precise. However, in actuality, the result was the opposite. The increase in time performance is more significant with the wider targets, that is, for actions with lower precision.

Regarding self-control, clearly the task with two wide targets needs a lower level of self-control, which may be described as an automatic level of self-regulation. However, when we change the task from two to four wide targets, the task becomes significantly more difficult and needs a higher level of self-control, which would require conscious self-regulation. The task with the narrow targets needs a high level of self-control. The switch from a low level to a higher level of self-control significantly increases time performance of actions. Thus, time performance increases gradually within the automatic level of self-control, but increases sharply, when self-control changes from automatic to conscious.

It is also interesting to compare T_1 and T_2, where the distance between the targets is changed, for tasks with narrow targets and tasks with wide targets. If targets are narrow, the time for the larger amplitude action is $T_2 > T_1$ significant at the ($p < .05$) level. If targets are wide, the difference is not statistically significant. The experimenter concluded that if the amplitude of a low-precision action increases, it does not significantly change performance time in the work environment. However, if the amplitude of a high-precision action increases, it significantly changes performance time. These results should be taken into consideration for the design of efficient manufacturing work processes and equipment.

When subjects switched from two narrow targets to four, errors jumped from an average of 2.0 to 3.6. Subject strategies also changed when they transferred from two to four targets. When a subject switched from two to four narrow targets, the target area used expanded more significantly than when wide targets were used. This means that the subjective standard of success changed: Subjects used almost the entire width of the target. When they used four wide targets, only part of the target width was used, which means that with four narrow targets, they used a risky strategy, which is why the precision of their action decreased. These data suggest that precision depends on the operator's correct selection of strategies and the significance for them of the various components of the task.

This research indicates that a person does not merely react to external stimuli. Subjects select their own goals and actively interface with the environment, changing their strategies of activity, and their subjective standards of success based on self-regulative processes. Finally, results indicate that positioning actions cannot be seen as independent and uncorrelated. This contrasts with Fitts' (1954) law, which may only be used when an operator reacts to a single signal with maximum speed and performs an isolated act.

Another study (Bedny, 1985) suggested that while performing two actions consecutively, the subjects, depending on goal and conditions, continually change their strategy to redistribute information while performing similar external actions.

2.5 FUNCTIONAL STRUCTURE OF SELF-REGULATION

The concept of self-regulation is based on the fundamental notion of feedback. Feedback is necessary to provide the operator with information required for system operation. Russian psychologists believe that the concept itself, outside the overall context of self-regulation, is not especially fruitful. Without psychological "machinery," such as goals of activity, subjective criteria for the evaluation of success, and strategies, the concept is incomplete.

As an example, a study was performed (Stepansky, 1974) in which subjects had to stop a rotating arrow at a designated position by pushing a button. When this was done, the arrow stopped rotating. Subjects could react either before the arrow reached its designated position (early reaction) or after the arrow passed that position (delayed reaction). The subjects could immediately see their error and correct it during the next trial. In most cases, the number of early and delayed reactions was similar.

In the next stage of the experiment, the subjects received an electrical shock if they stopped the arrow in advance of the designated position. As a result, the number of delayed reactions sharply increased. In both stages of the experiments, feedback was the same. In the second stage, however, the significance to the subject of an early reaction error increased. As a result, their strategy of performance changed. One can see, therefore, that feedback alone (without its interaction with other factors) may provide misleading conclusions.

There are two types of self-regulation process in human beings: physiological and psychological. They are interconnected, but each of them has an individual dynamic. The physiological self-regulation model is based on homeostasis. There is a tendency for physiological disturbances to be reduced and for departures from an optimal state to be corrected in order to restore balance. Many physiological imbalances are corrected automatically. The structure of physiological self-regulation processes is wholly predetermined.

The psychological self-regulation system can change its own structure based on its experience. Such a system can form its own goals and subgoals and its own criteria for activity evaluation. Self-regulation provides the integration of cognitive, executive, evaluative, and emotional aspects of activity. Given the complex set of variables involved, people exhibit an infinite diversity of activity. Internal changes in the psychological aspects of self-regulation emerge not only from external influences, but also are based on those changes that are prefigured in previous experiences. Thus, in systems of self-regulation, memory assumes a fundamental importance.

It is important to take into consideration that during any individual self-regulation process, the program of self-regulation, criteria of evaluation, and even the goal of self-regulation may be changed. Self-regulation functions in time; it is based on past experience, evaluation of the present situation, and the anticipated future. From a psychological point of view, self-regulation may be considered as a process that supplies coordination among the various psychological functions in accordance with a specified goal. The Russian understanding of self-regulation differs from that prevailing in the current research literature. Thus, we cannot agree, for example, with Heckhausen (1986), Kanfer (1996), Kuhl (1992) who considered self-regulation as motivational or volitional processes. Self-regulation cannot be explained without considering human information processing.

Strategy is fundamental to self-regulation. Broadly speaking, strategy is a method of taking into consideration information during planning, as well as throughout the processes of achieving the goal. However, between plan and strategy, there are significant differences. A plan is something that is stable, even rigid, because it describes a more specific situation than does strategy, which has a more dynamic character. Strategy implies flexibility, plasticity and variability of means, the capacity to change a program based on task outcomes, changing conditions, internal state of the human.

The major problem facing the self-regulation system is the process of continuing reconsideration of behavior strategies when internal and external conditions have changed. Sometimes this results not only in changes in the methods of achieving the goal, but a change in the goal itself.

When we talk about action, operations, motions, and so on, a major consideration is temporal-spatial localization of the different elements of activity. All this refers to the morphological description of activity, in other words, its construction. When we talk about self-regulation, we describe activity functionally. In this case, during the analysis function blocks or specific systems of actions and operations resulting from coordination and integration emerge, through which one can achieve specific goals. Any function may be realized by different mental operations and motor actions. Function blocks can be defined in an unvarying manner, but their content may and, indeed, will vary. Task context is also extremely important in determining how function blocks are used, their degree of development, and how they perform. The content of the function block can change but the purpose of each function block in a self-regulation model will remain the same. The meaning of function blocks in any specific activity can be understood only in relation with other function blocks. Hence, the morphological description of an activity should be linked with the functional description of that activity. Here we encounter the issue of stages of activity analysis.

The concept of function block was used in the first section of this chapter when the self-regulation of motor actions was described. In cognitive psychology when one tries to describe psychological microprocesses, the notion of function block is also used. In this case, it describes psychological microprocesses such as iconic memory, mechanism of scanning information, and so forth. However, when Russian psychologists try to develop a block scheme of self-regulation of activity as a whole, the function block has a much more complex architecture. Under these circumstances, the function block represents a coordinated system of subfunctions that have a specific purpose in the structure of activity. For example, some function blocks can be responsible for creating an image of a situation, or a program of execution and complex function of control and corrections. The function block at the later stage of analysis may be decomposed into more detailed subfunctions. Below we will describe the model of self-regulation of activity developed by Bedny.

A glance at Fig. 2.6 shows that a fully elaborated model of self-regulation has the following characteristics:

1. All of the function blocks are interrelated.
2. Each block is a functional subsystem directed to achieve specific subgoals of activity.
3. Each function block is part of a flow of activity with multiple entry points and exits.

Theoretical and experimental data presented in preliminary sections of the book describe different psychological mechanisms that can be presented as function blocks of process of self-regulation. In this section, we present some additional information that describes these function blocks. First, we list the function blocks of the process of self-regulation:

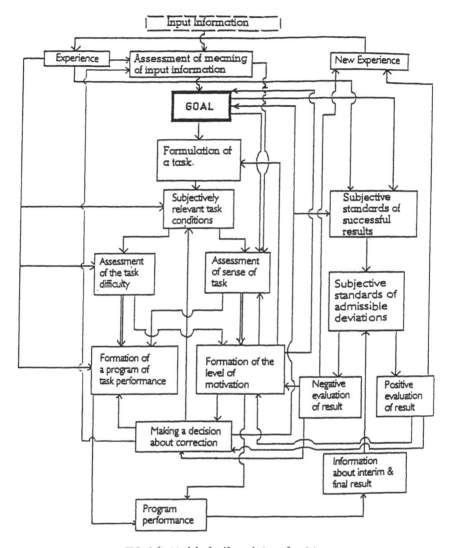

FIG. 2.6. Model of self-regulation of activity.

- Past experience.
- Assessment of the meaning of the input information.
- Goal.
- Formulation of the task (goal-conditions).
- Psychological representation of the subjectively relevant task conditions.
- Assessment of task difficulty.
- Assessment of the sense of the task.

- Formation of a program of task performance.
- Formation of a level of motivation.
- Making a decision about correction.
- The program performance.
- Information about interim and final result.
- Negative evaluation of results.
- Positive evaluation of results.
- Subjective standards of admissible deviations.
- Subjective standards of a successful result.
- New experience.

The function blocks in this model represent processes that occur over time in a series of sequential stages and that have inherent in each of them a number of more molecular subfunctions. For example, the block of "past experience" assumes a memory subfunction; "evaluation of results" implies a subfunction of comparison. The model, therefore, is in a sense a top-level summary of integration concepts that have been discussed previously. At the same time, the model is a combination of different elements. For example, some function blocks are very broadly descriptive of the past (like "past experience") whereas others are more narrowly descriptive of the immediate present (like "information about interim and final results").

Some of the function blocks have been discussed previously; others require further discussion. The block "formation of task" is an important component of the self-regulation process. Perceiving and understanding the same task can be accomplished by different people in different ways. The subjective representation of the task is a complicated component of activity that becomes obvious in creative processes. An example that demonstrates the importance of the aforementioned function block is the "four points" task shown in Fig. 2.7. The objective formulation of the task is as follows: "The requirement is to connect four points by using three lines without raising the pencil from the paper in such a way that the pencil returns to the starting position."

Solving the problem is possible only when the subject imagines that he or she can leave the space restricted by the four points. From this example, we can see that the subjective formulation of the task and its objective representation are different. In the process of subjective formulation of the task, both conscious and unconscious elements are present (Ponomarev, 1976).

The next function block that operates in close interdependence with "formation of task" is "subjectively relevant task conditions." However, they constitute distinct stages of activity. "Subjectively relevant task conditions"

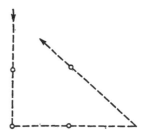

FIG. 2.7. Four points task. From
Ponomarev (1976).

is a dynamic mental representation of individual components of the task
relevant to a specific stage of task performance. This function block includes
not only conceptual but also imaginative components of activity. When the
imaginative components of activity predominate, one can use the notion
"operative image of task" introduced by Oshanin (1977). How the individual
imagines the subjectively relevant task conditions is vital for the development
of the correct approach to the problem. A practical example illustrates the
point (Bedny & Zelenin, 1988). A major catastrophic shipping accident,
involving a collision between a freighter and passenger ship in the Black
Sea, took place in 1984. The Russian coauthor was assigned the task of
investigating the psychological factors involved. The captain had been examined by psychiatrists and found not to be suffering from any major dis-
orders. He himself could not explain the reasons for his refusal to heed the
warning of his mate, who alerted him to the danger of a passenger ship
cutting across his course, when it was still 5 miles away. Though the captain
had clear visual contact with the ship, and had been warned by his mate,
he failed to take any emergency actions. As a result 400 people perished.

Examination showed that although he was an experienced seaman, the
captain's models of subjectively significant task conditions were inadequately
developed with respect to mapping from the spatial to the temporal domain.
In other words, the captain's model of subjectively significant task conditions
did not balance space and time components of the task. The distance of 5
miles between the ships was perceived by the captain as substantial, though
the facts were that the physics of the vessel movement required over 2 miles
to halt the ship's momentum. Accordingly, it was incumbent that the captain
act well in advance of the time that any contact would be made by the two
vessels. As a result of this investigation, recommendations were made to
include more focused training on cultivating the transformation of the spatial
information about the distance between ships into temporal information.

Next, it is necessary to discuss a very important function block, "assessment of task difficulty." The essential purpose of this block is the evaluation
of task difficulty by the person who performed it. Evaluation of the difficulty
of the task is a cognitive process that involves consciousness of the objective
complexity of the task. Evaluation of task difficulty may be performed in

different ways. For example, the individual may under- or overestimate the objective complexity of the task. Individuals may assess the task as being so difficult that they will reject the task, if this is objectively possible. On the other hand, individuals underestimate the difficulty of the task and as a result select an inadequate or inappropriate strategy for goal attainment, thereby failing to solve the problem.

Evaluation of task difficulty is a complex process that involves comparison by the person of his or her own capability with that of other people to solve the same problem. There may be different decisions. For example, a person may conclude that a task is very easy for others but very difficult for him or her because of individual idiosyncrasies. On the other hand, the task may be objectively very complex, but because of an individual's experience and competency, she or he may conclude that the problem can readily be solved. The evaluation of task difficulty affects task strategy, task performance, and most prominently, motivation, which energizes the individual to reach the goal. Evolution of task difficulty connected with a person believes that he or she does or does not possess the necessary abilities and experience to solve the problem. At the same time, the evaluation of the task difficulty does not predetermine the motivation process and the success of solving this task. Because the function block "assessment of task difficulty" has complex relations with the other function blocks (e.g., "formation of level of motivation"), no simple derivation from the evaluation of task difficulty to outcome may be made. Incorrect assessment of difficulty can result in inadequate personal sense or motivation to sustain efforts for completing the task.

"Past experience," as it says, reflects the past experience of subjects. The interaction of past experience and new input information results in assessment of the meaning of the immediate input information. When the input information has an optimal of complexity and is sufficiently clear for the subjects, then anyone may infer the same meaning from the identical input information. Here objective input information is transferred into similar subjective meaning for the subjects who have similar past experience. When input information is very complex and vague for the subjects, then substantial variations in subjective meanings will be observed among diverse subjects exposed to that same information. Subjective meaning is therefore an interaction of input information and individual past experience available to the subject at the present time.

It is necessary to discuss more explicitly the emotional and motivational components of self-regulation. Earlier, we considered these processes solely in connection with the formation of a goal and evaluation of the significance of the tasks (see Fig. 2.6). Information about the external situation interacts with the present need that governs motivation as well as with set and instructions (immediate intentions). This refers to the emotional state of the operator immediately before task performance is initiated. We refer to the motivation that precedes the stimulus information to the dominant (primary) motivation. This initial emotional state is associated with the function block "assessment of the sense of task." The function block "formation of level of motivation" can be treated as secondary, action-oriented motivation. Because of feedback, the activity is tuned to goal formation and goal directedness. Because the function block "formation of the level of motivation" is connected with the block "making decision about correction," it became possible to correct the goal or even change the goal.

Past experience (emotional as well as cognitive) also influences the sense of task. The functional relationships between "past experience" and "assessment of the sense of the task" are mediated through the function block of the "assessment of the meaning of input information."

The previously described function blocks and their interrelationship can be considered as part of the more general system of self-regulation presented in Fig. 2.6.

What we have described is the first subsystem of self-regulation. This subsystem may be called "goal formative and orientational components of self-regulation." At this state the subject orients to the situation, develops goals, formulates a subjective image of the task, and singles out the salient components of the image of the task. These cognitive components and emotional components act as "advanced organizers" (Ausubel, 1968) that preliminarily orient the subject. These have a dynamic character that permits ongoing revision.

As was mentioned previously, motivational processes have a leading role in the formation of the goal. People do not merely solve a task or problem, but they also have a bearing or "attitude" toward the problem. The greater the sense of task for the person and its associated value, the greater the effort expended and the greater the motivation.

Motivation changes in different ways depending on whether a task has or has not been successfully accomplished. For example, if a task presents a risk of danger for an individual, the sense of the task will exhibit greater intensity. The motivational aspects of self-regulation can be decomposed into two blocks represented by "assessment of the sense of task" and "formation of the level of motivation." The first block is the emotional-evaluative component of motivation. The second is the inducing component of motivation. Both are interdependent. Sometimes these are not harmonious. Through the interaction of these two blocks and their complex interactions with the other blocks, the structure of motivation is subject to alteration. This means that the motivational blocks affect executive and cognitive components of activity. In turn, motivational blocks are subject to feedback from negative and positive interim and final results. Motivation in the self-regulation model functions in the following way: Based on an accepted or formulated goal, the subject formulates a task, singles out task relevant conditions, evaluates difficulty and significance of task, and based on that, develops a level of motivation.

The motivational components of self-regulation often unfold automatically. For example, a person can understand the importance or danger of a specificaction, and in response to this importance or danger, an emotional reaction will be immediately elicited. In a new task, a different situation may emerge. A person may not be aware of the danger or importance of the situation. This may result in an inadequate emotional reaction to the situation. This, in turn, may cause an inadequate motivational process, which is connected with the mobilization of his or her efforts. Here one can see the influence of cognitive activity on emotional-motivational components. The negative evaluation of an outcome acting through the function block "making a decision about correction" can influence selection of subjectively relevant task conditions and cause reevaluation of task significance, which in turn alters the level of motivation. Obviously, accidents may occur due to failure to evaluate the significance of the task appropriately.

Accordingly, the function block "assessment of sense of task" is distinct from the function block "subjectively relevant task conditions." The latter may be well defined consciously in general terms, but not transformed in a manner that relates to personal significance.

In the same way, one can demonstrate a relationship between motivational processes and evaluation of the difficulty of the task. People may single out certain task conditions as being important, others as being less so. These conditions may be evaluated as either "difficult" or "easy" depending on the individual features of the person and his or her past experience. Increasing the difficulty of the task causes intensification of information processing, increasing thereby the workload on memory and thinking processes.

The significance and complexity of the task are correlated. The difficulty of the task influences, first of all, the function block "formation of program of task performance," discussed later. The prominence of this causal relationship is designated with a double-line arrow in Fig. 2.6. The function block "formation of a program of task performance" is an informational function. "Assessment of the task difficulty" also affects the level of motivation. The more difficult the task, the more energetics components must be mobilized.

Thus, the relation between "assessment of the task difficulty" and the "formation of the level of motivation" is depicted with a single-line arrow in Fig. 2.6. On the other hand, we depict the causal relationship between "assessment of the sense of task" and the "formation of the level of motivation" with a double-line arrow.

"Assessment of the sense of the task" also affects information processing, thus affecting the "formation of a program of task performance." The relationship among these function blocks is depicted with a single-line arrow in Fig. 2.6.

"Formation of the program of task performance", "program performance," and "making decision about correction" involve the executive (implementation) components of activity. The function block "formation of the program of task performance" involves the development of the program of execution of actions directed to achieve the accepted goal. This function represents information regarding the method to be used in achieving the task goal and may or may not be conscious. This program is developed prior to task performance and may also be modified during that performance. The program depends on "subjectively relevant task conditions" and "assessment of task difficulty." At the same time, "program performance" often affects the function block and assessment or sense of task. Where the task is simple, it will be performed automatically and the individual may be conscious only of task outcomes. If task conditions change or if unanticipated outcomes occur, the program of activity or task performance may reenter consciousness.

Self-regulation requires evaluation of outcomes and decisions about corrective processes. Performance of the task enables the individual to obtain new data regarding deviations that the individual can take into the next performance cycle. In some cases, changes of "Goal" of activity, "Subjectively relevant task conditions," or "Assessment of sense of task" can be so significant that it can result in changes of "Assessment of meaning of input information." This can happen because of feedback influences from functional block "Making a decision about correction" to functional block "Assessment of meaning of input information." In this case, a situation can emerge that the person can perceive and interpret the information incorrectly. We can encounter this situation when we investigate serious accidents and catastrophes, when the person who has caused these accidents cannot correctly explain his obvious irrational actions. For example, the people who were responsible in Chernobyl of committing a sequence of grave mistakes could not correctly explain their behavior. One of these people, before he died in a clinic, ". . . kept repeating the same agonized thought: I did everything right. I don't understand why it happened" (Medvedev, 1991, p. 61).

The nature of the self-regulative process is that it forms a gestalt, every function affecting every other function, the degree of influence of one function on another depending on local task conditions, past experience, and idiosyncratic factors.

Social learning theory has emphasized the significance of an individual's setting of a subjective standard of success. This standard depends on standards that were established by other people through a process of social comparison (Bandura, 1977, 1982). Bandura showed that during the process of creating this standard, the individual's evaluation of the standard established by other people who had performed the same or similar tasks is important in self-regulation theory. Development of subjective standards of success has a dynamic relationship with the goal and past experience. Such standards can be modified during the goal formation and decision-making process. In some situations, a subjectively accepted goal may be utilized as a subjective standard of success, although by itself the goal does not contain sufficient information for the evaluation of the results of activity. Accordingly, ongoing experience with attempts at goal attainment and past experience may establish a subjective standard of success. Sometimes the formation of this function block occurs almost immediately; in other cases, its formation emerges as a complicated process associated with evaluation of interim and final results of task performance.

In the literature, more attention is paid to proving that a subjective standard of success exists, than to how this standard is formed. Konopkin, Engels, and Stepansky (1983) developed an experimental method that made it difficult for the subjects to develop such a standard. The subjects had to measure the length of an object by using special instruments. Because the precision of the measuring instruments was not very great, the subjects could not conclude anything from one trial. Thus, multiple repetitions of the measurement was required. Based on observation and verbal protocol analysis, the experimenters could determine how the subjects changed, what they considered a "good" standard. It appeared that they used various strategies to discover the best standard. A common strategy was to develop a preliminary standard based on the frequency of obtaining similar results in measuring. Progressively, the subject refined the value of the standard by discarding the results that sharply deviated from the standard because this result was considered as erroneous.

In the other study of positioning actions similar to one described in section 2.4, Bedny also used two conditions. In one condition when subjects missed targets, they received information about only erroneous actions. In other experiments, subjects additionally received a shock in the left hand. It was discovered that in neutral conditions without shock, the subjects preferred use of most of the width of the target as the parameter of success (risky strategy). They also did not often correct their strategy after erroneous actions. Subjects in the experimental "shock" condition changed their strategy of activity, narrowing the acceptable width within the target (unrisky strategy). At the same time, they immediately corrected their own action when they approached the edge of the target. As a result, the speed of action decreased under the experimental shock conditions.

In Morosanova and Stepansky's (1982) study, subjects had to stop a pointer moving in a circle, at a specified objective stop position. Errors varied to the left and the right of the stop position. Errors made in one area (left or right) were considered by subject only as chance variations and the subjects did not correct their subsequent actions if the pointer stopped in this area. This area was considered by subjects as admissible deviation. Errors in the other area (with more significant deviation) were considered by the subjects as critical and they corrected their actions if the pointer departed into this area.

Morosanova, Stepansky, and Sablin (1980) studied sharpshooting and found much the same thing. It follows from these studies that the subjective standard of success while necessary, is not sufficient for self-regulation. Personnel must define which errors are significant and which are not. This definition derives three function blocks in the model of self-regulation: "subjective parameter of success," "subjective parameter of deviation," and "information about interim and final result." As was pointed out previously, these criteria are, influenced by objective criteria but are also affected by past experience, individual differences, and level of aspiration. This process may occur either consciously or unconsciously.

Any output that varies from the "subjective parameters of a successful result" and exceeds "subjective parameters of admissible deviations" will be evaluated negatively or positively. Of course, if the output is exactly what is required, it will be evaluated only positively. As part of this process, individuals must define which deviations are significant and which are not.

Most approaches to self-regulation establish criteria for the correction of influences that register only when the person obtains negative results. The model presented in this chapter, on the other hand, also enables modifications to be made in response to positive evaluations of interim or final results. For example, "positive evaluation of result" may increase achievement motivation through connection with the block "formation of the level of motivation" and be reflected at a cognitive level in the functioning of "making a decision about correction." This psychological process where success motivates further success marks the distinction between psychological and homeostatic self-regulation.

In self-regulation there are sequential stages of motivation. The first stage is associated with the formation and acceptance of the goal of activity and of task. The next stage of motivation is associated with evaluation of the task (difficulty and significance). A third stage of motivation is connected with the performance of task. The fourth stage is associated with the evaluation of the result of the activity. These stages are intimately interconnected. For example, positive or negative evaluation of result of activity has meaning for a person only in a situation where the person is motivated to achieve the goal. Similarly, motivation for achieving a goal maybe unrelated to the motivation for the processes required for executing the activity. Often, one encounters this in both work processes and learning activity. Positive motivation for goal attainment may be combined with negative motivation for task performance. This may be considered as a contradiction or conflict within a motivational process. By the same token, positive motivation for goal attainment may combine with negative motivation attached to the assessment of the task difficulty. This may lead to the rejection of desired goal. Negative evaluation of one's own results may be uncoordinated with positive external sources. In this situation, external reinforcement cannot be effective. At the same time, overestimation of one's own result can lead one to ignore external evaluation, thereby losing valuable information.

In the model, there is a function block for "new experience." The result of the self-regulation process is not merely achieving a goal, but also the formation of an experience of goal attainment.

The goal in the self-regulation model has informational or cognitive components that have a complicated relationship with its motivational ones. This contrasts with goal-setting theory in which the goal has two primary dimensions: content and intensity. The most important attribute of the goal content is difficulty. The basic postulate of goal-setting theory is that difficult or unattainable goals, if accepted, lead to greater job performance than do easier goals (Lee, Locke, & Lathman, 1989).

However, Russian psychologists would say that goal-setting theory merely substitutes complicated motivational problems with the phrase, "If the goal has been accepted." They feel that this phrase subsumes a number of motivational processes of which goal difficulty is only one important cognitive characteristic.

According to the model of self-regulation, a complex relationship exists between difficulty and motivation. Increasing the difficulty of the task does not always produce an increase in the level of motivation, as stated in Locke's goal-setting theory (Locke & Lathman, 1984). Locke and his colleagues used a static explanation: Increasing the level of motivation always follows from increasing the difficulty. However, the level of motivation depends on a complex relationship between the function blocks of "assessment of task difficulty" and "assessment of the sense of task."

These two function blocks can be depicted graphically as bars "difficulty" and "significance" because significance is linked to sense of task. Task difficulty and significance can be changed from a very low level to a very high level. If that difficulty is very high and significance is very low, as designated in Fig. 2.8, a person will not be motivated to perform the task because he or she does not have any reason to waste a great deal of effort on a task that has no significance for him or her. In Fig. 2.9 where the "difficulty" is very low and the "significance" is very high, the task is easier to perform and very significant; this produces an optimal level of motivation.

If "difficulty" and "significance" are low but not extremely low, the person is still motivated to perform the task, but with a very low motivational level. This kind of work is usually perceived by a person as monotone and boring (see Fig. 2.10). If "significance" is more sharply reduced, in spite of the task being easier to perform, the person stops working (the work is very easy, but also very boring and is, therefore, not important).

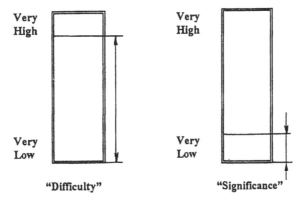

FIG. 2.8. Very high difficulty and very low significance result in the absence of motivation.

If the task is optimally significant to the person and is reasonably difficult (see dash lines on Fig. 2.11), and if task difficulty is increased, motivation also is increased. The higher the initial level of significance, the more the level of difficulty results in increasing motivation.

Here we consider the construct of self-efficacy developed by Bandura (1987). According to Bandura, the stronger the belief in self-efficacy, the stronger the person will pursue the desired result. He suggests that people with high personal efficacy set more difficult goals and show greater persistence in the pursuit of them, while people with low efficacy set lower goals and often abandon them in the face of adversity. It follows from this that all motivational manipulations are effected through self-efficacy.

From the self-regulation point of view, if a person evaluates a goal as very difficult due to his own low self-efficacy, the resulting negative influence

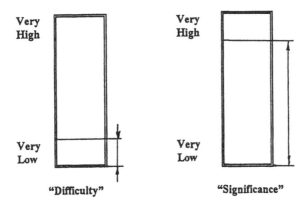

FIG. 2.9. Low difficulty and high level of significance stipulate an optimal level of motivation.

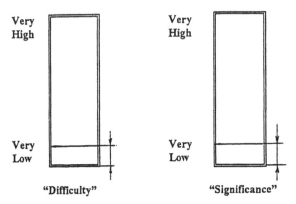

FIG. 2.10. Relatively low level of difficulty and significance result in a low level of motivation. (The work is monotone and boring.)

on motivation increases the probability that the goal will be avoided or abandoned. On the other hand, if the particular goal is significant or highly desirable, those with low self-efficacy can nevertheless be motivated to strive for the goal. In certain situations, high self-efficacy can have negative effects. For example, athletes that regard their self-efficacy as very high may under-estimate the strength of their opponents, diminishing their motivation, and consequently their performance. Sports lore is replete with tales of the "stronger" team losing to an "inspired" underdog.

The model of self-regulation also enables the development of a criteria of success that may not match the objective criteria. Thus, for example, an athletic sports team that tends to evaluate its own talent as inferior to a given opponent may develop as a criteria of success the goal of keeping

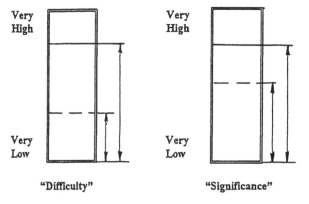

FIG. 2.11. Increasing difficulty of the task when the level of significance is high enough results in increasing the level of motivation and significance of the task.

the score respectably close, losing by a dignified rather than a lopsided score, but losing nonetheless. In accepting this lower criteria of success, the team may even accept a lower quality of performance on their own part than they are actually capable of. Conversely, as noted earlier, because the perception of self-efficacy is based on subjective goals and criteria of success, the weaker team may be highly motivated in competition by "rising to the challenge" presented by a team perceived to be superior.

These examples show that self-efficacy derives from complex self-regulative processes and exercises variable influences—both positively and negatively. We can see that self-efficacy beliefs are not enough to explain motivational processes. Motivation can be explained as a dynamical process which depends on complex relationships between different mechanisms of self-regulation. Thus, the static explanation of motivation as a solitary mechanism is not correct.

Let us consider some examples that demonstrate how we can consider mechanisms of self-regulation while conducting experiments in engineering psychology. It is well known that the apparatus most frequently used in the study of human visual perception in cognitive psychology is the tachistoscope. This is a device that presents visual information for a very brief period of time. Subjects should react in different ways according to the information presented. It is in this same way engineering psychologists conduct experiments while designing visual displays and instrumental panels. They use reaction times and errors as major criteria for evaluation of the pilot's instrument panel or different display apparatus. However, this procedure is often insufficient, because it ignores mechanisms of self-regulation.

In an experimental study conducted by Dobrolensky and colleagues (1975), scientists tried to determine how pilots interact with the visual display when it is showing a failure signal. An instrument presented a failure signal, at random, between other instruments. There was a delay in the reaction to this signal. The scientists first concluded that the delay occurred because the instrument itself was deficient. However, comparing different data about eye movement, response time, and the pilots' subjective opinion proved that there was a considered reason for the delay. The pilots voluntarily delayed their responses to these kinds of signals because they had higher subjective significance in comparison with other signals. Delay responses to emergency signals increases the precision and reliability of the pilots' actions. This means that the delay response cannot be considered as evidence that the display itself is deficient.

Consider this next brief example. A correctly designed emergency instrument is very important to increased flight reliability. If the flight regime is disturbed, the pilot must change the regular sequence of actions, distributing his attention between the ongoing task and the emerging situation. The pilot should also extract the more important components of emergency problems.

However, very often in this situation this does not happen. This can be explained, first, by insufficient functioning of the self-regulating mechanisms "goal," "assessment of meaning of input information," and "subjectively relevant task conditions." An insufficient functioning of this mechanism provokes inadequate response actions. These data were proved in special experiments (Beregovoy, et al., 1978).

Very often engineering psychologists recommend paying attention to the modality or physical intensity of the emergency signal. However, this recommendation is insufficient. In some cases, for example, increasing intensity of signals can elicit the opposite result, worsening the operator's performance.

Let us consider another example. It is important that in different conditions the flight speed should not surpass particular limits. Therefore, while approaching the critical speed parameters, the pilot must be receiving correct information about the flight speed. Sometimes, engineering psychologists recommend for these purposes additional acoustical or visual signals. But in special experiments, it was proven that introducing these signals and regulating them according to their intensity is not correct for this particular emergency situation, because the pilots shifted their attention from the ongoing task to the emergency signals. The time for interpretation and comprehension of emergency signals significantly increased, and the transformation of attention from one goal to another resulted in the pilots' forgetting information about the ongoing task.

For the purpose of the efficient functioning of the mechanisms "assessment of meaning of input information" and "subjectively relevant task conditions," the goal of task performance in an emergency situation should not be ignored. Based on these data it was recommended that the designer place emergency signals directly into the speed indicator. When the pilot approaches the critical speed, a particular area of the scales lightens with a red color. The intensity of light and the area of the light increase as speed increases. In this situation, the pilot is more likely to act quickly and precisely to emergency situations. Basically, it was determined that transformation from one goal to another has negative consequences, particularly in those situations when the ongoing task is more difficult and dynamic. From this experiment we can see that understanding the mechanisms of self-regulation permits the scientist to more correctly interpret data obtained from experiments and more efficiently make design decisions. It is also obvious that we cannot consider the operator a device for processing information. A person actively selects information and interprets it according to goals and significance of activity.

We should also note the main difference between the model of self-regulation described above and the model of self-regulation developed by Lord and Levy (1994) as the base of the control theory. There exist two models of self regulation, the homeostatic models, and the goal-directed models.

The homeostatic self-regulative system is directed toward maintaining the constancy of certain parameters of the system. Many technical and physiological systems operate according to this principle. For example the concentration of blood sugar, blood pressure, etc., are maintained within fairly narrow limits on the basis of homeostatic principles. The model of self-regulation developed by Lord and Levy works according to the same principle. In their work they wrote: ". . . control systems serve to compare sensed information to standards and act to keep resulting discrepancies within adaptable limits" (Lord, Levy, 1994, p. 338). This principle, which is the main principle of their model, contradicts the psychological principle of the theory of regulation of activity. This principle cannot explain why a person consciously creates various deviation or discrepancies from a standard course of action, analyzes these deviations, and forms new goals or strategies toward the achievement of these goals.

Functional blocks of the model of self-regulation developed by Lord and Levy are not fully supported by psychological data, and are not based on general theoretical principles. For example such blocks as, "Standard," "Comparator," "Deaccession Mechanism" are deduced using a functional principle. However blocks such as "Sensor," and "Effector" correspond to the morphological principle of description. While blocks such as "Task or Activity" and "Disturbance" cannot even be presented alongside mechanism because they themselves are not such. We will not go into the analysis of this model any deeper since it has been criticized in detail by such authors as Locke (1994), Hacker (1994), etc.

The goal-directed, self-regulative model is useful in the analysis of volitional processes since it helps to explain "the freedom of choice in goal selection" and "freedom of action." In our work we examine will in its narrowest definition as the support mechanism that sustains activity toward the obtaining of a goal insight of obstacles or contradicted motivational tendencies. It is in this narrow definition that the concept of will can be of interest to human factor specialists.

We should note that according to our model of self-regulation when studying volitional processes, blocks such as "Goal," "Assessment of the task difficulty," "Formation of the level of motivation," "Assessment of the sense of task," and "Making a decision about correction" are of great importance. For example when obstacles are present on the path to a certain goal the difficulty of the particular task increases, this in turn increases the motivation, and is accompanied by volitional efforts for the reaching of the goal. If however these volitional efforts are too great or the goal is not significant enough for the subject, then through the decision making block the decision is made to drop the goal, or to adopt a new goal.

Mechanisms of self-regulation are not developed evenly. Multiple interactions engender various rates and synchronicities within the development of various function blocks. Some of these may be developed first, other later, other simultaneously. Initially in the development of particular functional systems of self-regulation, the function block of "Assessment of Meaning of Input Information," "Goal," "Assessment of the Sense of the Task," and "Formation of the level of Motivation" assumes greater importance. Due to feedback and feedforward they become more precise and differentiated. Function block "Assessment of the Sense of the Task" evaluates the subjective significance, not of the task (which does not yet have an objective existence), but of the significance of the goal and the meaning of input information. At later stages, other function blocks begin to develop.

In particular cases, some function blocks are not developed in precise form, nor are they all of equal weight. For example, dynamical representation of task may be reduced in self-regulation process. In this case we may only attend to the function block "Formation of Task." Thus, it may be seen that in practical application professionals may emphasize certain function blocks and their relationships, as well as neglecting or treating as background other function blocks and their relationships.

2.6 CONCLUDING COMMENTS

The models of self-regulation presented in chapter 2 are intended to explain the self-regulation of activity as a whole, including those kinds of activity that cannot be determined by external reinforcements. Of course, in regulating this kind of activity, motivational components also have an important meaning. Self-regulation appears as a result of the existence of a complicated functional system involving different function blocks. Each function block has a particular function and includes a complicated combination of different psychic processes. Self-regulation can be realized unconsciously. An example of this type of model is the model of self-regulation of condition reflex presented in chapter 1.

A more complicated kind of self-regulation, which involves the existence of a conscious goal, is known as the conscious self-regulation of activity. The model describing self-regulation is presented in Fig. 2.6. Specific psychological processes and phenomena that are involved in different function blocks are not considered in this model. The content and specificity of each function block can only be considered in the context of a particular activity. The process of conscious self-regulation is subordinated to the accepted goal. At the same time, it provides flexibility and adaptivity in a wide range of human behavior. A person actively and creatively selects and interprets information about his environment and himself in order to optimize the strategies of activity performance.

The conscious level of self-regulation was even observed during the performance of the simplest reactions. The study demonstrated that even the simplest sensorimotor activity cannot be considered as an activity that is directly determined by a stimulus. A person reacts to this stimulus with different strategies depending on the stimulus' characteristics, the specificity of its interpretation, environmental conditions, and the person's inner state. The more complicated and undetermined the activity is, the more important role the conscious level of self-regulation plays in achieving the goal. In this chapter, it was demonstrated that it is possible to intentionally influence the strategy of activity performance, motivation, and goal formation through the influences of different function blocks of self-regulation.

What most differentiates the self-regulation theory from other theoriesis its comprehensiveness and level of detail. Other theories also deal with goals, tasks, feedback, input information, experience, and meaning (among other components), but as one can see from Fig. 2.6, self-regulation theory makes us aware of the tremendous complexity of the regulative process. For example, where many theories assume the existence of fixed goals, self-regulation places much emphasis on the formation of goals. Again, Russian theorists place great emphasis on the premise that the motivational components of the self-regulation theory are differentiated from the cognitive components, although they interact, of course.

As can be seen from the examples provided in this chapter, self-regulation theory can be efficiently utilized in experimental ergonomic design. This issue is discussed in more detail in chapters 4 and 5. In chapters 6, 7, and 8, we consider how this theory can be applied in analytical design and learning.

Information Processing

3.1 OVERVIEW AND INTRODUCTION

3.1.1 Information Processing

It follows from the definition of activity that activity is a system that integrates cognitive behavioral and motivational components. Accordingly, psychological processes are studied in the context of the theory of activity. Knowledge about psychological processes is required for describing mental actions. This chapter provides some new information about mental processes from the perspective of the theory of activity. There are three levels of information processing: sensori-perceptual, imaginative, and verbal-logical.

In creating new images, two subprocesses exist: productive imagining and creative imagining. Operator activity occurs in four stages: receiving information, evaluating and updating information, decision making, and implementing the decision.

No matter how simple and basic a task is, such as those studied in psychophysics, the entire activity is involved: goals, motivations, and strategy. Sensory discriminations vary according to the criteria used by the subjects. Idiosyncratic and contextual factors are also important determiners.

Operative or working memory is distinguished from short-term memory. Memory productivity is dependent on a number of parameters: information, structural-spacial, modality, and temporal. Memory does not exist on its own: Memory influences activity, but activity also influences memory. Short-term memory is sensitive to goals, task characteristics, and motives.

In indeterminate tasks, the process of task formulation may be more difficult than task solution. Russian psychology distinguishes a special type of thinking, operative thinking, which is found in solving practical problems. It has three functions: planning, control, and regulation. There are three sequential stages of task solution: (a) manipulation of single elements of the situation, (b) integration of these elements, and (c) development of a general principle leading to task solution.

3.1.2 Decision Making

There are two types of decision-making theory: rational decision theory and psychological decision theory. The latter is of greater interest to ergonomics, because it corresponds with the decision making performed in the "real world." In the latter, task conditions may be unclear and dynamic; the individual often has incomplete knowledge, and lacks criteria to evaluate consequences. An important variable in such decision making is the level of aspiration.

There are two classes of the decision-making process: decisions on the sensori-perceptual level and those on the verbal-thinking level. Components of the second level may include: lack of information, restricted time to decide, situational uncertainty, and the individual's strength of will to overcome uncertainty. Decision types can be divided into: impulsive, risky, balanced, cautious, and inertial. Actions taken as a result of decisions can be either reactive (largely based on the objective situation) or proactive (largely based on idiosyncratic factors.) Risk is interpreted in different ways in Russia and America. In Russia, it is associated with danger (hazard); in America, it is associated with the possibility of failure.

3.1.3 Attention

Attentional behavior depends, like other behavioral processes, on goals, motives, and strategies. Several models of attention are discussed: One considers attention as mental effort, which may involve single or multiple resources; another treats attention as an information-processing mechanism. Studies demonstrate that stimulus recognition requires only a low level of attention; decision making requires a high level.

The Russian model of attention builds on the work of American psychologists, but adds to their models a complex function block called a "regulative integrator," which creates a goal to which attention is directed and compares one's expectations (anticipations) with input information. The Russian model includes three separate subsystems: informational and energetic substructures, and coordination mechanisms.

3.1.4 Images

In Russian psychology, one of the important aspects of the study of work activity is research on imaging and imagination during the worker's task performance. Although perception of an external object determines the character of the image, the perceptual image is not a copy of the external stimulus; it is affected by subjective factors, especially when the object presented is ambiguous. These are primary images; secondary images appear in the absence of real objects.

An image, being the outcome of an activity, performs not only a cognitive, but also a regulative, function. The most important component of an image is the goal. A number of types of images are discussed: conceptual model, conceptual image, operative image, image goal, image object, and psychic image. There are both similarities and differences between the operative image and situational awareness. The use of images in analyzing the pilot's flight task is discussed.

In summary, individual processes and mechanisms like the reception and processing of information and mechanisms and attention must, in the Russian theory of work activity, always be viewed in the context of the total activity.

3.2 RECEIVING INFORMATION

In research on cognitive processes from the theory of work activity viewpoint, Russian psychologists postulate three levels in processing information: sensori-perceptual, imaginative, and verbal-logical (Lomov, 1984).

The sensori-perceptual level describes the situation in which objects and environmental phenomena directly affect a person's sense organs.

The imaginative level describes the process of image formation. The image emerges purely internally, without being influenced by a sense organ. In the Russian language, two concepts are distinguished: *predstovlenie* (imaging) and *voobragenie* (imagining). *Predstovlenie* are those processes that are reproductive in nature and are linked to the memory process. For example, having once seen an actual apple, one can imagine an apple without actually viewing that apple. In contrast, *voobragenie* describes the creation of totally new images that do not depend on prior viewing of a physical object. The latter process is closely related to intuitive thinking.

When a person performs a problem-solving task, two systems of processing information are involved one being conceptual and the other imaginal. The more complicated and indeterminate situation, the more the mechanisms of imagination are involved (image manipulation.) This process has an intuitive character and a person is often not consciously aware of the processing.

In a situation where a person creates new images, the term imaging is used. In imagining, two subprocesses exist. In the first, one creates images based on a description of objects or phenomena (productive imaging). In the second, one creates new images, ideas, and objects as in creative thinking. The second subprocess is called creative imaging.

The third level of information processes is verbal-logical. In conceptual thinking, a person operates with abstractions and generalizations framed in symbols. The most developed symbol system is human language, although there are other symbol systems, such as mathematical or graphical.

The activity of an operator in the human-machine system occurs in four consecutive stages: receiving information, evaluating and updating information, decision making, and implementing the accepted decision. In this section, we briefly discuss some aspects of an operator receiving information. Receiving information requires sensory and perceptual processes.

From the activity point of view, it is important to specify what the stimulus is and the experimental procedure in studying it psychophysically. Zabrodin (1985) defined "the stimulus" as a specific message that has motivational coloring and an inducing force in relation to an individual's action.

The essence of a psychophysical experiment consists of an analysis of how a subject solves the tasks presented. The key word is *task*. We not only study isolated psychic processes, such as sensation, but the subject's whole activity, including the goal, motivations, and strategy. Although sensory processes predominate in this task, the sensory processes work with other psychic processes. The psychophysical task, like other tasks, requires problem solving.

A task that is objectively presented to subjects during a psychophysical experiment and the subjective representation of the task are assumed not to be the same. Because of this, Zabrodin, for the purpose of describing their relationship more precisely, divided the task into different components (Zabrodin & Lebedev, 1977). These include the goal (the requirements of the task) and the situation (the conditions in which the goal is presented to the subject).

Based on these, the subject forms his or her own subjectively accepted task. Because the tasks in psychophysical experiments are sensory ones, the components of the task require of the subject sensori-perceptual and imagery responses. Accordingly, the subject extracts the image goal and the image of the situation as basic components of the task. Very often the subject's image goal is vague and contains multiple alternative meanings of the objective stimuli and alternative possible ways to achieve the goal (even though the experimenter tried to present precise instructions to the subject).

During task performance, the subject develops one or more strategies to achieve the goal and criteria for evaluating the success of his or her task performance. In consequence, the task and the manner of its performance have probabilistic features. Because of this, Zabrodin (1985) also introduced the notions of "sensory space" and "space of decision." Sensory space is comprised of alternate possible psychic images that can adequately represent the situation. The space of decision includes alternative possible responses that may accomplish the goal of the psychophysical task. From this it follows that the basic problems of psychophysics are the analyses of cognitive functions: the study of the structural and dynamic features of the psychic image and sensory space, and the study of regulative functions of that image and its influence on the decision-making process.

Figure 3.1 describes the structure of psychophysical problems and, obviously, it involves more than simple sensation and perception. The essence of signal detection theory (Swets, 1964; Tanner & Birdsall, 1955) is that much more is involved in apparently simple psychophysical tasks than sensation and perception. Similarly, the fundamental assumption of the activity approach is that, even in the simplest situation tied to the extraction of weak signals, a problem exists that must be treated from the perspective of problem solution. The characteristics of the image of the same reality will be different in different tasks (Zabrodin, 1985).

Determining sensory thresholds is the basic task of psychophysics. This requires the solution of a problem by the subject. The problem is often solved differently in a laboratory experiment and in the real-world situation. Data describing absolute and differential thresholds determined by psycho-

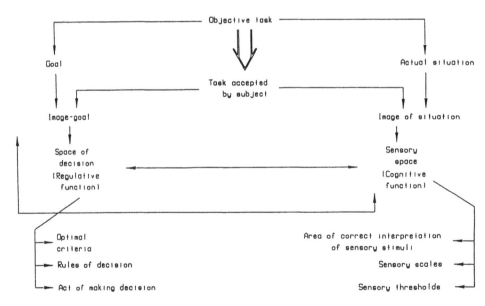

FIG. 3.1. The structure of fundamental problems solvable by psychophysics. From Zabrokin (1985).

physical methods under laboratory conditions very often cannot be used in realistic and practical situations, because these situations are not reproduced in the laboratory. This is why the activity approach to determining sensory thresholds makes it necessary to model the real-world conditions of an operator's activity.

Threshold data represent the extreme performance of which the individual is capable, which means that she or he must exert maximum effort to perform. Although this information is needed, it is not expected that in practical ergonomics tasks, such as designing an appropriate display, one would require of the operator the same maximum effort, because this would lead to undue stress. Because of this Dmitrieva (1964b) introduced the notion of the "operative threshold" (Fig. 3.2), which can be considered a practical, working threshold.

The relationship between stimulus differences, as an independent variable, and the accuracy and speed of the discrimination process as the dependent variable, is a nonmonotonic function. Up to a certain level, increases in the differences enhance the accuracy and speed of discrimination. Beyond that level, the accuracy of the discrimination will deteriorate. When the differences between two stimuli become so profound, a new adaptive process is required.

Assume that an operator must discriminate the brightness of two visual stimuli to a certain point; greater brightness differences between two bulbs result in more rapid and precise discrimination (Fig. 3.2, Point A). Further increases do not enhance the accuracy or speed of the discrimination. If one continues to increase the differences between these two stimuli beyond a certain point (Fig. 3.2, Point B), the discrimination process deteriorates due to the need to readapt to these more extreme differences. Thus, Point A may function as an operative (working) threshold. According to Dmitrieva

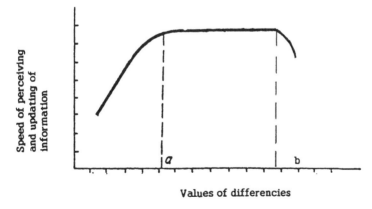

Values of differencies

FIG. 3.2. The operative threshold of discrimination between two stimuli. From Dmitrieva (1964).

(1964b), the operative threshold is 10 to 15 times greater than the differentiation threshold. These phenomena are often applied in practice. For example, the operative threshold may easily be calculated by multiplying by a factor of 10 to 15 the initial differential threshold values from standard tables of thresholds.

Varying instructions and information in the psychophysical task lead to a change in the psychic image of a task and determine the selection of information from a sensory space (see Fig. 3.1). From the position of the self-regulation model of activity, the specificity of the selection of information depends on such function blocks as "goal," "subjective image of task," "subjectively relevant task conditions," "evaluation of the difficulty of task," and "evaluation of the sense of task." Changing the components of self-regulation leads to different strategies in performing sensory tasks.

Signal detection is determined not only by the sensitivity of the sense organ, but by the criteria used by a person. This is why in psychophysics we have the notion of the "threshold region." In other words, the threshold is not a constant value.

Researchers who applied signal detection theory to psychophysical problems considered that the observer's response can be presented as a point in a sensory space (region). In this region, values can be viewed as one-dimensional if the physical variables associated with them can be shown along one axis (Swets, 1964).

On the contrary, Zabrodin (1985) posited the idea of a multidimensional sensory space. It is assumed that sensory space can be evaluated in terms of multiple criteria. Each criteria can be represented by its own axis. The set of those axes represent a multidimensional space.

Bardin et al. (Bardin, 1982; Bardin, Sadov, & Szen, 1984; Bardin & Voytenko, 1985) began to implement this concept in their studies. They demonstrated that the observer, while performing the task of signal detection in noise, can use additional criteria corresponding to different axes of sensory space. They derived this from the fact that "the same changes of physical characteristics of a signal can lead to different changes in its psychic image" (Zabrodin & Lebedev, 1977, p. 95). For the sake of simplicity, Bardin et al. restricted themselves to the use of only two axes' criteria in the multidimensional space.

For example, for signal detection in a noisy environment, when two signals are very difficult to discriminate on the basis of loudness, the subjects began to use additional criteria. They began to perceive previously unnoticed qualities in the acoustical stimuli and used these as discrimination criteria. For example, they reported that the sound seemed to become dimmed, brilliant, resonant, dull, and so on. These features were named "additional sensory features" (qualities) by Bardin. They may be not only acoustic but may also possess other modality qualities more or less difficult for subjects to verbalize. It was hypothesized that they very often appeared involuntarily from the subject's need to solve the psychophysical problem. Specific major acoustical qualities of stimuli are very difficult to verbalize. The additional qualities can vary in their intensity and specificity, which causes them to have their own axis of measurement in sensory space.

The results of one experiment when the subjects used additional qualities in the discrimination of two sound stimuli are presented in Fig. 3.3.

The distributions of $f(x/s_1)$ and $f(x/s_2)$ are illustrated on the X-axis (loudness) and the additional sensory features or criteria $f(y/s_1)$ and $f(y/s_2)$ on the Y-axis. The set of sensory attributes as shown in Fig. 3.3 may assume various positions in relationship to both axes. It is important that these two sets are separated more along the X-axis than the Y-axis. This is explained by the fact that the Y-axis is an additional axis.

A subject will use the criteria along the Y-axis only if detection along the X-axis is to difficult. For example, Fig. 3.3a shows the situation when $d'_x >$ d'_y. In this situation, the subject can easily detect the signals according to the level of loudness. However, the difference in the intensity of the compared signals S_1 and S_2 begins to decrease; after a certain length of time d'_x $\approx d'_y$ (Fig. 3.3b). If one continues to decrease the difference between the signals along the X-axis while preserving the differences along the Y-axis, Fig. 3.3c results. In this situation, $d'_x < d'_y$, the subject will go from using the criterion of loudness (X-axis) to using the other criteria along the Y-axis.

When the operator cannot detect a signal according to sensory criteria, she or he attempts to use additional criteria, such as sharpness or brightness. Bardin (1982) and Bardin et al. (1985) demonstrated that the use of these additional criteria is determined by idiosyncratic factors. If an individual cannot differentiate between stimuli according to the specified criteria, they will develop their own.

One of the most important conclusions in contemporary psychophysics, based on signal detection theory, is about the independence of sensory sensitivity from nonsensory factors (e.g., decision making). The influence of nonsensory factors is restricted by the shifting of criteria along the axis of observation. During this shifting, the fraction of correct and incorrect

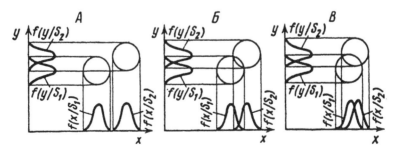

FIG. 3.3. The model of two dimensional discrimination. From Bardin (1985).

answers either increases or decreases. However, the sensory sensitivity factor d' does not change (D. M. Green & Swets, 1966).

Bardin and Zabrodina (1988) attempted to see whether d' would change if the differences between the stimuli related to thinking problems. The control task initially was performing a sensory task (discriminating between straight lines that differed in length on some trials). The experimental task involved the same discrimination, along with the solution of thinking problems. The problem-solving tasks varied in complexity: Some were imaginative, others numerical. In the imaginative problems, subjects had to perform geometric tasks whose solution depended on whether or not the lines were equal in length. The numerical problems required subjects to transform mentally one numerical sequence into another.

Bardin and Zabrodina (1988) found that a minimal level of complexity in a thinking problem does not affect sensory sensitivity. Beyond this level, however, an increase of complexity in thinking problems did change sensory sensitivity in a statistically significant manner (the value of d' changed). Subjects used various strategies connected with distribution and allocation of attention between sensory and thinking task. Bardin and Zabrodin's studies showed that detection is closely linked with cognitive processes and that cognition can determine sensory sensitivity. Hence, the premise widely accepted in psychophysics, based on signal detection theory, that sensory and cognitive functions are independent, is not always the case. Such an independence observed in simple cognitive tasks, but does not obtain in more complicated tasks that involved sensory functions.

The next stage of receiving information connected with perceptual processes can be divided into two different types. One type of perceptual processes is associated with perceiving unfamiliar objects. Subjects construct images of objects and meaningfully interpret these data. Another type of perceptual process is recognition, which can be considered as the ability of subjects to relate the presented object to previously exposed objects. Persons match current stimuli with what is stored in the memory. In recognition, perceptual processes are integrated with the function of memory.

The process of recognition should be distinguished from identification. Identification always suggests dividing all presented stimuli into two classes: those that are identical according to all features to templates stored in memory (positive identification) and those that are not identical to templates according to at least one feature (negative identification). Sometimes a template can be presented externally. Identification can be considered as a reaction performed according to relation of similarity or differences between two stimuli or stimuli and templates already stored in memory. Recognition is distinguished from identification because it additionally includes categorization (classification) and its verbal designation (Shekhter, 1967; T. P. Zinchenko, 1981).

From an activity point of view, the perceptual process can be considered as a system of perceptual actions that performs different functions (Zaporozhets & Zinchenko, 1982). For example, it can be orientational, explorative functions or construction functions. Actions that perform construction functions are particularly important when the subject perceives unfamiliar objects and perceptual processes require creation of perceptual images. The perceptual process includes different levels of perceptual actions: detection, selection adequate for task informative features, identification, and recognition.

According to microstructure analysis, the perceptual process can be divided into sequences of very short subprocesses that can be performed in a sequence or simultaneously. They can be presented as function blocks. T. P. Zinchenko (1981) developed a model of visual recognition process according to microstructure analysis (Fig. 3.4).

From this model it follows that the recognition process includes not only visual operations but also verbal operations. These operations suggest selection of verbal templates that can correspond to their visual equivalents. Success in verbalization depends on the number of verbal templates, compatibility of verbal and visual templates, capacity of working memory, past experience, and training. The aforementioned is important for developing the method of training and of presentation of information to the operator. The decision-making process on perceptual and verbal stages later integrates into a unitary decision-making process that at the last stage is expressed by verbalization. Successful interaction between these two systems results in an effective recognition process. It would seem, then, that the decision-making process on the sensori-perceptual level is an important component of an operator's receiving information.

In the operator's activity, the decoding process is also important. If identification is the comparison of the sign with the template, decoding is relating the sign to the object being described by the sign and always requires preliminary identification. The more complex the code, the more time is needed to memorize its relationship with real objects. T. P. Zinchenko (1981) found that for one-dimensional visual codes, decoding time was 1 minute, 30 seconds; two-dimensional codes required 2.5 seconds for decoding; and three-dimensional codes took 6.61 seconds. Decoding is the transformation of the signal's image into the image of an object. That is why an abstract code makes the process of decoding more difficult. The speed, precision, and reliability of such a process depends on how the "alphabet of signals" and the "alphabet of the object being described" are related. In abstract codes, the association between the signal and its meaning takes longer to establish and requires more effort.

The sensori-perceptual process includes decision making on the sensori-perceptual level. When making these decisions (about whether a stimulus is one thing or another), it is important to consider previously obtained

104

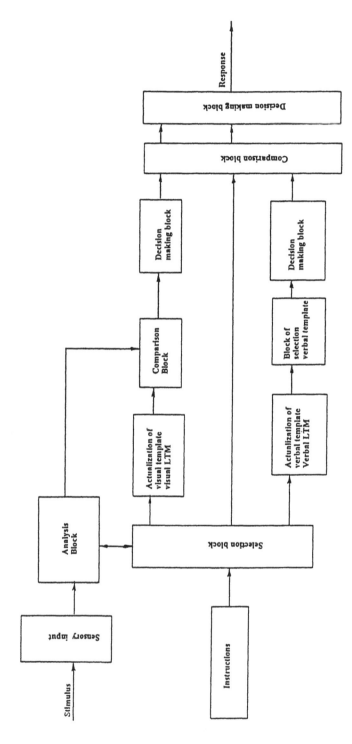

FIG. 3.4. Interaction of visual and verbal system in recognition process. From Zinchenko (1988).

information, whether qualitative or quantitative. Quantitative data involve
the probability of object appearance, whereas qualitative data describe an
object's surroundings. A study by Rubakhin (1974) demonstrates the importance of quantitative information. Twenty subjects, divided into two groups,
one experimental and one control, were shown 16 photo images of differing
clarity at varying intervals between 3 seconds and 1.5 seconds for 2,000
trials. Clarity was determined by the number of lines per millimeter. Subjects
in the experimental group were informed of the probability of different
categories of photos being presented, whereas the control group subjects
were not. Knowledge about the presentation probability of the pictures led
to a significant increase in correct identifications.

Rubakhin (1974) also conducted experiments related to the process of
deciphering army reconnaissance codes. Two groups of average skill were
chosen to decipher the codes. The first group was told that they were
determining the position of a "strong" enemy, whereas the second group
was informed that they were to locate a "weak" enemy. Deciphering information about the weak enemy resulted in a decrease in correct identification,
but deciphering the strong enemy resulted in a sharp increase of false alarms.

The structural scheme of perceiving familiar signals is shown in Fig. 3.5
as a sequence of specific stages; although in reality, the demarcations between stages are blurred. Requirements for the presentation of information
can change depending on what stage of information reception is involved.

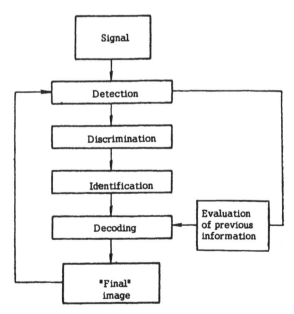

FIG. 3.5. The structural scheme of the process of perception of signal. From
Rubakhin (1970).

3.3 PROCESSING INFORMATION

The theory of activity assumes that information received by an operator goes through several stages of processing.

The work of Sterneberg (1969) was the catalyst for the development of various Russian models of short-term memory, one of which is shown in Fig. 3.6 (V. P. Zinchenko, Leonova, & Strelkov, 1977). As far as information reception and updating are concerned, Russian psychology (e.g., Strelkov, 1972) has been greatly influenced by American and British work and accepts the basic concepts of that work (e.g., B. F. Anderson, 1975; Atkinson & Shiffrin, 1971; Baddely & Patterson, 1971; Lindsay & Norman, 1992; G. A. Miller, 1956; Tversky & Kahneman, 1973; Wickens, 1992) as it deals with sensory memory, short-term memory, long-term memory, and rehearsal.

The model of memory shown in Fig. 3.6 presents a sequence of function blocks that correspond to different psychological operations as well as determine the path of information from its entrance to the operator's response. In Russian psychology, operative or working memory is distinguished from short-term memory.

V. P. Zinchenko and Munipov (1979), with their colleagues, suggested the separation of two kinds of work activity. The first (short-term memory) is associated with the presentation of small amounts of information per unit of time to an operator with immediate decision making and performance. This kind of activity was named "informational search with immediate performance." In this situation, the working memory is not strongly involved.

The second kind of activity is characterized by the fact that the informational search does not end with the immediate performance of action. For example, if an item of information is presented to an operator, he or she evaluates the importance of the elements in the item, then puts the elements in a specific order, and finally keeps this order in the working memory during the task performance. This kind of activity was named "informational search with delay of performance." Putting elements in a specific order requires an operator to keep the defined order and alter it according to the changing situation.

For determining how much time is required for this kind of task performance, Repkina (1967) suggested the following empirical formula:

$$t(x,y) = .174xy + .056x - .28y + 1 \qquad\qquad (3.1)$$

where $t(x,y)$ = time spent for putting in order different objects (time per one object), x = number of objects, and y = number of attributes of the object.

In determining the number of attributes, the dependency of t on x and y can be expressed by a straight line (Fig. 3.7). The angle of the straight

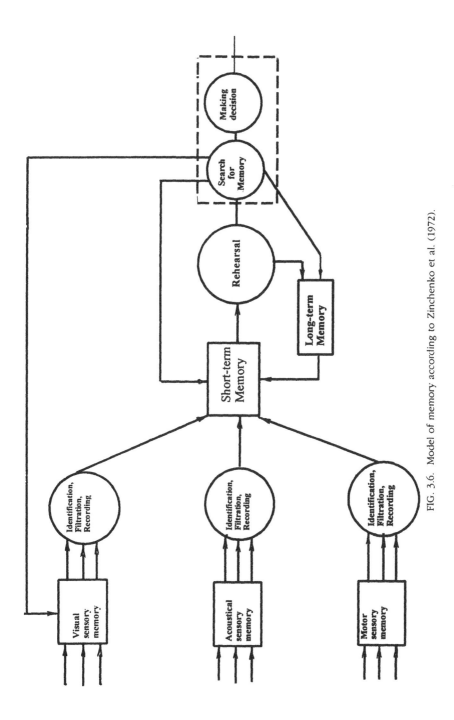

FIG. 3.6. Model of memory according to Zinchenko et al. (1972).

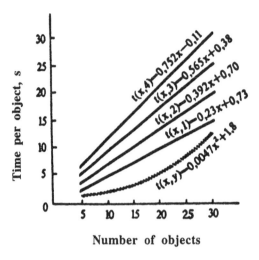

FIG. 3.7. Relationship between time (t) spent to put objects into a specific order and overall number of objects (x) and number of attributes (y) for each of these objects. From Repkina (1969).

line changes if the number of attributes changes. During skill acquisition, subjects develop more economical and efficient ways of searching, putting objects in order, and automatizing memorization actions. Yntema (1963) found that an operator performs better when working with a few objects that have many attributes than with many objects with only a few attributes.

Mel'nik (1976) developed a special questionnaire that helped him to estimate probabilistically the influence of various memory productivity parameters that are shown in Fig. 3.8. These attributes and the degree to which they influence memory are informational (.48), structural-spatial (.15), modality (.25), and temporal (.12). Informational parameters include the quantity of information and method of coding and informativeness of signal (ability of signal to transmit information). The quantity of information, as viewed subjectively, is affected by its regularity, degree of organization, and the operator's memory capacity. The study showed that redundancy had a positive influence on memory by increasing probability of stimulus occurrence. Informativeness in the statistical sense derives from the probabilistic features of the stimuli presented. Nevel'skii (1969) discovered that stimuli with a higher and lower probability of occurrence were memorized more efficiently.

The method of coding is also related to informational parameters. The selection of a particular code depends on the characteristics of the task performed by the operator. For example, the principle of similarity means that symbols should recall a real object. If, however, these objects are in reality very difficult to discriminate from others, the principle of similarity cannot be used. This is because symbols also become similar to each other. In these cases, it is better to use an abstract code or combination of abstract codes with symbols, which is to use the principle of similarity (Litvak, Lomov,

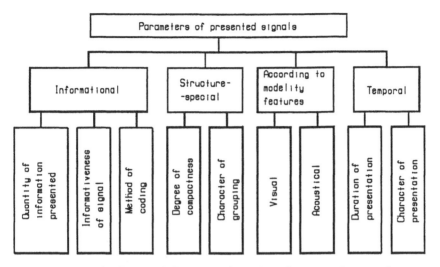

FIG. 3.8. Classification of objective attributes that influence productivity of memorization. From Mel'nik (1976).

& Solveichik, 1975). For example, in Fig. 3.9a, a submarine and a launch are shown on the left, and on the right symbols for these vessels are presented. They were developed based on their similarity to the actual vessels. The symbols are very similar to each other, and the operator can confuse them. In Fig. 3.9b, an abstract symbol is used for coding the same launch. This symbol is less similar to the launch, and is less likely to be confused with the symbol designating the submarine.

The other principle is called "use habitual association." For example, a red signal means danger, a yellow is a warning, and green means a normal state.

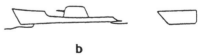

a

FIG. 3.9. The method of coding of different objects. Adapted from Litvak, Lomov, and Soloveichik (1975).

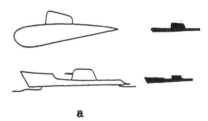

b

The method of organizing presented material is related to structural-spatial parameters (see Fig. 3.8). Memorization productivity in these cases is influenced by the compactness and grouping of stimuli. Memorization efficiency depends on where a symbol is place in the sequence of stimuli. It is especially important in cases when increasing lengths of sequences stress short-term memory capacity. Symbols at the beginning of the sequence and at its end are reproduced with fewer errors (Fig. 3.10; Dushkov, 1981).

One way of improving the compactness of presenting information in order to improve memorization is to use special tables, formulae, or equations. In these, a specific symbol conveys specific information about the object referred to.

The method of coding according to modality is also important to memory. Information retained in working memory can be represented in two forms or codes: verbal (words) and spatial (visual images). Wickens, Sandry, and Vidulich (1983) described the best format for different memory tasks. For example, speech or text is more efficient for a task that demands a verbal working memory. It is more effective to use special spatial-visual codes for a task that demands a spatial working memory.

When one presents the operator with visual information, which should be kept in the working memory, it is important to distinguish between two kinds of errors. One type is connected with recognition and the other with recall (Lindsay & Norman, 1992). For example, if one presents the operator with different letters under conditions that make it difficult to identify them, she or he may report P instead of R, O instead of C, or vice versa. These errors are connected with visual features that are common to a number of

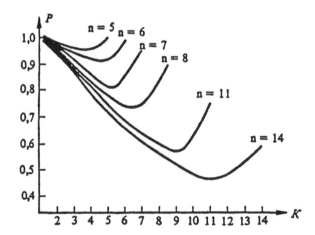

FIG. 3.10. Relationship between correct reproduction and position of a symbol in presentational sequence of symbols. From Dushov and Smirnov (1986).

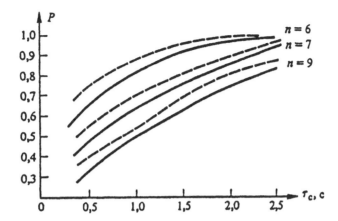

FIG. 3.11. Relationship between probability of correct reproduction (p_{cr}) and time of presentation of the symbol (t) and number (n) of symbols. From Krilov (1986).

letters. If the operator keeps letters in the working memory, acoustical errors can occur. The operator may report D instead of C, X instead of F, and so forth. This can be explained through the rehearsal process. Errors of recollection from the working memory suggest that visually presented information is translated into an acoustical form.

With regard to the temporal parameters of stimuli, the length of the stimulus sequence and the character of its presentation (simultaneous or consecutive) have the most effect on the memory's productivity. The probability of the correct reproduction of information (Pcr) from these factors is shown in Fig. 3.11 (when the number of symbols equals 6, 7, or 9) (Krilov, 1972). The solid lines represent the value Pcr when the information is presented sequentially and the dashed lines represent the value Pcr when the information is presented simultaneously.

3.4 THINKING

Another important component of work activity is thinking. An operator may have to perform cognitive tasks that are ill-defined. Based on the information available to them from the work situation, the worker may not be able immediately to formulate a task. This makes the task a problem and so it must be formulated as a problem. The process of task formulation can often be more difficult than the actual task solution. The situation may be complicated because of time constraints. During the transition from problem situation to task formulation, the relationship between instrumental and noninstrumental stimuli is very important.

Stimuli presented by equipment displays are instrumental stimuli; stimuli that are not presented by equipment displays are noninstrumental stimuli, internal to the individual. The latter are of two types: those based on instructions and those based on learning. The ability to use noninstrumental stimuli is important in mastering skills and in intuitive thinking because they influence the task formulation process. Instrumental and noninstrumental stimuli can contradict each other, producing an incorrect task formulation. For example, a fire warning light may illuminate but the pilot may see no evidence of fire.

Russian psychologists tend to distinguish among different types of thinking processes, depending on the type of work involved. Thus, for example, they distinguish between theoretical and practical thinking, the latter being more important in examining task performance.

Practical thinking can be categorized as manual-manipulative, operative, and supervisor thinking (Zavalishina, 1985). Manual-manipulative thinking is considered more suitable for blue-collar workers. When an operator works in an automatic and semiautomatic system, operative thinking processes dominate. In this situation, the operator uses cues from instruments and manipulates controls. Sensori-perceptual and imaginative processes are extensively involved in this type of thinking. Supervisor thinking is directly linked to its influence on subordinates. Here, logical components and planning become very important. Because humans are involved in supervisor planning, the task has a social involvement. One of the components of this type of thinking is the formulation and distribution of goals among personnel and the coordination of these goals (Kitov, 1981). For operative and supervisor thinking, a time limit is usual; because time limits in Russian concepts can produce emotional elements, operative and supervisor emotional components are also involved in operative and supervisory thinking processes.

These are strange concepts to Western scientists. Different tasks obviously involve different types of thinking processes, but the emphasis is on the task, not worker distinctions. A mathematician or physicist obviously does not utilize highly proceduralized thinking, because their tasks do not require such a type of thinking. A worker performing a repetitive manual or semiautomatic task may have to follow very formalized procedures, although as has been pointed out previously, the need for such work is being reduced because of increasing automation. What is interesting here is that the relationship between types of thinking and types of work is made at all.

The most important features of operative thinking are (Pushkin, 1965):

1. Operative thinking is directed toward the solving of practical problems.
2. A practical action is formed on the basis of operative thinking and is immediately analyzed in practice. From this it is possible to immediately correct and change the thinking process.

3. Operative thinking is often performed under time constraints and can be accompanied by stress.

Operative thinking has three functions: planning, control, and regulation. In some cases of operative thinking, diagnostic functions are important.

The following components of the thinking process are important: creation of structure (the development of enlarged meaningful units of thinking based on the connection between different elements of a situation), dynamic recognition, anticipation of the effects of the final situation, formation of algorithms (developing principles and rules of task solution), and sequence (development determining the sequence of actions required to perform the task).

For example, an experienced chess player does not manipulate the individual figures; he or she manipulates groups of chess pieces as separate units that are part of a total problem. Depending on his or her strategy and the stage of the chess problem, he or she can extract different relations between the chess pieces and use these relationships as units of the thinking process. The process is quite different with an inexperienced chess player. The experienced player can recognize the final solution of the chess problem in an early or intermediate configuration of chess pieces without conscious awareness of intervening logical steps.

Three stages of task solution can be identified. In the first stage, the individual tries to manipulate individual elements of the situation. In the second stage, she or he attempts to integrate these elements. The third stage involves the development of a general principle of task solution based on the correct interconnection of the elements of the situation. This last enables one to perform the task.

The study of operative thinking may involve various methods. Subjects can be asked to solve typical tasks and verbalize as they perform. If visual information is important, eye movements can be recorded. It has been found (see V. P. Zinchenko & Vergiles, 1969) that when visual stimuli are presented, if the strategy of task solution changes eye movement also changes. Another method involves the analysis of operator errors, difficulties encountered, and, naturally, task duration.

Rubinshtein (1973) suggested that the thinking process is first of all analysis and synthesis. When trying to understand a problem situation, it is necessary to decompose elements of the situation; this is analysis. However, because objects and their features do not exist in isolation, it is necessary to discover their interrelationships; this is synthesis. Analysis and synthesis are always finally integrated.

Various mental actions are performed during the thought process: comparison, classification, systematization, generalization, abstraction, and concretization.

When eye movements are recorded in studying cognitive task perform-ance, it is necessary to distinguish between perceptual eye movement and thinking eye movement. The latter often is unconscious. Such actions are explorative in nature and directed to extracting the meaning of a situation. It may seem to the reader that objective differentiation between perceptual and thinking eye movements would be possible to make, but Russian re-searchers make the distinction based on a comparison of the physical eye movement data with qualitative analysis (subject self-report and examination of task characteristics).

With the help of sequential fixations the subject can extract from identical objects distinct essential characteristics that are germane to the solution of a particular problem. These features are not always available to consciousness or verbalization. According to Tikhomirov (1984) these thinking eye-move-ments provide for the extraction of "nonverbalized operational meaning," a notion that is close to the "situation concept of thinking" developed by Pushkin (1978). From this it follows that the thinking eye-movement is involved in the performance of inner mental operations. The deduction of the interaction between the elements of the situation gains significance in this case. The same external situation is constantly changing in the mind of the subject. This continual change of the situation in the mind of the subject in light of its external constancy has been called the "gnostic dynamic" (Pushkin, 1978). At the base of gnostic dynamics lies self-regulation process, which provided the mental transformation, evaluation, and correction of the situation. This process is carried out using nonverbalized, unconscious thinking operations. They can be transformed into conscious, thinking actions, that can be verbalized. In this case thinking actions are directed toward conscious goals. Therefore con-scious and unconscious components of self-regulation can be extracted from the thinking process, and they are interconnected with the verbalized and nonverbalized aspect of thinking. The thinking process can be presented as the interaction of conscious and unconscious strategies and their mutual transformation. Thinking is analyzed not simply as a process but as activity which consists of mental operations and actions.

Tikhomirov (1984) also studied how the blind play chess with only the sense of touch. In this situation, he also discovered the difference between percep-tual and thinking actions. The chess players were allowed to use only two fingers of the right hand (the thumb and the index finger). A ring with a light attached to it was put on the index finger. The movement of the finger with the ring was recorded by video camera. The movements of the finger were compared to the combinations of the chess board. The verbal reactions of the subjects were also analyzed. In this study, it was also shown that sensorimotor actions directed at extracting nonverbalized operative meanings exist.

In this study it was shown that one of the most important purposes of these actions was to provide tactual functions. However, these actions should be distinguished from the perceptual tactual actions. Motor actions that are involved in perceptual processes are directed toward analyzing the information regarding shape, position, the distance between objects, and so on. This type of action is important for the reception of information by the operator. In contrast, motor actions described by Tikhomirov (1984) are directed to interpret the situation, analyze the interaction of the elements of the situation, anticipate the dynamical elements of the situation, and the like. These actions extract nonverbal operative meaning of the situation. The aforementioned actions are not perceptual but are thinking operations. They also differ from executive-motor actions directed to transform the situation. The thinking actions just described are closely connected with intuitive processes and are very important in tasks involving sensori-perceptual features. It thus appears that verbalization is not the only component of the thinking process. At a later stage of thinking, actions can be partially verbalized by an operator. This is in essence the process involved in concurrent verbal protocol analysis (Newell & Simon, 1972).

The solution of problems may be treated as a sequence of cycle processes the purpose of which is the reduction of differences between current state and goal-end state. The problem solver is working on sub-goals as a means of reaching the ultimate goal. Means-ends analysis that was introduced by Newell and Simon (1972) in the theory of activity is treated as a self-regulative process. Failure to achieve the formulated goal, disappointment of proposed hypotheses constitutes feedback information for modifying the formulated goal, changing strategies, etc. Therefore, thinking is considered as a cycle process where the notions of goals, sub-goals, feedback, strategies, etc., plays an important role.

In this self-regulation process exploration has profound significance. Exploration may be performed in external, motor form, on the one hand, or in mental plans, on the other. Usually these two kinds of exploratory activities are intimately inter-connected. This is evidence for the unity of external and internal activity during the thinking process. Based on the evaluation of these exploratory activities, it is provided formation and correction of mental actions and strategies of thinking. Of course, during the solution of practical problems, external explorative activity is not always feasible. In such cases exploration may be explored only through a mental plan. In this self-regulation process emotional and motivational factors have very important implications. Due to self-regulation of thinking activity it is provided not merely the acquisition of new knowledge, but the production of new knowledge.

Based on the goals of the activity and the problem-solving process itself, the individual develops operative images of the task that reflect the dynamics of its solution in subtle ways. The actual ongoing situation is compared with these operative images. Based on this, the method used in performing the task can be corrected. The process of performing a task, therefore, depends not only on information presented to the operator, but on a system of internal operative images created by the operator. These act as filters that select information during task solution (Hacker, 1980).

The analysis of the transformation of the goal has great importance in the study of operative thinking. Changes in the internal operative image of the task, and the reformulation of the task, can lead to changes in the goal of the task. This creates progress toward the solution of the given problem (Brushlinsky, 1979). Operative thinking has been studied in the activity of pilots and air traffic controllers (Dmitrieva, 1964a; Pushkin, 1965; Zavalishina & Pushkin, 1964).

Russian ergonomists (e.g., Galaktionov, 1978) have discovered that in operating complex process control systems, it is useful to present the control/display mechanisms in a manner that "mimics" the organization of system components. They refer to this control/display arrangement as a mnemonic scheme, because it presumably facilitates the operator's memory. In effect, they see the mnemonic control panel as improving the development in the operator of a conceptual model of the system.

In studying the process of utilizing such a control panel, Galaktionov (1978) utilized a number of procedures, including the favored eye-movement recording, but also observation, interviews, and most important of all (according to Galaktionov) laboratory and field experiments using what he called a "blind" information search.

In these, an experimental control panel was developed that permitted presentation of alternative versions of the graphic mnemonic scheme. The special characteristic of the mimic control panel was that displays could present information requested by subjects only after the subjects requested the correct information. The problem task was one of troubleshooting a malfunction in the system. The experimental procedure worked in the following manner. If, in requesting information about a parametric deviation, the operator threw an incorrect switch (incorrect for the problem as presented to them) or did so in an incorrect order, the displays would not present the requested information. If the operator operated the switches correctly, they would receive the requested information. This procedure allowed the experimenter to observe the operator's strategy of gathering information.

Eye-movement recording indicated that while solving various problems, the operator often looked at an area of the control panel that was not linked with the information required by the problem. This was not an erroneous search for information. The operator often observed instruments unrelated to the particular problem she or he were trying to solve, for the purpose of determining that the parameter values displayed in these instruments were, in fact, unrelated to the problem. (This is entirely in accordance with what nuclear reactor operators trying to solve a nuclear power plant problem typically do.) Because of this, use of the eye-fixation recording method was determined to be incorrect, unless such data were interpreted properly.

The results of the study indicated that the "best" layout of instruments on the control panel depended on the past experience of the operator and his or her skills in solving problems. Because of this, it was suggested that, during the design process, ergonomists should take operator experience into account (although how one does this in the actual engineering design process is unclear, unless the past experience of future operators of the new system is determined before design begins). Galaktionov (1978) used a procedure in which the best subjective version of the graphic mnemonic scheme was drawn by subjects at different stages of learning.

Whether or not the Galaktionov procedure is feasible in design, the results suggest that the operator develops different conceptual models of the system during learning and this will influence his or her strategies of solving malfunction problems.

In summary, studies of thinking can extract the more important aspects. First of all, thinking may be treated as an activity where thinking is described in terms of actions and operations—not merely as a process. Thinking actions and operations may perform distinct functions within the structure of the thinking process. They may be directed toward the promotion of different hypotheses, formulation of goals, mental transformation of situation, etc. Mental actions are not only directed to analyzing existing situations, but deliberating upon past situations, and conjecturing future situations. Logical organization of mental actions develops under the guidance of self-regulation mechanisms.

3.5 DECISION MAKING

Two kinds of decision-making theories have emerged over the years. The first is rational decision theory, epitomized by subjectively expected utility (SEU; Edwards, 1987), and the second is psychological decision-making theory, which is the framework preferred by Russian researchers.

Rational decision theory, which makes extensive use of Bayes' theorem and other mathematical mechanisms, was created by mathematicians and economists. This theoretical approach attempts to discover how people make decisions in highly structured situations of a laboratory type. Psychological decision-making theory, on the other hand, tries to find out how people make decisions in reality, where problem situations are much less structured, ambiguous, and subject to many more influences than those permitted to function experimentally. Decision making has a number of attributes that must be considered. Decision-making tasks may be static or dynamic, well or poorly defined, risky or not risky, involving or not involving diagnosis.

Of greatest interest to ergonomics are dynamic, risky, and poorly defined (open-ended) tasks, because it is these that occur in the real world.

In these cases, task conditions are not clear for an operator. The operator does not know all the alternatives, the situation is very dynamic (may change during problem solutions), and he or she may not have enough information to evaluate the consequences of alternative decisions. Because this kind of task is very difficult to formalize in the laboratory, qualitative analysis of what the subject does is very important. One important variable in this analysis is aspiration level (a^*), which interacts with the objective scale of achievement. For example, if an athlete jumping the high bar overcomes 7 feet, she or he tries to increase the goal by 2 inches. According to this concept, the aspiration level sets the standard used by people to evaluate the outcome of task performance (Kozeleski, 1979). When an individual initially attempts to solve a problem, she or he creates a level of aspiration. From the self-regulation point of view, as the first stage of task solution, a person very often performs exploratory actions, the purpose of which is to use a sequence of trial-and-error actions to analyze their consequences. After that, the individual creates a hypothesis about the situation and then formulates preliminary goals about his or her activity. This goal can be considered a level of aspiration. The aspiration level is developed as a result of the individual's evaluation of the situation, his or her view of what he or she is capable of doing, and alternative goals.

Aspiration level, as it relates to decisions, can be analyzed from the point of view of utility or $u(a^*)$. If a decision is considered to have a lower level of success probability than $u(a^*)$ [i.e, $u(a_s) > u(a^*)$], then the individual discards it. If a task is perceived as very complex, then the person may lower his or her aspiration level. Here potential success or failure can correct the aspiration level.

Kozeleski (1979) presented the following example. In Poland, the government wanted to increase the efficiency of the railroad system to accommodate more trains. Specialists discussed several alternatives and accepted one that included the use of 30-ton wagons. Kozeleski noted that the specialists chose this alternative not because it maximized the goal of more trains, but because it corresponded to the aspiration level of the decision makers.

In well-defined, highly rational closed-loop tasks, one chooses alternatives that maximize utility. In open-ended tasks, one tends more often to choose alternatives that correspond to the decision maker's aspiration level. This aspiration level is multidimensional and is produced by many factors; it may also be changed during problem solution.

When psychologists study ill-defined tasks, they may employ various methods such as observation, questioning, protocol analysis, and examination of erroneous actions and decisions during problem solution. From the activity point of view and the process of self-regulation, aspiration level is only one important component of goal formation.

In this discussion, we emphasize those aspects of the problem described in the Russian language not familiar to English readers.

As pointed out earlier, according to Russian concepts, there are two classes of the decision-making process. The first involves decisions on, the sensori-perceptual level, whereas the second involves the verbal-thinking level. Because the first was considered in section 3.1, the second is discussed here.

The situational components of decision making on the verbal-thinking level include relative paucity and ambiguity of information and time constraints that may produce conflicting hypotheses, uncertainty inherent in the problem situation that can provoke a conflict in the person's motives, and the individual's strength of will which enables him or her to overcome uncertainty.

Kotik (1987) defined risk as action executed in dangerous conditions, where a decision can result in a worse situation than the present undesirable one. The process of making a decision in an indeterminate (hence, by definition, "risky") situation occurs as follows. The operator who is monitoring a system (of any size) determines that the system is exceeding predetermined limits and that a malfunction (an undesirable situation) either exists or will shortly occur, destabilizing the system. The problem to be solved is what to do and the goal is (a) to stabilize the system and (b) to determine the cause of the malfunction. If the situation has occurred in the past, the appropriate decision may be well known to the operator, and he or she recalls that decision and applies it immediately. In such a situation, there is practically no gap between stimulus and the decision response. If the situation is unfamiliar to the operator, and available mental tools are not adequate, the operator will have to consider the alternative possible meanings of the stimuli and generate alternative hypotheses based on those meanings. If there is a conflict between the two goals, the operator will have to decide which one to emphasize. She or he will have to test the alternative hypotheses to decide on one that will determine the decision. The testing process can be performed in one or both of two ways: imaginally and conceptually, utilizing a mental model of the system architecture to determine which hypothesis has the highest probability of being correct; and by physical "tweaking" of the system (e.g., manipulating a physical component and seeing what results).

The idiosyncratic aspects of decision making involve the emotional willing and motivational spheres of personality and their influence on the decision-making process. The decision-making process includes will actions, which are actions connected with overcoming difficult tasks and obstacles, and are accompanied by the emotionally motivated stress. Kulyutkin (1970) suggested a classification of decision making based on the emotional will process. In this situation, he considered the relationship between the process of creating an hypothesis (P) and evaluating effects of the hypothesis (C). Depending on the relationship of these two, he distinguished the following decision-making formulae:

- P >> C is an impulsive decision (the process of constructing an hypothesis overwhelms the forces that constrain decision making).
- P > C is a risky decision.
- P = C is a balanced decision.
- P < C is a cautious decision.
- P << C is an indecisive decision (a contradiction in terms, perhaps, but the individual postpones decision making excessively and when he or she does make a decision, lacks confidence in it).

It is possible to classify decisions in terms of the informational processes involved and the nature of the situation in which decisions are made. For example, Fogel (1961) classified decisions as deductive (e.g., following learned operating procedures with little uncertainty), abductive (determining causal factors and how to proceed in a situation involving at least some uncertainty), inductive (requiring the integration of multiple parameters and utilization of a mental model of a system), and prognostic (decisions requiring the forecasting of transformations in the initial situation.)

Actions resulting from decisions can be divided into two groups: reactive and proactive (Reykovski, 1979). Reactive actions depend on features of the objective situation; proactive actions, which are determined largely by personality characteristics, usually take place in nonstructured, complex, and indeterminate situations.

Kotik (1987) claimed that risk can be interpreted in different ways. In Europe, for instance, risk, or rather the elimination of risk, is considered the desired goal, the achievement of which is associated with physical danger. In America, risk is considered as the possibility of selecting in uncertain situations a decision that may not produce desired consequences; failure is associated not with danger but with not achieving desired results.

The first approach emphasizes physical danger and is oriented to analyze cases where a subject selects more potentially dangerous goals or more dangerous methods of goal achievement. This approach is useful in studying work safety. In the American orientation, risk is connected with choice difficulty and the danger of not achieving goals because of poor decision making. When a person chooses a more difficult goal, she or he sacrifices reaching an easier one where the achievement is guaranteed (J. Atkinson, 1957). The two orientations, although different, are of equal value.

From the standpoint of self-regulation, risk can be considered as a specific kind of self-regulation activity where goals are achieved in a dangerous environment, or as the possibility of not achieving a significant goal. Danger can be considered as the possibility of failure associated with negative physical and psychic consequences.

The difference in interpretation can be disregarded if we assume that risky goal activity is not only the possibility of failing to achieve a desired result but achievement without physical and psychic damage.

The components of danger that affect objective task conditions are: (a) potential rewards of success, (b) the potential for failure, which includes physical danger and the possibility of injury or loss of life, and (c) probability of successful goal accomplishment or avoidance of danger with any individual decision.

The behavior of a person in a dangerous situation from the position of self-regulation depends on the individual's representation of the task, his or her perception task relevant conditions, his or her evaluation of task difficulty, and the motivational components of activity. The risky behavior is a specific method of self-regulation, accompanied by strategies of achieving a goal in a situation of risk.

F. Burkardt (according to Kotik, 1987) developed a schema of the worker's activity strategy in risky situations. From Fig. 3.12 we can see that the tendency for caution depends on three factors: extent of the possible risk, predominant motivation, and work experience. The tendency for caution increases when a person achieves the desired results without risk or when risky behavior leads to accidents. However, the experience of accomplishing positive task outcomes following the performance is risky; behavior weakens the likelihood of caution. Risky behavior is shaped by a person's experiences.

Work, educational, and social activities are the three main kinds of activity. Each of these activities has its own specifics, but they all share two general aspects: process and outcome. The role of these aspects in different activities is not the same. In work and education, the most important factor is outcome; in social activity, process is more important. We see this in sports like gymnastics or ice skating, when athletes achieve greater status by performing more complex actions.

Russian theorists suggest that some individual risk taking occurs not for the achievement of a specific production goal, but rather to obtain a gratifying emotional state, associated with overcoming danger. Such behavior can, of course, have undesirable consequences. According to the model of self-regulation (Fig. 2.6) increased difficulty leads to increased motivational levels, and an increase of task significance rooted in the function block of "sense of task." With repeated experiences of task performance (learning), task difficulty decreases thereby diminishing the "sense of the task" as well as "motivation," in turn depressing in the worker interest in job performance (presumably the worker becomes bored with well-learned tasks). The worker

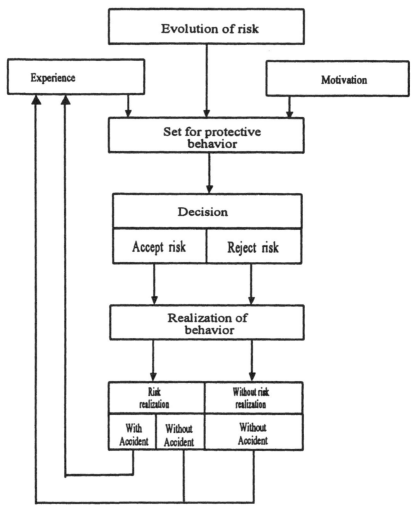

FIG. 3.12. Scheme of strategy of risking and protective behavior according to Burkaradt. From Kotik (1987).

tries to maintain his or her interest in the work performance through acquiring new methods of work, increased productivity, and so forth. As a result, task difficulty will then begin to renew its challenge through increased difficulty. However, the cycle recurs again with experience and skill acquisition. Again, the worker will be led to more complex methods or higher performance goals and increases his or her interest in the task. In sum, diverse emotional tendencies emerge and are connected not only with unconscious but also with conscious factors. In this way, the worker realizes a broad array of self-esteem and satisfaction of self-expression.

On the other hand, these dynamics present the danger of subordinating the task at hand to the emotional gratification associated with risk taking. Successful performance of actions with risk combine both the emotional states of accomplishment and that of having survived the risk. Further complication of the task increases the risk, which may lead to an accident or other misfortune.

In chapter 2, it was pointed out that the quality of performance after a sequence of successes can sometimes decrease. After a short period of failures, the quality of the task performed may increase again. This is explained by the fact that after a continuous successful performance of the same or like task difficulty decreases. Repetitive achievement ceases to have positive motivational significance. The reduction of difficulty and task significance further diminishes motivation directed toward goal achievement. However, after a short-term failure, the significance of the task rises once again and further improvement in task performance is noted.

3.6 ATTENTION

The concept of attention is important not only in cognitive psychology but also in engineering psychology and ergonomics. People do not analyze all the information that is available to them at any one time. One can select individual items of information and still be in a more or less attentive state. Considerable effort has been expended in search for an effective model of attention. In this section, we consider this problem from the point of view of activity theory.

Russian ergonomists make the point that one cannot realistically and validly study attention without consideration of the goal, the individual's motivation, his or her strategies, and the total work context, including the specificity of work requirements. They reject the experimental approach that removes attention from the work situation and in effect fractions it.

Currently we can extract two classes of resource models. One considers attention as mental effort, the other treats attention as an information-processing mechanism.

Kahneman's (1973) model is the best example of the single-resource theory. He assumes a single undifferentiated pool of resources available to all tasks and processes. The performance of multiple tasks presents few difficulties unless the available attentional capacity is exceeded. If task difficulty is increased, or the person must perform two tasks simultaneously, it requires more resources, as well as allocation of these resources. Because the resource pool is limited, the more difficult a task is, the fewer resources are available for the second task. One limitation of the single-resource theory is, for example, in interpreting the well-established empirical finding that when concurrent tasks are in a different modality or use different codes (e.g., spatial or verbal) allocation of resources becomes much easier. Again, however, the single-resource model cannot predict how the various components of information processing (e.g., significance of information, operator strategy) can affect attentional performance.

The multiple-resource theory (Navon & Gopher, 1979) argues that instead of a single, undifferentiated resource, people have several different capacities with distinct resource properties. For example, it is easier to perform two different tasks that require a different modality (visual, aural) than two tasks that require the same modality. In this situation, time-sharing tasks will be more efficient.

Another problem that is also important in the study of attention is the relationship between peripheral and internal processes. A deficiency of the multiple-resource theory is its inability to determine whether the advantages of cross-modal tasks over intramodal tasks are attributable to central or peripheral processes. For example, time sharing may not in fact be the result of central resources, but rather the result of peripheral factors that put two intramodal tasks at a disadvantage. For example, two visual or two auditory task may mask one another (i.e., the cause is peripheral rather than internal processes). Needless to say, at present no one theory of attention completely explained the phenomenon that people do not react to stimuli as a purely reactive wall, but rather create complex strategies of activities directed to achieve a conscious goal. This has significant implications for a theory of attention.

Attention is embedded in a total system of activities and cannot be understood outside the scope of these activities. It is possible, however, to create a model of attention that can integrate conceptual and energetic components and incorporate features of single- and multiple-resource theory.

An information-processing model that comes closest to how Russian ergonomists view attention is Lindsay and Norman's (1992) model, although it too ignores energetics (motivational) components. These authors introduced the concept of "data-driven" and "conceptually driven" processing. Data-driven processing, which is an automatic process, depends on input information. This process is similar to the Russian concept of involuntary attention (Dobrinin, 1958), which through certain physiological mechanisms is connected with the "orienting reflex," which is known in the West through Sokolov (1963).

Conceptually driven processing involves expectation of the generation of hypotheses about the nature of sensory signals, conceptualization, and past experience. Remembered information is combined with information from sensory data. The Russian equivalent is the concept of voluntary attention (Dobrinin, 1958).

One procedure for studying attention is to use two simultaneous tasks, so-called time-sharing tasks. For example, through one earphone a message is given to the left ear; through the other earphone, a message is presented to the right ear. Typically, the instructions require the subject to track one or another or both of the channels.

A new approach was utilized by Bedny (1987). A second task varying in complexity was presented immediately after the first task, which also varied in complexity. The more complex the second task, the more resources it required. By analyzing how the time performance on the first and second tasks changed depending on their complexity, one can determine how the individual allocates attention between the two tasks.

The first task required a response to acoustical stimuli by the left hand. The second task required a response by the right hand when visual stimuli were presented. Because the two tasks were presented sequentially, the peripheral factors of attention in a time-sharing task were eliminated.

When attention is studied, experimenters use either intramodal or cross-modal procedures. In intramodal procedures, the stimuli or messages used are the same modality (two auditory or two visual stimuli). In this situation, they can mask one another; in other words, peripheral factors influence task performance. In a cross-modal time-sharing task, one stimulus or message can be visual and the other can be acoustical. In this situation, peripheral processes are eliminated. This is why cross-modal time sharing can be more effective, not because of separate attentional resources, but because peripheral factors place the intramodal situation at a disadvantage (Wickens, 1992). The advantage of the cross-modal time-sharing task is that it permits researchers to study the hypothesis that, for messages that are addressed to different modalities, different resources of attention exist. This is one of the basic arguments for using the multistore model of attention. If one wishes to prove the advantage of the single-store model of attention, one should conduct an experiment in which peripheral factors are excluded. If, in a time-sharing task, peripheral factors are excluded and, at the same time, both tasks will interfere with each other depending on the complexity of the task, central processes become important. In Bedny's (1987) experiment, cross-modal procedures were used when two tasks were performed sequentially.

In this situation, any delay in the time to perform the second task, depending on the complexity of both tasks, can be mediated only by central mechanisms of attention, which have limited resources or capacities. This strengthens the single-store model of attention. The advantage of cross-modal sharing, in comparison to intramodal sharing of the task, which is performed simultaneously, is that peripheral factors also play a role in the attention process.

However, this does not discount the advantage of the single-store model of attention, if two tasks interfere with each other depending on their complexity. The single-store model of attention, with one undifferentiated pool of resources, can explain interfering in the time-sharing task because of central mechanisms that are connected with single resources, and also peripheral interference of messages.

In Bedny's (1987) experiment, the complexity of the first and second tasks was varied based on changing information conveyed by stimuli. In a second experiment, we also varied the complexity of the first task based on the discriminative features of the stimuli. This permitted a more precise determination of the interrelationship between informational and energetical components of activity.

In Bedny's (1987) study, a meter recorded the right-hand reaction, another, the left-hand reaction. The number of auditory stimuli varied from zero to four; the number of the visual stimuli varied from one to eight. Four switches were available for the subject to respond with the left hand, eight for the right hand. The number of possible choices for the left hand and for the right hand varied from one series of experiments to another. Four male university students were the subjects.

In the first control situation, only visual stimuli were presented to the subjects, and the subjects had to react only with the right hand. Response time was measured. In the second experiment, reaction time with the right hand was measured after a previous left-hand reaction. The number of both aural and visual stimuli was progressively increased, for example, one-two (one sound and two visual stimuli), one-four (one sound, four visual stimuli), and so on until the program was completed with four-eight. Only time was recorded, not errors.

Data obtained in the control situation were compared with the experimental programs, using analysis of variance. It was found that the more complex the stimuli for the left hand, the greater the delay in reaction time for the right hand. The more complex the stimulus for the second reaction, the greater the delay when the stimulus complexity for the first reaction increased. The complexity of the first and second tasks influenced the time performance of the second task.

It is well known that increasing the complexity of a task requires more resources of attention. The more attention was required by the first task, the more difficult it was for subjects to switch their attention from the first task to the second task. The more complex the second task was, the more difficult it was for subjects to allocate attention between tasks. From this it follows that in developing a model of attention, it is necessary to postulate a coordination block that is responsible for allocating resources of attention in accordance with the goal of the activity. This coordination block is called the "regulative integrator." Subjects also displayed indecision in connection with the visual stimulus, indicating that they recognized the digits presented, but could not move their arms.

According to Lindsay and Norman (1992), stimulus recognition is performed by passive automatic processes using a low level of attention whereas making a decision is linked with active processes requiring a higher level of attention and is thus slower than passive automatic processes. This is why in experimental conditions there was a break between identifying the digit and decision making about performance of second action.

Construction and revision of the strategy of attention when sequentially unconnected items of information are presented is the same as when the subject shifts attention from one item to another item of information when they are simultaneously presented. In the experiment just described, a subject cannot keep all of the information about two actions in short-term memory. It is, therefore, necessary to use information from long-term memory. The search for information in long-term memory can be started at any point in a structure of the information base of long-term memory (Lindsay & Norman, 1992). Hence, during the process of extracting the information from long-term memory, the alphabet used by a subject (the number of memory storage units of information) constantly changes; that is, this alphabet is a dynamic. As a result, the information processing changes too.

This means that information theory can be used for predicting operator performance time only if the amount of information used by an operator is not larger than the capacity of short-term memory. Misunderstanding of this leads to erroneous estimations of the quantity of information accepted by an operator over time. For instance, the following formula is suggested to estimate the amount of information taken by the operator from a display (Nikolaev, 1973):

$$H = \log_2[(X_{max} - X_{min})/S] \qquad (3.2)$$

where X_{max} and X_{min} are the maximum and minimum of the gauge's readings and S is the absolute error of reading the apparatus indications.

Such an approach, used to estimate the amount of information, assumes that the operator works as an automatic recognition device, and that the alphabet used to make a choice remains constant and depends on the number of scale intervals.

In reality, the operator can select different areas of the scale depending on the requirements of the task. This means that the alphabet used to make a choice during the perceptual process is continually changed and depends on the goal and system of expectation of the operator. Under these conditions, informational theory that uses bits as the units of measure is useless.

The mechanisms of attention play an important role in the process of integrating individual components of activity into a holistic system. The studies performed by Bedny (1987) indicate single reactions or actions performed sequentially cannot be considered as independent, noninfluenced acts. When tasks are repeated, they are combined into groups, and complex activity strategies are developed. These strategies can be conscious or unconscious. Subjects developed strategies in order to optimize their activity. In the experiment, increasing the speed of the reaction to the sound stimulus led to delay of the reaction to visual signal. Subjects tried to choose that optimum of their reactions, which allowed them to respond quickly to both signals. Allocation of attention between these two tasks plays the more important role in developing these strategies.

During experimental procedures, subjects shifted attention among different tasks, allocating attention and effort among them, attempting to perform one task more quickly, slow down other tasks, or vice versa, correcting errors and attempting to develop a more suitable strategy of activity. Separate actions are integrated into an holistic structure based on self-regulation. In developing these strategies, a particularly important functional mechanism is the regulative integrator.

Building on the work of Lindsay and Norman (1992) and the preceding experimental data, Bedny developed his own model of attention; based on his experimental study of interaction complex reactions (Bedny, 1987). Only that part that deals with cognitive task performance is graphically described in Fig. 3.13.

Following Lindsay and Norman (1992), the model contains the following function blocks: "sensory analysis," "analysis of features," and "short-term memory." These blocks are tightly linked with automatic processing of information. Information goes mostly to short-term memory; only some of the unattended information goes to long-term memory. Again, this is called involuntary attention in the Russian concept of activity. This kind of attentional process does not involve goal-directed activity, but is instead an additional channel of receiving information. Data flowing along an additional channel are attenuated and only partly enter long-term memory. Due to this the attention model has function blocks called attenuators. In this type of attention, the most important attenuators are features of external stimuli instead of internal willing processes that link with our consciousness and goal-directed activity.

Lindsay and Norman (1992) suggested in their model of attention that a function called conceptually driven processing implements switching from one channel of information to another, using attenuators. This function block continually constructs and revises expectations and controls and corrects sensory messages. It is a conceptually driven analysis that is necessary to distinguish it from data-driven analysis. Conceptually driven analysis has a limit to the number of analyses it can conduct in any given unit of time.

In the present model of attention, conceptually driven analysis is included in a regulative integrator. This block creates the goal to which our attention is directed and compares the individual's expectations with input information. Based on this comparison, this function block generates feedback and

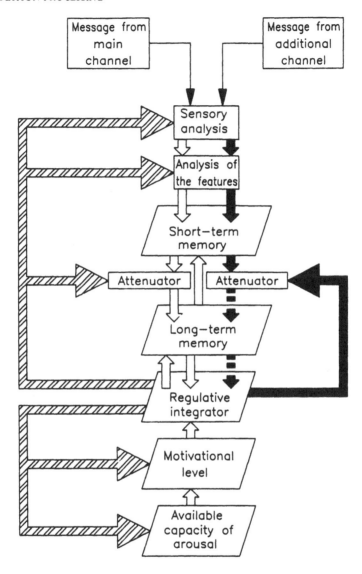

FIG. 3.13. Model of attention for cognitive task performance.

coordinates energetical and informational processes. It is also responsible for developing attentional strategies.

Integration between central and peripheral processes is provided by feedback connections. Thus, attention functions as a closed self-regulatory system that provides a complex strategy of attention directed to attain established goals.

The individual's expectations are included in a broader context of anticipation. Systems of expectation and anticipation are linked with the operator's set and goal of activity (Uznadze, 1961). When these last are altered, expectations are also altered. Initially, the individual has an unconscious set that may later be translated into a conscious goal. Goal and set determine the expectation and are tied with the regulative integrator that is responsible for these functions. Based on comparison derived from feedback influences, every function block pertaining to attention process may be subject to correction. The regulative integrator performs a much more complicated function than the conceptually driven process suggested by Lindsay and Norman. Formation of the goal of observation and the subjective image of the task, both extremely important, are implemented by the regulative integrator.

There is a negative relationship between attenuators. If one of them becomes more active, the other one becomes less active, and vice versa. In this case, the feedback shown at the left in Fig. 3.13 should be reversed. This happens when information secured through an additional channel becomes more meaningful to the individual.

Every information process must have an appropriate level of energy support. This is implemented by the function blocks "motivational level" and "available capacity of arousal." The latter can be seen as a physiological level of energy; the motivational block provides psychological energy support. It directs the person's emotional and activation states toward achieving the goal. By means of feedback the regulative integrator activates these function blocks. Because of the closed self-regulatory system, coordination between peripheral and central process is achieved.

This model of attention allows one to integrate single-resource theory and multiple-resource theory. The concept of "resources" is linked with the energy aspects of activity. The regulative integrator regulates both informational and energy processes, thus manifesting the unity of these two processes. However, even if two tasks exert the same demand for energy resources at the same time, they may have different demands of allocation of these resources to coordinate them with informational processes. The allocation of attention depends, therefore, not only on energy resources but also on the complexity with which informational processes themselves are coordinated and the coordination of informational with energy processes, when there is an undifferentiated pool of such resources. Hence the suggested model can be related to the single-source theory.

It appears then that central processes play an important role in a time-sharing task. It also follows that cross-modal time sharing is better than intramodal, not because it requires different resources but because it simplifies the functioning of the regulative integrator.

In conclusion, this model of attention, which is based on the theory of activity and cognitive psychology, differs from others in that it includes three separate subsystems: informational and energetic substructures and coordination mechanisms. The existence of forward and backward interconnections among function blocks allows the formation of a strategy of attention directed to achieve the activity goal. The central mechanism influences the process of time sharing. This model of attention allows for the explanation of single- and multiple-resource theory data from the same point of view. At the same time, performance in the time-sharing task cannot be explained merely as an attentional process. The performance of any task has its own motives and strategies of activity directed to goal achievement.

3.7 IMAGINATION IN WORK ACTIVITY

One of the most important aspects of the study of work activity is research on the imaginative process during the worker's task performance. This is again a concern that is manifested primarily in Russian ergonomics; one finds hardly anything of this in the West.

For example, in the studies of Pushkin and Nersesyan (1972) and Vekker (1964), it was shown that manipulation of spatial images is one of the central components of railroad dispatcher work. Similar data were gathered by Dmitrieva (1964a) in her study of railroad dispatchers.

The imaginative reflection of reality may be realized through a sensori-perceptual process or a mnemonic (memory) process (*predstavlenya*). During perception of an object individuals develop an image of the object. External information determines the character of the sensori-perceptual image. Nevertheless, the sensori-perceptual image is not a "copy" of the external stimulus; it is affected by subjective factors. The subjectivity of the sensori-perceptual image becomes most apparent when the presented display is ambiguous. Russian psychologists call these "primary" images.

Unlike the primary images, secondary images are those that appear in the absence of real objects. The meaning of these images derives from imaginative memory. They can be divided into reproductive images (*predstavlenya*), and productive or creative images (*voobroshenya*) (Petrovsky, 1986). The transformation from primary images to secondary images results not only in a loss of information, but is in many ways relevant to solving the task problem. This is not always clearly understood. For example, J. R. Anderson (1985) stated that the images tend to be fuzzy and less detailed compared to the originally perceived objects. However, according to Lomov (1966) considering the images as a "vague shadow of sensation and perception" is not useful. The studies of Anan'ev (1960), Rubakhin (1974), and Lomov (1966) demonstrate that the more important specific features of images are selectively retained, including features associated with a group of objects belonging to a specific category. During the transition from perception to imagining the structure of the image changes. Some features are strengthened whereas others begin to weaken and fade away. The specifics of the transformation of a perception into an image depends on the characteristics of the task being solved by the individual. The more important components of the situation, from the point of view of the goal of the activity, stand out, whereas the less important become secondary or are even dropped completely. For example, Lomov found that when a subject was asked to compare similar objects, in his or her imagination process differences emerged between the objects. If the requirement was to compare different objects, similar features were emphasized.

When people imagine their own body movements, the process is connected with visual movement and motor activity (ideomotor action). During visual imaging of the object, our eyes re-create, to some extent, the outline of the object (V. P. Zinchenko & Vergiles, 1969). This principle was used in a study of ideomotor training examining in the activity of three groups of skydivers performing acrobatic stunts. The first group contained experienced skydivers who had not trained for the previous 2 years, a second group included experienced skydivers who had maintained their training, and a third group were inexperienced. They were asked to close their eyes and to imagine the stunts. The time from the beginning to the end of the imaging performance was timed. The first group had difficulty imaging at first, but as time passed, their accuracy increased and they performed precisely. The second group demonstrated precision and accuracy right from the beginning. The last group, which included those who were in training for a minimal amount of time, had the least accuracy. This study was used to develop ideomotor training for soldiers who were performing complicated assembly operations with specialized equipment (Bezdyuk & Tsiganov, 1985).

A person manipulating images performs different mental actions, such as successive perception, rebuilding of images, and their mental rotation. The experiments conducted by Lazootkin and Magazannik (1984) showed that the effectiveness of visually imaging actions depends on the spatial organization of the information and on the specifics of the memorization of the visual information. The manipulation of images becomes significantly easier if there exists an external functioning model of the object. However, in some cases the external support for the manipulation of images can interfere with internal mental manipulation.

An image, being the result of an activity, performs not only a cognitive function but also a regulative one. The information input is analyzed and interpreted based on the image of the situation. The efficiency of an image as a regulator of activity is to a great extent determined by how well the individual anticipates. Lomov and Surkov (1980) showed that the process of anticipation is involved with different levels of processing information. On the sensori-perceptual level, the process of anticipation is limited by the action that is being performed. On the imaging level, there emerges the ability to anticipate potential actions. On the level of verbal-logical thinking, anticipation supplies the ability to plan activities as a whole entity.

The involvement of the thinking process in the manipulation of images leads to the creation of totally new images, which is characteristic of the activity of the scientist or inventor. For the equipment operator productive imagination is more typical, because it is based on machine stimuli and instructions. Nevertheless, in complex tasks, such as malfunction diagnosis, creative imagination may be needed.

The interaction of thinking and imagination leads to the formation of two methods of information processing and forecasting the outcome of the operator's activity. The first method is based on the development of images, the second on conceptual thinking, although both these methods are closely integrated. At any given moment a person is aware of only a small part of the objective situation. In the change from one action to another, the conscious part of the image changes. The image can be compared to an iceberg—at any given moment only the tip is visible above the surface of the water (Zavalova, Lomov, & Ponomarenko, 1986). The major role in forming conscious concepts belongs to the verbal-logical process. The leading role in imaginative reflection belongs to the visual system. The components of the image that are ordinarily unconscious in certain conditions may become conscious. Research (Ponomarenko & Zavalova, 1981) has shown that keeping the image of an object in the operator's consciousness is significantly difficult. The relationship between the reality of what is presented by equipment displays and the image of the situation based on this information is of great interest. The image may not match with the display input and the operator may lose awareness of the real situation. In some cases there is a great imbalance between perceptual data and imaginative data. An adequate image allows the operator to see mentally beyond the equipment display into the actual situation.

The image can be analyzed from two aspects, as a tool for comprehension of reality (cognitive function) and as a regulator of activity (regulative function). In the individual's activity, the image is not only a specific stage of cognition but also a regulator of his or her actions. The most important component of an image is the goal. In this case, the goal appears as an image of the future outcomes of the activity. Interpretation of events and plans of activity are formulated on this basis. In the process of activity, a person can manipulate the image of the object and the situation.

In particular, the notion of image-goal was introduced by Lomov (1977). The imagery components of activity in operator task performance are extremely important. The requirements of the task are embodied in the goal and these requirements can be presented not only verbally, but are also represented in the form of imagery within the individual. Traditionally, imagery is considered only as part of task conditions.

In engineering psychology a number of concepts are associated with the image, for example, conceptual model, operative image, image-goal, image action, image situation, and image of tasks. They are similar but not identical. Most general is Welford's (1961) concept of the conceptual model. He viewed the conceptual model as a global image that forms in the operator's mind. The concept of the conceptual image was analyzed in the framework of the theory of activity by Gordeeva et al. (1975). The imaginative conceptual model of activity is itself a multidimensional reflection of reality described in sensori-perceptual, symbolic, and verbal languages. In this model imaginative processes are dominant.

Research has shown that, on the conscious level, the essence of a situation can be abstracted from the situation, but not its specifics. For example, in experiments performed with chess players, using a tachistoscope, the chess players were shown complex chess positions for very short intervals and then were asked to re-create them. The chess players could evaluate accurately the strength of a position, but could not accurately re-create the actual positions of the figures (Zarakovsky, Korolev, Medvedev, & Shlaen, 1974).

The conceptual model is characterized by great informational redundancy. At any particular time only those image components are abstracted that are tied to the operator's specific task. The conceptual model is available to the operator even before he or she begins an action, based on verbal or written instructions and experience.

Underlining the anticipation aspects of the conceptual model, Oshanin (1977) introduced the concept of the operative image. In contrast to the conceptual model, which is global in nature, this is the specific image of the object formed in performing an action.

An operative image is produced during the process of performing a task. The task determines the structure and contents of the image. An operative image contains information that is necessary to the operator for performing specific actions.

An operative image is laconic. It ignores an object's specific details, including the essential ones, if at a given moment they are not important to the solution of a problem.

An operative image is affected by the operator's set and those aspects of the task that appear to the operator to be most significant. This last is related also to the significance of the task.

The characteristics of the image suggest ways of presenting information to the operator. For example, from the laconic feature it follows that only that information that is really necessary should be presented to the operator. The tendency for any image to deteriorate over time requires that we emphasize only the important aspects of any stimuli we present.

The features of the conceptual model are relatively constant over time, because they are related not to immediate, temporary objects, but to relatively enduring aspects of the system and environment. The content of the conceptual model changes slowly, but may actually increase in precision as the operator gains experience. In contrast, the operative image is dynamic and can be changed quite quickly (Morosanova, 1974). This enables a flexible transition from the perception of one feature of an object to the perceptual perception of a different feature. The dynamic flexibility of the operative image provides for the efficiency with which the image can perform its regulative function. Because of these dynamics, one can perceive different features of an environment. For example, in one study it was found that when the goal of a listening task was changed, the apparent loudness of the stimulus also apparently changed, even though its intensity was actually the same (Oshanin, 1977). This reflects the selectivity of the perceptual process.

The image-goal determines the specifics of selecting information by the operator (Lomov, 1977). There is precise awareness of the goal as an image. The image-goal as a construct is a more stable phenomenon as compared with the operative image. The operative image, however, is one of the most important components of the function block "subjectively relevant task conditions."

This function block also includes logical components that are connected with the thinking process. The specificity of this function block is that it reflects the dynamic characteristics of task performance. The operative image is an important component of this function block. It reflects not only the present and future, but also the past. This image is involved in regulating particular ongoing actions. The content of this image also depends on the significance of different components of the task for the performer. This means that the operative image of the activity also depends on the emotional and motivational aspects of activity regulation. The operative image is more tightly connected with the goal of an activity and its conscious components. However, not all the components of an activity connected with an operative image are conscious. The image includes both conscious and unconscious components of activity. During performance, it is possible to have transitions from vague imprecise conscious components into precise conscious ones and vice versa (Gellershtein, 1966).

The function block "relevant task conditions" is responsible for developing the operative image of the task and extracting the task's relevant logical conditions. An analysis of this function block permits one to conclude that this function block is more tightly connected with those phenomena that are considered to be situational awareness in American psychology.

An analysis of the phenomenon situational awareness is conducted from the position of cognitive psychology (Endsley, 1995; Smith & Hancock, 1995). In this case, the function of memory is more important. From the activity point of view, any function block, depending on the specificity of the task, can involve different psychic functions, in particular, thinking. The operative image is not simply a function of the memory, but rather a creative process.

In the operator's activity, the imaginative components are very important. The operative image performs the role of a filter based on which a person selects, actualizes, and organizes information important at a particular moment. The operative image emerges as the regulator of actions and an instrument of the thinking process.

On the basis of the conceptual model we form an image-goal and an image object (Ponomarenko & Zavalova, 1981). The state of a physical object at any given moment is dynamically perceived through the image of the object. The formation and functioning of the image of the object are performed on the basis of a continuing input of information about the object. All components of the image are closely interconnected. For example, with the change of the image goal, there occurs a change in the image object.

The imaginative and logical components of activity have a complex interrelationship. For example, the choice of action can be based on logic, but the method of performing the action may be based on imagination. Very often an operator receives information in conceptual-logical form, which is then transformed into imaginative form. An image reflects the real world as it is subjectively perceived by the individual and reflects idiosyncratic factors. The image is, thus, independent of the objective reality of that world, although this subjectivity is limited by the parameters of the real world (Zavalova, Lomov, & Ponomarenko, 1986).

Although an image is formed on the basis of integration of inputs from all sensory modalities, the leading role belongs to vision. During the process of activity, different sensory functions are integrated that support the development of the image. The image contains both rigid and flexible components. Because of the existence of rigidly programmed components, these provide single meaning connections between sense organs: Consequently, the individual's recognition of his or her environment does not require the individual to maintain conscious control of the imagery. An example is the system that provides orientation in space. In some circumstances there may be disagreement between sensory modalities and the information they supply, which can lead to distortion of the image content (e.g., perceptual illusions).

Imagery in the pilot's activity was studied by Golubev (1953), Demersky (1955), Platonov (1970), Beregovoy (1981), and others. To better understand the specifics of the image, we can briefly analyze it using the activity of the pilot as an illustration. The pilot, using both instrument and noninstrument signals, creates an image of the flight. This image is more of a general concept than that ordinarily employed in aviation psychology. A pilot uses not only abstract-logical and conceptual, but also imaginative, information. A creation of an adequate image becomes more complicated when instrument and noninstrument signals contradict one another. The reliable control of an aircraft on the basis of equipment data alone has limitations. Inputs from other sense organs are also useful. As a consequence, an image of the flight is created. The concept of an image of flight is comparable to the notion of conceptual model, but specialized for flight. An image of flight includes the goal, image of the task, images of various actions, and a system of knowledge about the object under control.

Even before takeoff the image of a goal for the flight is created. This a relatively stable image. At the same time an image that reflects the dynamics of the flight is created. This dynamic image can be compared with operative images that arise in the course of flying. The structure and contents of the image of flight according to Zavalova and her colleagues (1986) is shown in Fig. 3.14. As one can see, in addition to the conceptual model and the goal, there are additional components of the image, which are discussed next.

The image of the aircraft in space regulates the spatial orientation of the pilot and helps him or her to be aware of the position of the aircraft in relation to the land. During the analysis of this component, we must take into consideration that a great number of strong noninstrumental influences can be a cause of space illusion. Contradictions between perception and thinking, and between external, visual and internal receptor signals can emerge. In some cases, a pilot tries to repress false sensations by forcing his or her own will and using logical components of regulation activity.

"Instrument image" is simply an image that is created by an operator based on perceiving information from instrument displays. This image regulates the motor control component of pilot activity and produces automaticity in the activity. An instrument image is laconic in that it is not redundant. This permits speed and accuracy of performance of individual task components, but it may also lower the reliability of task performance as a whole.

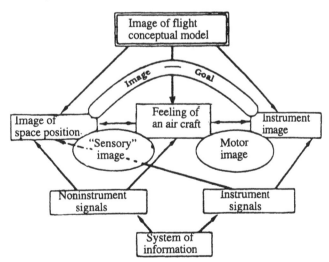

FIG. 3.14. The basic components of the pilot's image. From Zavalova, Lomov, and Ponomarenko (1986).

The pilot does not simply perceive the output of the instruments but interprets them in accordance with the existing image.

Another component of the image is the "feel" of the aircraft. Its formation is based on the influx of noninstrument signals, that is, vibrations, noises, resistance of the controls, and the like. Although these signals may cause errors, they are important for the feel of the aircraft. The sense of kinesthetics and equilibrium is also important. The lack of muscle feeling in the process of piloting increases the role of the visual sense. For example, the turning on of an automatic stabilizer of the aircraft's position increases the duration of the visual fixation on the displays of the aircraft (Ponomarenko & Zavalova, 1966). This is explained by the fact that the use of a stabilizer disrupts the level of resistance of the controls, effecting the pilot's feel of the aircraft. The controlling actions of the pilot change the position of the aircraft, which leads to the changing of instrumental and noninstrumental signals. Such signals give information not only about the position of the aircraft, but also about the result of the pilot's actions. Based on the consequence of the pilot's own actions, she or he can change his or her strategy of task performance and correct the image of flight. Here we can see that the pilot works in a self-regulative system.

A complex relationship exists between the perception of information and behavior. Based on perceived information, the operator creates within him or herself an image of the situation. This image performs two functions, one designed to ensure the "correctness" of the situation, the other to regulate motor actions. If the internal image is distorted, not only is information processing distorted, but also the motor actions required by that information. From this, Russian psychologists conclude that it is impossible to separate the study of motor activity from cognitive processes. Analysis of the interrelationship between input and output of parameters in human behavior is not enough in the studying of the activity of an operator. The internal world of the operator and especially his or her imaginative process is of great importance.

3.8 CONCLUDING COMMENTS

According to the theory of activity, psychological processes are instrumental in information reception, orientation to the environmental situation, and regulation of behavior. An operator receives information presented to him or her according to particular rules, called the "informational model." These provide interaction between the operator and the object being observed. The object is not observed directly, but rather, via information presented by different instruments and signals. Based on this informational model, the operator creates his internal conceptual model, which presents itself as a system of images, logical statements, a system of knowledge, emotional reactions to different signals, and so on, and can be considered as a dynamic internal image of external situations.

Here we have three problems: How does the informational model correctly represent external events? How does informational model interact with the internal world of the operator and help him or her create the conceptual model? And how does this conceptual model regulate the behavior of the operator? In all these cases, the psychological processes of operator are crucial.

In this chapter, psychological processes considered from the position of theory of activity are described. What we see in this chapter is that psychological processes and mechanisms, like perception, thinking, memory, and attention, must in the Russian theory of activity always be viewed in the context of the total activity, which is energized by will and motivation, even in an extremely simple psychophysical task. Psychological processes depend on goals, motives, specificity of task, and strategies of performance. They are a logical sequence of mental actions and operations, which are interconnected by feed-forward and feedback influences, and provide active tuning and regulation of different stages of processing of information from the point of view of goals of activity. From this it follows that mental activity can be presented as a system of mental actions and operations.

Also very important to the operator's performance are the imaginative components of activity. Image, like any psychological process, provides not only a reflection of the situation, but also regulates behavior of operator. The role that images played in the activity of the pilot and the structure of the pilot's images are described. The study of operator activity demonstrates that, in human information processing, there is a complicated, dynamic relationship between logical and imaginative components of activity. The study of the dynamic features of imaginative processes is very important for understanding principles of self-regulation of activity. For instance, experiments demonstrate that a lack of coordination between sensori-perceptual and imaginative-conceptual levels of processing of information can cause, in flight conditions, the emergence of a pilot's illusion. To overcome this illusion, the pilot should be trained to actively select the required information and to correctly interpret it.

Design of the
Human-Machine System

4.1 OVERVIEW AND INTRODUCTION

4.1.1 System Design

Russian ergonomists generally have accepted Western concepts of how to perform behavioral design, because these concepts are logical and meaningful in the context of system development. What engineering psychologists in both Western and European contexts prefer to do is to research "fundamental" behavioral processes. The Russians view system design in the context of work activity. The Russian approach leads to a rejection of what they term the *systemic-technical framework*, which is viewed as a method of optimizing input-output or the stimulus-response relationship. Russian theorists feel that the so-called "black box" orientation is inferior to the anthropocentric approach, which assumes that the human and the machine function in some sort of synergistic relationship. This leads to the concept that system design must conceive of the human as the central component in the design of work activity. This presents the problem of how one can design a system to be responsive to changing motives and goals. The "human" approach is to determine the operator's model of the system, which includes all the variables described in preceding chapters. The basic tool to create this model is task analysis, but in Russian terms this is much more tightly linked to the structure of the operator's activity. Russian psychologists conceptualize three types of models: conceptual, mathematical, and physical.

Design begins with functional specifications of the system, which is a symbolic model. Nested in this is the work activity model design. The system model developed by the designer must convey the same concept to the operator. In developing such a model the specification of system requirements is compared with the overall model of work activity and relevant elements of the latter are extracted to be the basis of behavioral design. These are then broken down into more detailed mechanisms. The resultant model is tested both analytically and empirically throughout development and is revised as necessary. Russian ergonomics would describe behavioral design as the process of describing an operator's activity, contents, structure, and the parameters of system components. The Russian approach to design is much more analytical than the Western.

4.1.2 Temporal Characteristics of the System

Russians use time as an evaluated criterion for the system as well as a means of studying information processing. In their research they have conceptualized such notions as reaction to a moving object and a switching reaction. The significance (importance) of a stimulus affects reaction time. Other factors studied in a temporal context include whether the individual is responding to an external or an internal command, training, the distribution of time values, and the amount of information presented by the stimulus. Temporal responses are also affected by varying operator strategies in response to multiple stimuli.

Russian psychologists have also developed the concept of "reserve time," which is the surplus of time the operator has over that minimum in which she or he must respond. This reserved time is both objective and subjective (operator estimated), the latter being affected by both technical (system) and idiosyncratic factors. A reduction in available time to perform leads to operational and emotional tension. Excessive time to response leads to monotony.

The very detailed analysis by Russian psychologists of microactions leads to an interest in predetermined time systems such as MTM-1, although not all Russian theories maintain this opinion.

4.1.3 Precision

Russian psychologists distinguish operator errors from what they term as operator failures, based on the criticality of the error. An operator action that affects work efficiency negatively but does not shut the system down is an error; an action that shuts the system down is an operator failure.

Russian theorists have developed error taxonomies based on (a) general characteristics of the error, (b) the position of the error in the system structure, (c) its position in the structure of the operator activity, (d) the causes of the error, and (e) its consequences.

Accidents are related to errors, although they may also be caused by purely technological factors. The process of accident investigation follows Western procedures, including reconstruction of the events leading to the accident. Russian theorists also distinguish between positive and negative significance of the situation, which affects how the accident situation is viewed.

With increasing automation the operator's role in system operations has changed and so has the definition of what constitutes an error.

Just as there is reserved time, so the concept of reserved precision is introduced. The latter suggests that a certain number of errors is permitted to the operator and so long as he or she makes fewer than the allowable errors, he or she has a reserve that permits him or her to make further errors.

Russian psychologists believe that the complexity of a task can be increased by limitations on time and precision. The significance of the task and the goal also determine the operator's work strategy. Accuracy and speed criteria are also influenced by stimulus intensity.

4.1.4 Operator Reliability

Russian psychologists have formulated two approaches to evaluation of human performance reliability: systemic-technical and anthropocentric, the latter based on the activity concept.

The systemic-technical approach divides the operator's activity into hierarchical levels, determines error rates for lower levels, and then combines these mathematically to arrive at a numerical value for total performance reliability. This is essentially the same as Western procedures, which have similar difficulties in lacking an appropriate database; it also assumes independent units of activity and fails to recognize variations in operator strategy, all of which violate activity concept assumption.

The activity approach to determination of operator reliability utilizes time to perform and intensity of the failures resulting from errors. The former requires that time to perform individual tasks be specified, and that necessitates either empirical data or expert judgment.

Russian psychologists have suggested certain psychophysiological qualities that influence operator reliability: endurance, resistance to distraction, continuous attention, attention diversion, and the ability to switch from one work activity to another.

One method of determining the impact of the operator's performance on system reliability is to measure that reliability with and without including the effects of operator performance. Time limitations increase the significance of the task and thus affect human performance reliability. Russian psychologist also suggest that physiological factors affect human performance by increasing arousal mechanisms. They also consider the operator's self-control to be an important mechanism of self-regulation; self-control is the individual's checking on the correctness of his or her actions, which requires feedback.

4.2 THEORETICAL APPROACHES TO SYSTEM DESIGN FROM THE STANDPOINT OF WORK ACTIVITY

Like Western engineering psychologists, Russian ergonomists view the operator as a component of a more or less integrated human-machine system. This is because, starting in the 1950s, the concepts of American human factors theorists pervaded the former Soviet Union. Morgan, Chapanis, Cook, and Lund (1963), Fitts (1951), Fleishman (1975), Meister and Rabideau (1965), and Meister (1976) all played an important role. Systems engineering and operations research (from which the so-called system concept was derived) were also important (Bertalanffy, 1969; Hall, 1962).

To view the human as a system component is not, however, to equate human with machine processes. For example, B. A. Smirnov and Samoshkina (1976) differentiated three aspects of design: technical (i.e., as related to equipment functioning), ergonomic (that equipment design that relates to the machine operator), and artistic (something that may be roughly equated to what in America is called industrial design, or the aesthetic aspects of system design).

Some specialists would prefer to view system design as a relatively simply linear process; it is none of that. Manifestly the design process varies, depending on whether it creates a single, relatively simple design, or whether the system contains a number of integrated, interactive equipments or subsystems, such as are found in parallel-processing computer systems. Meister (1971) provided a list of the factors affecting system design and the attributes (e.g., indeterminacy, uncertainty, hierarchy) that the system manifests.

However simple the equipment under design is, though, the design process is iterative, with many forward and backward loops, much like the processes found in the human's work activity. In fact, the process of system design has been viewed as essentially a problem-solving task, with all the attributes that problem solving manifests.

Russian behavioral design is divided into four stages (Munipov, 1983): (a) developing technical proposals, (b) developing technical suggestion (preliminary design), (c) developing technical design prototypes, and (d) developing detailed designs and producing prototypes to test them. These stages are essentially the same as those described in Western documents (e.g., Meister & Rabideau, 1965). Consequently only a brief description of these stages is given here.

The development of technical proposals determines the purpose of the system, as well as the principles and conditions of the application of a new system. Based on these, requirements specifications are developed in which are articulated the various conditions in which this system will be used, as well as, a system profile involving the temporal sequence of actions with which this system achieves particular goals.

The development of technical suggestions is based on previous data and a search for prototypes and related systems. It includes what in the West is called function allocation. A preliminary algorithmical description of the operator's activity is performed. Mockups may be built and tied into computers. Preliminary experimental data are gathered to support the analytical description of the operator's activity.

The development of technical designs involves final ergonomical decisions. Conclusions about personnel requirements are reached. The list of operator tasks is defined, as well as training programs, shift work provisions, selection criteria, and simulator development.

The final design leads to a test program, which involves determining that the behavioral design goal have been achieved.

In accordance with the principles that have guided the theory of work activity from the beginning, system design must be viewed in the context of work activity as a whole, that is, in terms of the operator's goals, motives, and work strategies. System design must therefore be responsive to these. Obviously, the extent to which one can think of system design in terms of the principles of work activity depends, among other things, on the size of the unit that is being designed. For operation of a single simple device with highly preceduralized operating instructions, the principles of work activity to be incorporated in design are highly circumscribed. The larger the system, the more complex its operations, the more automated it is, the less determined it is, the more behavioral principles of work activity apply and must be incorporated into system design. Needless to say, Russian ergonomists are aware that the system of the 21st century will increasingly be of the second type.

Lomov (1977) suggested that there are two approaches to system design: systemic-technical and anthropocentric. The basic task of the systemic-technical approach is to develop a method that optimizes the system's input–output relationships. For Lomov, one of the weaknesses of this approach is that the human is treated as a simple system chain. The basic goal of ergonomics research is determining the input–output characteristics of individuals. The core of the method resides in the S-R approach to psychology.

It should be noted that there is nothing intrinsically wrong with attempting to determine individual input–output characteristics, because this knowledge can be of value. If, for example, we find that there are certain stereotypes in the way in which people treat information, this knowledge will be of assistance if we endeavor to design systems according to information principles.

The point, however, is that the input–output or black box approach to engineering psychological research is incomplete and must by utilized within the context of the anthropocentric approach. In Russian ergonomics there have been important initiatives around the anthropocentric approach to system design. In this approach the human and the machine do not function together in an undifferentiated manner. Rather, machines are treated as tools for the human. Technology does not exist solely to improve productivity and the profit margin of entrepreneurs, but also, and even more important, to allow the worker to perform his or her job more effectively, more easily, and more comfortably. If this increases productivity and profit, so much the better. In other words the anthropocentric approach does not consider technology alone, but makes the human the centerpiece of technology.

Humans do not function in the same way as do machines; this must be self-evident, because to date artificial intelligence scientists have been unable to incorporate goals, motives, and strategies in even the most complex of systems.

Proponents of the anthropocentric approach conclude that the design problem is not one of designing the system as such (this is too restricted) but rather the design of "work activity," and the technical components of a system. In the latter approach the human emerges as the central concern. The anthropocentric approach makes it necessary to determine the technical components of the system that affect human activity within the system.

This raises the question of how to optimize the relationship between human activity and the technical components of the system. This poses a particular problem for Russian ergonomist because the theory of work activity emphasizes the internal state of the operator, his or her needs, goals, strategies, and so on. It is easier, for example, to develop invariant lists of principles that constitute "good" human engineering, when these principles apply only to the input–output aspects of the system.

For example, the principle that states that controls and displays that are used in conjunction with each other in equipment operation should be colocated on a console is probably one that, applying to the input side of operation, will never change.

Anthropocentric and systemic-technical methods of ergonomics design differ, but do share important similarities in that both rely on methods of systemic analysis. Some anthropocentric ideas can be found in American sources. For example, Meister (1979) formulated a basic principle of ergonomics: Different system design structures imply different methods of operator work. This principle is fundamental to anthropocentric design. Of course, the way in which this principle is realized in the West and in the former Soviet Union is sometimes different.

The implementation of work activity design involves linking its analytic description—the analysis of the structure of work activity—with the internal and external tools of that activity. By internal tools we understand such things as motives, goals, criteria of evaluation of outcomes, habits, skill, knowledge, and motivational and physiological resources of individuals. External tools refer to the workplace, control panels, information embodied in equipment signals, and instructions and mnemonic schemes used by operators during work activity.

Many experts in ergonomics use the term *design* incorrectly. That is perhaps because the discipline lacks a viable theory of design. There are consequently many design definitions. Attempts were made to develop a standard definition in the former Soviet Union. According to GOCT 22487-77, design may be considered a description of a system whose implementation entails iterative refinement of initial descriptions, optimization of the required features of the objects, or algorithms of its functioning. The initial tentative description becomes the point of departure for subsequent reformulation in different symbolic languages. Accordingly, we create abstract models of the original object. The central meaning of design in Russian ergonomics revolves around the notion of "model."

Often the problem of design is treated as one of physical modeling (mockup studies) of equipment and observation of how people perform in terms of these physical models. This kind of activity is erroneously called a "model of real activity." However, only the equipment has been modeled; the activity has been imitated rather than modeled. Deitrich (1981) argued that design should be seen as human activity operating within the information, but not activity in the field of physical objects (p. 57). Neumin (1984) offered the following definition of design: "Design is the creation of an abstract image of artificial objects in accordance with some requirements and characteristics with the purpose of materialization of these objects" (p. 141).

If the notion of model is critical to design, it is necessary to explore this concept more fully. Some theories (Bedny, 1975; V. P. Zinchenko & Munipov, 1979) suggest that the designer has or develops a model not only of how the system will function, but also of how the eventual operator will perform with the new system.

It is apparent that the same reference object may have a number of different models, and the more they match, the better off we are. The descriptive language employed for a model is a critical factor in its development. The engineer describes his or her model of how the system will perform in hardware and software terms, in terms of, for example, motors generating power. The model of work activity is much less precise, because it is less concrete; it talks of goals, motivations, actions, and so forth. We see later that the general model of work activity (as it has been described in this book) must be translated into a more specific model of the operator's functioning with the new system. The process of performing this translation—from the general the specific model—is performed by means of a body of methods called "task analysis."

The whole process of design can therefore be conceptualized as one of translating one descriptive language into another. If, as an ergonomist, I want the designer to include certain motivational aspects in his or her design (because these are important to the operator functioning with this system), then I have to translate those aspects into hardware/software equivalents.

Any model is only an approximation because every object, event, phenomenon has many features and as a practical matter it is almost impossible to include them all in any one model.

If we want to talk in quasi-philosophical terms of something called "reality," it is possible to create a number of different models of reality for the same object. The choice of model derives from the problem at hand (Montmollin, 1967).

The advantage of a model consists in reducing the potentially infinite number of system features to a manageable number. We do this by determining what will affect what, what physical system features will affect work activity and how.

There are three types of models: conceptual (thinking), symbolic, and physical. Conceptual models are images about systems, events, objects, and phenomena that can be created in the mind. Symbolic models describe the object or phenomenon verbally, graphically. One type of symbolic model is the mathematical model, with equations or algorithms representing relationships that occur in real objects. Finally, there are physical models that reflect some of the physical and functional features of real objects (e.g., simulators).

When addressing a problem experimentally, the choice of the model depends on the experimenter. However, when we address design problems, such freedom in the choice of models is not available. In technical design, for example, in contrast to in experimentation, the model is used not only by the designer but also by the production personnel.

As a general principle, there must be concordance between the specialist who created the model and those who must interpret it for production purposes. This may be attained in part by standardizing procedures and language for the description of the model during the design process. For example, without the knowledge of specific drawing principles design engineers and production personnel cannot understand each other.

Viewing the design process as a problem-solving task, we have to ask the question: Which of the postulated features of the system will influence work activity? In this very general form, this is a relatively easy question to answer. In the automatized system, for example, it is the information display. Now comes the hard question to answer: How will the information displayed affect the operator's information-processing activities? To break it down more concretely, how much information is available; how much should be displayed to the operator; if there are multiple subsystems, how should the interaction of these subsystems be combined and presented?

It must be emphasized, however, that this is only the beginning of the model development process. At the beginning we have only rather general behavioral requirements elements. For example, it becomes apparent in the highly automatized system that a major behavioral function will be monitoring to determine whether the system is within parameters. This implies the functions "recognition" and "identification" are out-of-parameter conditions. What is involved from a design standpoint in the cognition function? Is it more effective to give the operator a continuous gauge or a digital indication? Should recognition be enhanced by use of auditory and/or visual alarms (as they are used in many automatized systems like nuclear power plant control rooms and ship propulsion control rooms). Should there be an auxiliary expert system to help stimulus interpretation? This decomposition of general functions into more detailed ones, with their design implications, is called by American human factors specialists the top–bottom approach to design, although in the iterative processes of design there is also a complementary bottom–top approach based on feedback from tests during the design process.

The development of a design model is largely an analytic process for which empirical methodology (e.g., observations and questionnaires) is not useful. However, research results achieved empirically can be very useful in providing design answers to questions by the model. In the previous example of the recognition of a potential malfunction situation, research is available that will help answer the question of continuous versus digital information, auxiliary alarms, and expert systems.

The initial model, once developed, is not "cast in concrete." That is, it is tested during the design process and is subject to revision if tests reveal an inadequacy or new design ideas occur to design personnel. Research models are or should be tested formally through experimentation for validity (although most are not). There are, however, few or no formal tests specifically for design models, because they are system specific, and the criterion of their value is not validity but utility. If the model helps the engineer's design processes, it is utilized; if it does not, it is revised, or ignored.

The design model, insofar as it pertains to operator performance, will be tested indirectly during design whenever an operator-involved test is performed. There are comparatively few such tests, but mock-up tests and part-task and full-task operational tests in which real or simulated operators must function in relation to equipment will provide such tests.

Many designers are retrograde and cannot even conceptualize the notion of behavioral design model. Nevertheless, despite what the designer does not do, a design model is created in the designer's thought processes. In other words, the absence of a formal design model does not mean that there is no design model. There is always a design model, even when it is sorely inadequate, overly complex, and insensitive to the operator's needs.

Mention has been made of using task analysis as a means of helping create a behavioral design model. Unfortunately, the general term *task analysis* subsumes a number of techniques, and one cannot talk of a theory of task analysis, although the procedure assumes a stimulus-response decomposition process. The original task analysis methodology developed by Miller in 1953 has gradually been expanded and modified, and one of the variants is called "cognitive task analysis." However, because of the S-R orientation, which in Russian eyes is still current in Western ergonomics, certain modifications of cognitive task analysis would probably be required to coordinate it with the theory of work activity.

There is, moreover, a discrepancy between the task analysis level of detail and the detail level implied by the theory of work activity. For example, Fig. 4.1 describes a function flow diagram. The first function specifies "monitor and check out onboard subsystems." The Russian ergonomist would ask: What is the operator's structure of activity described by this function? The Russian ergonomist assumes that the most minute detail is needed to reveal the operator's structure of work activity, the assumption being made that knowledge of that structure is necessary to develop a behavioral design model.

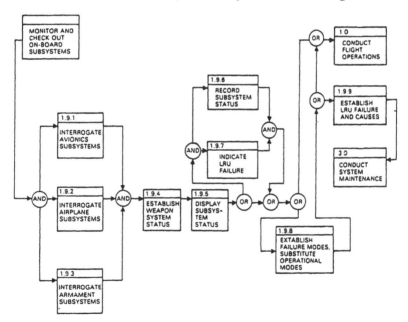

FIG. 4.1. Functional flow diagram. From Geer (1981). *Human Engineering Procedures Guide* (Rep. AFAMRL - TR—81-35. Wright-Patterson AFB, OH: Aerospace Medical Division.

We have said very little about physical models, because behavior cannot be represented by a physical model.

Russian ergonomics places a great deal of emphasis on language for describing work activity processes. It insists that this language be standardized, hence the concern for taxonomies, (e.g., conceptual model, image-goal, image-object, motor and mental actions) which are specified almost to an agonizing level of detail.

Russian ergonomics accepts the iterative character of design, with its feed-forward and feedback loops, which are linked to changes in the nature of design models. Russian ergonomists also emphasize the use of mathematical models to analyze and compare different models of work activity; Western human factors specialists emphasize heuristics, with its pragmatic orientation. The actual accomplishments of Russian mathematical formulations have, however, been much fewer than appears warranted by the effort expended in their development.

In summary, Russian ergonomics would describe behavioral design as: the process of describing an operator's activity contents, structure, and the parameters of the technical components of the system. The emphasis appears to be on analysis of work activity parameters, and their connection with physical aspects of the system.

The goals of Russian ergonomic design are (a) the concretization of the functions that the person will perform in all regimes of system functioning, (b) the determination of indices and criteria of the quality of the functions being performed by the person who is the operator, (c) the determination of quantitative indices and criteria of human performance, (d) the development of the model of activity of a person, (e) the determination of the parameters of equipment that influences the activity of a person and provides his normal functioning, and (f) the determination of the parameters of the surrounding environment that affects the activity of a person and protects his or her safety.

4.3 THE TEMPORAL CHARACTERISTICS OF THE OPERATOR'S ACTIVITY

Any task that is performed by a person in the human-machine system requires some time. In work activity, the index of time emerges as one of the most important criteria of work productivity and efficiency. Failure to function within time limits is viewed as the failure of the person in the human-machine system. Russian ergonomists view time as one of the more objective and the easiest way to measure a person's performance. Time not only reflects the distinguishing features of external behavior but also the specifics of the internal psychic process. For example, chronometrical studies play an important role in cognitive psychology (Sperling, 1960; Sternberg, 1969; Velichkovsky, 1982; Zinchenko, 1978).

Reaction time can serve as one of the evaluative criteria for the human-machine system, as well as a method of studying a person's method of processing information. Time also is important in evaluating the training process, reflecting a person's individual differences, as well as his or her relation to the situation at hand.

Donders (1862) found two fundamental types of reaction: simple reaction time (designated as reaction type A) and choice reaction time (designated as types B and C). In the case of reaction type B, an operator identifies a stimulus from a series of stimuli and chooses a corresponding action. Type C occurs when two or more stimuli are presented and a person reacts to one and not the other. As described, these reactions are rarely encountered in exactly the same way in practical situations. For example, in a real situation an operator can react differently to the same stimulus. More often, the operator must evaluate the total situation and not just the isolated stimulus, and he or she does not always reacts with maximum speed during task performance. It is possible, for example, that many airline accidents occur not because a pilot's response actions are slow, but because she or he is rushed.

The psychophysiologist Zimkin (1934) identified a new type of independent reaction, which he designated as a reaction to a moving object (RMO). For example, a subject's reaction includes stopping a mark on a screen in a certain position, stopping a car in a precise spot, or a lathe operator cutting a piece of metal exactly by stopping the machine's cutting instrument in a specific position.

Platonov (1970) conceptualized a switching reaction that appears when additional stimuli appear; the method of reacting to the major stimulus changes.

As in America, Russian studies were performed directed toward the extraction of variables that influence reaction time. It will come as no surprise to the reader that stimulus modality, stimulus intensity, temporal uncertainty, stimulus-reaction compatibility, and so on, all affect reaction time. Following American scientists, Russian psychologists performed studies using information theory in studying choice reactions. Additional data about this are discussed later in this section.

Platonov (1970) discovered that an increase in stimuli intensity decreases the time of simple sensorimotor reactions. Nevertheless, the subject-to-situation relationship violates this law. Ol'shannikova (1962) demonstrated that a person reacts faster to a significant but weak signal, then to one less significant but more intense. In her experiment, the significant signal represented an accident. Similar data were obtained by Zarakovsky and Medvedev (1971). They studied the reaction time of drivers to an equally probable sequence of red, blue, and green signals. Reaction time and the number of mistakes were least when responding to the red signal, which commonly denotes "stop!" This can be explained by the fact that the green signal is less important to the driver. These data tie in with what was said in chapter 2 about the factor of significance.

It is well known that an operator's readiness to react is increased with stimulus predictability. Nevertheless, reaction readiness can be modified by self-command in response to the operator's plan of action or by external requirements (Kotik, 1978). His subjects were presented with red and white signals which they had to turn off by moving specific levers. The white signal was to be turned off in 3.0 seconds, the red in .9 seconds. In the first series of tests, the subjects were told the sequence in which signals would appear. This is how they could plan their activity and create the necessary readiness for their fastest reaction to the signal. The readiness to react was also detected in the recordings of physiological criteria (pulse rate, rate of breathing, galvanic skin responses). In the second series of tests, the signal sequence was not announced but a preliminary warning would be given before the appearance of the red signals (external command). The subjects' latent and motor components of reaction to the signal were recorded. The results of the experiment are given in Table 4.1. Note that choice reaction time is shorter to the external command than to the self-command. The differences are of statistical significance ($p < .05$). Reactions to external commands create greater mobilization of body responses before the reaction. Leont'ev (1977) wrote ". . . that it is easier for soldiers to rise to attack according to the command of the commander than according to self-command" (p. 209). Luria (1975) showed that a child forms the ability to perform an activity when an adult tells him or her to do so before the command is internalized. This suggest a genetic component to command responses that can in some cases cause them to be more effective than self-planned actions. This fact is well known in gymnastics. Very often a gymnast is afraid to

TABLE 4.1
Reaction Time According to Self-Command or External Command

Kinds of Readiness	Reaction Time		Frequency of Pulse Rate Before Reaction (beats per minute)
	ms	ms	
Readiness based on self-command	96.89	51.68	93.99
Readiness based on external command	90.37	47.16	97.46

Note. From Kotik (1978).

perform a dangerous stunt for the first time, and only after repeated commands by the coach does the gymnast perform it. All of this suggests some practical uses in ergonomics. In many cases it is useful to issue additional commands and "prompts" for the operators' procedures.

Platonov (1970) studied how reaction time changes as a result of training. The more complex the type of reaction the more important its training becomes; the simpler the reaction the faster stabilization of performance time occurs. Figure 4.2 shows that, as a result of training, the time of a simple sensorimotor reaction is also reduced. The data of the experiment must be taken into consideration in practice. The dynamics of acquiring a skill is an important criterion for evaluating its difficulty.

The more complex the reaction is, the more complex the skill needed for its performance. The more complex skills are, the longer it takes to acquire them.

Gellershtein (1958) showed that when a person receives accurate information about his or her reaction time after performance, the speed of future responses increases significantly. Also, the subjects acquire the ability to control the speed reaction. They acquire a "feeling of time," related to the ability to differentiate the microintervals of time.

When studying reactions to a moving object (RMO) the "preceding time" is quite important (Platonov, 1970). The term preceding time refers to the time from first perceiving the moving object to the time it achieves the stop position. For example, a subject has to stop the moving object in the stop

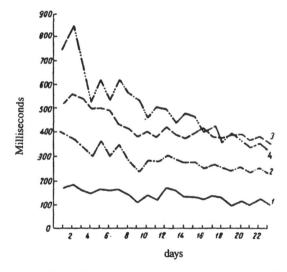

FIG. 4.2. The effects of the training on the reaction time of various levels of complexity. From Platonov (1970). 1—simple sensorimotor reaction; 2—simple sensorimotor reaction with the transition of finger from one button to another; 3—complex choice reaction; 4—switching reaction.

position. She or he can see the moving object only through the window. Preceding time "t" depends on the speed of the moving object and the distance "S" (see Fig. 4.3). When the preceding time is greater than 300 milliseconds, accuracy to moving target equals ±50–10 milliseconds. However, if the preceding time is 200–300 milliseconds, reactions are delayed. When the preceding time is greater than 400 milliseconds, the difference between the reaction of the hand and foot is absent. If the preceding time is 300 milliseconds or less, reaction accuracy by the leg is lessened. These data may be useful when choosing control display equipment. The decrease in the field of view of a moving object is reflected more negatively in the accuracy of the reaction than on the moving object's speed.

Sibulevsky (1962) determined that the variability of reaction time values follows a normal logarithmical distribution law. The mathematical mean of that distribution $m(t)$ and the standard deviation $\delta(t)$ have a correlation of .58. The greater the $m(t)$ the greater $\delta(t)$. The physiological minimal reaction time for different operators is approximately the same. The value $\delta(t)$ is, however, more variable for different people than the mean reaction time (see Fig. 4.4).

Many Russian studies were performed to study the effect of stimulus information on response time (see, e.g., Dmitrieva, 1964b; Glezer & Tsukkerman, 1961; Kossov & Rotanova, 1965; Nikolaev, 1973). Hick (1952) and Hyman (1953) were the first to discover the linear relationship between the average information (I) and reaction time (RT):

$$RT = a + bI \qquad (4.1)$$

where a and b are constant coefficients dependent on the conditions of the experiment. In section 3.4 in describing the model of attention we have shown

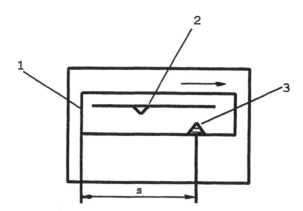

FIG. 4.3. Example of reaction time for moving object. 1—window; 2—moving object; 3—stop position; s—preceding distance.

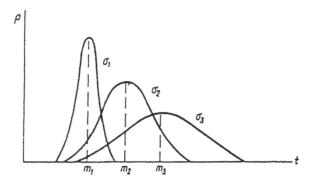

FIG. 4.4. The relationship between the average value of reaction time $m(t)$ and standard deviation $\sigma(t)$. From Sibulevsky (1962).

that information theory cannot be used to predict the speed of operators' work. Other studies also suggest the difficulty of applying information theory in practice. For example, Krinchik and Risakova (1965) demonstrated that the relationship between the average informativeness of a stimulus and reaction time to the stimulus depends on its significance (see section 1.6).

The effect on reaction time of such factors as significance, preliminary readiness, and training suggests that despite the automatization of such actions (skill-based behavior, according to terminology suggested by Rasmussen, 1986), these actions are performed according to self-regulation mechanisms. This was also demonstrated by the study of Konopkin (1980) described in chapter 2.

Similar data were obtained by Bedny (1987) in researching reactions performed sequentially as described in section 3.4. A person does not simply react to individual stimuli but builds different strategies of activity, strategies that change in accordance with changes in stimulus situations, his or her present mental and emotional state, and past experience. Such strategies include unconscious as well as conscious components. This is why unlike the approach in which a person is viewed as a reactive system, or in which a person is viewed primarily as an information-processing mechanism, Russian psychologists view the individual as a complex self-regulative system that constantly changes its strategies for achieving the goal. It is thus impossible to understand changes in reaction time without considering strategy changes.

Usually an operator reacts not to a single stimulus but to a whole complex of stimuli. This complex will be perceived both simultaneously and sequentially. Most research has been performed on stimuli presented simultaneously. Poltorak (1974) showed that stimuli duplicate each other by different sensory modalities; for example, sight and hearing increase the speed of reaction. However, if difficulty of task increases, effects of stimuli duplication decrease, and when the task becomes very difficult, the operator's reaction time becomes even longer than to complex stimuli. It was also discovered that in the simultaneous visual perception of complex stimuli, identification proceeds through certain hierarchical processes. First, brightness is extracted from the background, next, color characteristics, and only then is the shape of the stimulus formed. In complex stimuli certain dominating features are the first to be identified, determined by the physical characteristics of the stimuli, and their importance to the operator.

Experimental studies of reaction time do not unfortunately account for the solution of real-world tasks. It is important to know not only the times of individual actions executed at the appearance of certain stimuli, but also how much time is needed for performance of the total task and how this time correlates with time allowed for task execution. The task includes a system of stimuli and a system of reactions to these that are organized in accordance with a set goal. Russian psychologists consider that a task is successfully performed only if the human-machine system is transferred by the operator from an initial undesired state to one that is required and desired. Very often this process is affected by system malfunctions that, in the absence of necessary operator actions, will increase until the system fails or an accident occurs. The time during which the system is transferred from the initial to the required state is called by Russian theorists the "cycle of regulation time." Western scientists call this "task performance time."

The cycle of regulation time may be presented as the sum of time delays that are produced both by human information processing and the physical responses of the equipment. Thus,

$$T_0 = t_0 + t_1 + t_2 + t_3 \tag{4.2}$$

where T_0 is the cycle of regulation time, t_0 is the time during which the equipment presents information, t_1 is the time required for the operator's perception and processing of the information, t_2 is the time needed for the operator to perform his or her control action, and t_3 is the time required by the equipment to respond to the operator's control action.

Another important system characteristic is the reserved time it possesses. Reserved time is defined as the surplus of time over the minimum that is required for the operator to detect and correct any deviations of system parameters from allowable limits and to bring the system back into tolerance. Thus,

$$T_{res} = T - T_0 \tag{4.3}$$

where T is the time that cannot be exceeded without peril to the system, and T_0 is the cycle of regulation time.

For example, when a ship is following a certain course and an obstacle (e.g., iceberg, reefs) suddenly appears, the human-ship system needs, let us say, at least 8 minutes to effect a change course and avoid the obstacle. If the ship can make the change in 5 minutes, it has a 3-minute margin or reserved time. In an accident situation parameters quickly shift to the minimum allowable value and reserved time decreases sharply. Usually the equipment responds more quickly than the fastest operator action. In such cases the operator's delay as a component of the system can significantly effect the length of the reserved time. Nevertheless, there are systems in which the delays exceed the delays of the human components of the system.

From the self-regulation of activity point of view (see Fig. 2.6), it is necessary to differentiate between the objectively existing reserved time and the operator's subjective evaluation of that time. In many cases they are not the same. This may lead to an inadequate evaluation of the situation and the inadequate behavior of the operator in an accident situation. This was discussed in section 2.2, where an example was given of a ship captain who visually evaluated the distance between ships incorrectly and misjudged the reserved time he had.

A person often roughly evaluates his or her reserve time by making statements such as "I have plenty of time" or "I have a little time" or "I have no time." Such statements may reflect a sharply changed activity strategy, which suggests that the transition from one level of activity regulation to another has discrete features.

A decrease in reserve time can occur in relation to equipment as well as psychophysiological factors. Most prominent of these equipment factors are malfunctions. The psychophysiological factors include the deterioration of the physical (e.g., fatigue) and mental (e.g., stress) state of the operator and lack of adequate training. To a certain point, a reduction in available time for task performance may produce only an insignificant change in activity strategies. Zavalishina (1977) and Loos (1974) showed that such insignificant change even has a positive effect on performance. This results from mobilization of the organism's reserves, the speeding up of psychic and motor components of activity, and the combination of activity elements. Further reductions in available time cause additional changes in activity structure. The operator concentrates only on achieving what she or he considers to be the most important aspects of the activity structure. Thinking becomes less detailed and secondary components of activity drop out. If the available work time is reduced sufficiently, the activity may become totally disorganized. This disorganizing effect of time limits has been labeled *psychic tension.* Psychic tension is the state that arises in difficult activity conditions characterized by the temporal reduction of the stability of the psychic functions.

In Russian psychology this notion is considered to be a particular kind of stress that involves changes in the emotional-motivational, intellectual, perceptual, and behavioral components of activity caused by complication of task, complication of its performance, and/or increasing of the task significance. These complications may be accompanied by physiological changes. However, the same physiological reactions can be initiated by totally different factors. For example, the person who put his or her hand in extremely cold water can experience the same physiological changes as the patient waiting for a risky surgery. In the first case these reactions were initiated by physiological factors; in the other one they were initiated by psychological factors.

Nayanko (1976) distinguished between what he called operational and emotional tension. Operational tension is determined by a combination of task difficulty and lack of available task time. Emotional tension is determined by the personal significance of an activity to the operator. These two types of tension are closely interrelated, and under certain conditions one type of tension causes the other. For example, lack of available time for a significant task cause more tension than lack of time for an insignificant task. Russian theorists have developed something called the coefficient of tension (Lomov, 1982) in which

$$K_t = T_{comfort}/T_{real} \qquad (4.4)$$

where K_t is the coefficient of tension, $T_{comfort}$ is the time needed by the operator to perform the task, and T_{real} is the time that the operator actually has to perform the problem.

Siegel and Wolf (1969) presented a formula for the evaluation of the tensions (K'_{si}) in which

$$K'_{si} = T_i/(T_o - T_j) \qquad (4.5)$$

where T_i is the time needed by the operator for the performance of the remaining "i" essential actions, T_o is all the time that the operator has available to him or her, and T_j is the time used by the operator to perform j actions. If a reduction in time availability makes the task more difficult and causes stress, however, an unlimited amount of time may lower work productivity and create a feeling of monotony. As a result, a coefficient of workload was introduced (Vishepan & Vishepan, 1975). It is determined from the expression $K_{wl} = t/T_{sh}$ (where t is the time of processing of information by operator and T_{sh} is the shift time). This coefficient according to Vishepan is not supposed to exceed K = .75.

The informational workload of an operator is also affected by a queue of stimuli that must be updated. Such situations emerge when new information is presented before the completed processing of previous information. This causes tension for the operator, the amount of which depends on the length of the stimulus queue and how often it materializes. The time during which there exists a queue is designated as (T_q). This index divides the duration of work (T_w). As a result, one can calculate the coefficient of queue ($S = T_q/T_w$). The average length of the queue is designated by (k). This length cannot exceed the capacity of the operative (working) memory or the operator will miss some of the stimuli. The number of simultaneous stimuli that an operator can attend (keep in memory) when a dynamic sequence exists cannot be more than three or four. To calculate this index one should first determine the span of time during which an operator is presented with (n) stimuli. Then one determines the fraction of time during which these stimuli were in queue. From this one calculates the mathematical mean of the expected queue length.

Kotik (1987) conducted an experiment with pilots to determine the functional relationship between minimum time limitations placed upon the pilot in performing a given task and its importance to her or him. The pilots were presented with various verbal descriptions of deviations from the required flight regime. As experts, they had to determine the significance of each deviation on a scale of 1 to 7. The study revealed that the subjects considered a deviation more important if it was associated with time limitations. Previously, for each task a reserve time had been calculated. The results are shown in Fig. 4.5. The curves show the interconnection between the t_{res} (reserve time) of the task and the average significance (S^*). An analysis of the data reveals that the connection between the reserved time of the determined task and its significance can be approximated by the exponential relation of the type

$$S(t_{res}) = e^{-\lambda t_{res}} \tag{4.6}$$

where $S(t_{res})$ is the significance of the situation or task as affected by time limitations; λ is the mean arrival rate per unit of time, equal to the $\lambda = 1/\bar{t}_{res}$; and $\bar{t}_{res}$ is the average reserve time with which an operator works.

The theoretical curves are built according to the given formula shown in Fig. 4.5 by bold lines. The expert evaluation $S^* = 7$ points corresponds to the calculated value of $S = 1$ on the graph. The statistical comparison of experimental and theoretical dates gives evidence that curves fits each other ($p < .05$).

This research reveals that one can approximately determine task significance by the minimal time limits of these tasks. The formula has meaning only after a certain minimal time requirement is defined. For example, if I am late to my job, I will be fired. I live across the street from the workplace and can reach my job in 10 minutes. If I have an hour of reserve time, it does not matter how much time I have because I have plenty of time to

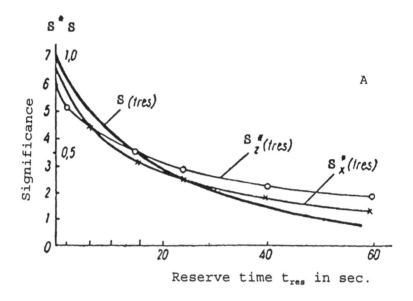

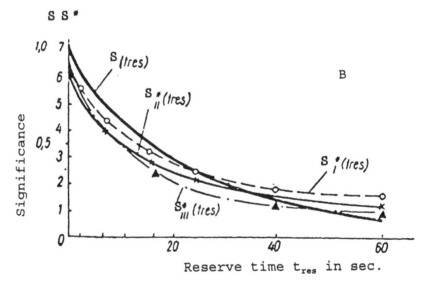

FIG. 4.5. The relationship between t_{res} and significance of task. From Kotik (1978). A—significance of information about deviation of aircraft in relation to its longitudinal axis (Z) and cross axis (X) as function of reserve time S^*_z (t_{res}), S^*_x (t_{res}), (experimental data). B—evaluation of significance of information, data of the first class pilot S^*_I (t_{res}), second class S^*_{II} (t_{res}) and third class S^*_{III} (t_{res}), (experimental data). Theoretical curves—solid bold lines.

reach my job. However, if my reserve time is reduced to 12 minutes any subsequent reduction in time becomes significant for me.

Significance and importance are not the same. The objective importance of a task is not significance. The subjective importance of the task is significance. Manifestly the first must be translated into the second. In most work situations objective and subjective importance (and hence significance) are similar. However, the same objectively important situation can produce different significance for different operators.

Reserve time is related to automation and something called the principle of "the active operator," which can be defined as the operator being actively involved in task performance. If reserve time is increased because of automation (e.g., a pilot spends more of his or her time being inactive rather than actively controlling the aircraft), the pilot's ability to take corrective action in an emergency decreases (Zavalova & Ponomarenko, 1968). Zavalova and Ponomarenko explained these results by claiming that nonutilization of a pilot's motor apparatus is connected with decreases in the flow of muscle impulses. A method of "joint" control (combined control, with the pilot and autopilot working together) was suggested. This lowers the pilot's workload and allows him or her to preserve his or her "activation." This problem is discussed in more detail later.

Russian psychologists also considered the concept of a logical filter-transformer that processes and redistributes information among the sensory organs (Akhutin, 1977). However, unless this concept is operationalized it merely restates the obvious. Zavalova and Ponomarenko (1970) demonstrated that to reduce reaction time to an emergency it is necessary to combine correctly instrumental (equipment-derived) and noninstrumental (operator-sensed) signals. To validate a noninstrumental signal about an aircraft malfunctioning, an instrumental signal must be presented to the pilot no later than .8 seconds following the noninstrumental one. For example, at the moment the pilot hears a change in engine noise or vibration in the resistance of the controls, an instrumental signal may not yet appear in the control panel. The pilot develops one or more hypotheses based in these noninstrumental signals. He or she immediately scans his or her instruments to validate the hypothesis. If the instrument information is unduly prolonged, the pilot may mistakenly reject a correct hypothesis and accept an erroneous one.

Konopkin (1962) studied the variations of time production operations produced by the change of a conveyor belt's pace. At a slow pace, the variability of the operation increases; at an optimum pace, it decreases. If the pace of a conveyor belt is greatly increased, then the variability of the operation increases again. Two types of production operations were studied. In the first, operation activity is performed based on external (equipment-derived) signals. In the second, operation workers performed mostly on the basis of prescribed instructions. Production operation regulated by instructions varied more significantly when the pace was increased than that regulated by external signals. These results should be considered when designing an operator's activity that is limited by time conditions. It is important to decrease the operator's memory load. Memorized instructions must be replaced by external signals. Such signal instructions must not only be easily differentiated and interpreted, but must advise the operator on how she or he must react.

Search for information is obviously important. It involves identifying on an instrument panel the characteristics of an object as denoted by the color, shape, position of the instruments, nature of the information shown on the instruments, and so on.

The most significant achievement in studying the temporal characteristics of work activity has been made by specialists in time-motion analysis (e.g., Gilbreth, 1911). One of the more widely known systems is MTM-1 (Karger & Bayha, 1977). The possibility of using this system in ergonomics is pointed out by the various authors (see, e.g., Eastman Kodak, 1990; Meister, 1985, 1989; Montmollin, 1967).

In the past few years some Russian ergonomists (e.g., Gordeeva & V. P. Zinchenko, 1982) have criticized this method because they believe that a predetermined time system is based on summation; that is, the total time for the performance of an activity equals the sum of time for elements of activity. However, this is incorrect. Sometimes, these methods are criticized because of their excessively detailed analysis. However, analysis of the structure of activity may require microanalysis, because some activity components are of very short duration, for example, pushing a button, throwing a switch. It is the belief of some Russian psychologists (including Bedny & Zelenin, 1980) that the MTM-1 helps with the task analysis of work activity and certainly there are occasions when it may be useful to have data of the sort found in Tables 4.2 and 4.3.

4.4 OPERATOR PRECISION

Every system sometimes deviates from its requirements. When this deviation exceeds permissible parameter values, the system either is proceeding to malfunction or has actually malfunctioned. When a system cannot function or when it cannot achieve its goal, it has failed.

Russian psychologists, like Westerners, assume that the accuracy and reliability of the operator's work influence the efficiency of total system performance. The two are obviously not the same. Accuracy refers to the precision with which a goal (e.g., a target) is achieved (how close one comes to some standard), whereas reliability refers to failures of a system and how probability of failure can change over time.

Because the operator is a system component, an error on his or her part can lead to system malfunction and breakdown. Hence any analysis of system

TABLE 4.2
Duration of the Visual Fixation

Task	Average Time of Fixation T_f
Search for simple geometrical figures	.18 - .20
Search for letters and numbers in table	.30
Search for letter-number lists	.31
Search for the target on the locator screen	.37
Orientation and navigation with the help of a locator	.64
Acquaintance with the situation denoted by conventional symbols	.63
Search for conventional symbols	.25 - .33
Detection of changes in a familiar situation denoted by conventional symbols	.55
Counting of the conventional symbols	.52

Note. From Lomov (1982).

failure must consider not only technical (equipment) factors but also the influence of the operator on system functioning. The more directly the operator determines that functioning (most often the case in manual systems), the more likely it is that operator error will negatively impact systems operations.

Traditionally, errors have been classified as errors of omission (failing to perform a required action), out-of-sequence errors (performing a required action at an inappropriate stage of a procedure), errors of incorrectness (performing a required action incorrectly), irreversible errors (once made,

TABLE 4.3
Temporal Characteristics of Operator Work Activity

Actions Performed	Average Duration (seconds)
Reading of digital indicators:	
Fluorescent light IN-1	.73
Optical projection board	.45
Seven-segment luminescnet	.58
Eight-segment luminescent	.63
Reading of pointer indicators:	
Damper	0.4
Average damper	1.0
Small damper	1.5

it cannot be corrected), or reversible errors (one that can be corrected without influencing the system).

Merely to classify operator errors in such a manner, however, adds nothing to our understanding of how and why the error occurred, because the underlying rationale for the study of error is to discover and eliminate the mechanisms that produce these.

Russian ergonomists distinguish operator errors from operator failures on the basis of the criticality of the error. An operator action that affects work efficiency negatively without shutting the system down is considered an error; an action that renders the operator incapable of further functioning or shuts the system down or causes the system not to achieve assigned goals is considered an operator failure. Western ergonomists would recognize the distinctiveness of the two effects of negative action but would probably use the single term *error* to describe them both.

Sometimes an increase in equipment precision can be produced only by increasing the complexity of the system, as by introducing new components, and this sometimes results in lower system reliability. Reliability can also be reduced when one begins to approach the limits of the system. For example, at low and average speeds of flight, firing an aircraft's guns can produce very precise results, but as the pilot approaches the maximum permitted speed, firing accuracy may be sharply reduced.

A starting point in the understanding of operator error is the development of various error taxonomies. In the former Soviet Union, the better known taxonomies were suggested by Zarakovsky and Medvedev (1979) and Bodrov and Orlov (1985). These taxonomies, including certain corrections and additions, follow: (a) general characteristics of an error, (b) the position of an error in the structure of the system, (c) the position of an error in the structure of the operator's activity, (d) the cause of an error, and (e) the consequence of an error. These criteria are considered in more detail in Table 4.4.

It is obvious that this error taxonomy is based on the concepts of work activity and self-regulation. For example, during the analysis of error in an activity's structure, it is important to analyze the operator's goal formation process and how he or she extracts relevant task conditions, as well as to determine the relationship between task difficulty and significance, task performance strategy, and subjective criteria of success. In analyzing experience, one must attend not only to particular knowledge and skills but also to how a person acquires efficient strategies of work performance and can successfully use them in practical situations.

When one analyzes erroneous actions, one should determine what are the action goals, what kind of information was utilized by an operator during action execution, what information is required, and what interfering factors complicated the performance of this particular action.

TABLE 4.4
Criteria for Analyzing Errors

Criteria	Parameter	Dimensions
General characteristics of an error	Relationship of errors to task in which they occur:	In what task do errors occur
	Detectability	Obvious, hidden
	Operator's awareness of error	Aware, unaware
	Existing causes	Predictability
	Probabilistic characteristics: Typical Expectedness Kind Quantitative	Usual, unusual Expected, unexpected Constant, variable Frequency of occurrence and number
	Functional characteristics	Reproductive or procedural, perceptional, motoric, symbolic, creative, or calculational
Structure referred to defines errors by the particular system or subsystem in which they occur, specific stage of system operation in which they occur, and how they are revealed in the system's functioning	Where errors happen and the time of their appearance	In what component of the system, at what stage of system operation, and on what work shift
	Character of the working conditions of the system in which error occurs	Good, bad, overloaded, underloaded, etc.
	External manifestations (error consequences)	Consequential, nonconsequential
Determines of the position of an error in the structure of the operator's work activity	Goal formation and evaluation of the meaning of a situation and task	Incorrect understanding and formulation of goals, incorrect analysis of task conditions
	Evaluation of specific characteristics of tasks:	
	Difficulty	Under- or overestimation of a task's difficulty
	Sense of task	Incorrect evaluation of the significance of task and its components
	Personal attitudes	Inadequate motivation

(Continued)

165

TABLE 4.4
Criteria for Analyzing Errors

Criteria	Parameter	Dimensions
	To what particular mental actions do errors belong	Errors resulting from sensory and perceptual actions, mnemonic actions, thinking actions, or decision-making actions
	To what particular verbal and motor actions do errors belong	Errors connected with discrete and continuous motor actions verbal errors, or errors connected with undesirable involuntary responses
Cause of errors	Error stems from idiosyncratic (personal) characteristics, from equipment, or from organizational characteristics of the system	Specific personal characteristics, technical or organizational factors; e.g., the aircraft goes into a spin when engine died during takeoff (error stems from technical factors); the aircraft goes into a spin when pilot incorrectly pulls hand wheel during takeoff (error stems from personal characteristics)
	Dynamic factors (factors that change over time)	Increased fatigue, boredom, monotony, decreasing vigilance; wrong and untimely selected strategies of activity, or incor-rect transition fro one strategy to another
	Stress-producing factors	Time limitations, dangers, and external interferences
	Technical factors	Incorrect distribution of functions between human and machine, inadequate information and instructions provided, inadequate design of equipment
	Organization factors	Interference from supervision, inadequate work/rest schedules, excessive or insufficient information transmission
	Operator experience	Insufficient training and experience, insufficient knowledge and skill, inability to select required strategies of activity, and to timely transfer from one strategy to another
	Idiosyncratic (personal feature)	Operator unsuited for work because of inadequate intelligence, physical fitness, or emotional stability

(Continued)

TABLE 4.4
Criteria for Analyzing Errors

Criteria	Parameter	Dimensions
Analysis of consequences of errors from three points of view	Influence on efficient of the system	System malfunction or shutdown, failure to achieve goal in assigned time, accident
	Influence on operator's activity	Performs required further actions incorrectly or out of sequence, or fails to perform further actions
	Influence on operator's state	Tries to correct error with or without success, fails to note error, maintains or loses self-confidence, becomes confused, continues or fails to continue activity

The analysis of erroneous actions from a self-regulation standpoint thus begins with the discovery of goals and motives. The cause of erroneous actions may be the incorrect formation and understanding of the goal, or the operator may incorrectly select a strategy of goal attainment. Other situations are also possible. The operator can correctly formulate the goal but accidently perform the wrong action. When the operator incorrectly formulates goal activity and as a result performs wrong actions, this according to Norman (1981) is a *mistake*. But, when the operator correctly formulates goal but accidently performs the wrong action, this is considered a *slip*.

According to Landa (1976), Zarakovsky (1987), and others we can extract a threefold level of regulation of activity:

1. Level of stereotypy or automaticity of performance.
2. Level of consciousness of the regulation of activity in terms of acquired rules and familiar strategies.
3. Level of regulation of activity based on general knowledge, principles, and heuristic strategies.

All of these levels have a hierarchical relationship. This classification can be useful when we conduct human-errors analysis. Errors more frequently occur when the level of regulation of activity and conditions of performance are poorly matched. For example, in the face of unpredictable changes during performance stereotyped methods of activity may result in errors. In conditions of low levels of predictability with complicated problems, the ability to use appropriate knowledge and skill, as well as adaptive new strategies assumes greater significance. As may be seen, the ability to exploit different levels of activity and transfer from one level to another is germane to controlling the number and gravity of errors.

Few Western ergonomists would seriously dispute any of the parameters and dimensions of the Russian error taxonomy but, except for special circumstances, the taxonomy is somewhat overambitious and idealistic. Manifestly, to answer the questions implied in the taxonomy would require a major investigation, and this would hardly be permitted except in extreme cases such as the Chernobyl disaster or where major errors of great significance to system function occurred with a high and continuing frequency. At the same time, the error taxonomy reminds one of all the factors that can contribute to error production.

It must be admitted that the study of error-causing mechanisms did not evolve in the former Soviet Union as it did in the West. For example, there is nothing in the Russian literature that corresponds to a formal theory of error production as that found in, for example, Reason (1990). On the other hand, despite energetic Western efforts to predict human reliability on the basis of error rates (see, e.g., Meister, 1993), the practical effects of attempts at error reduction and prediction have been quite few.

Related to errors are accidents, some of which may be caused by operator error, others by purely technological factors. Although the error is usually almost instantaneous, the accident in its physical manifestation is not; it has a definite progress, which creates a task for the operator (accident containment and elimination). That task also involves all the factors operative in routine operating tasks (e.g., goal formulation, image of the situation, task conditions, task significance and, above all, the strategy for containing the accident).

Obviously, it is necessary to determine whether or not the operator made an error that produced the accident. An essential part of answering that question is to reconstruct both the operator's and the machine's actions immediately prior to the occurrence of the accident and to try to link the two. The process of reconstruction has much in common with the walk-through used in design evaluation at an early stage to try to simulate the operator's activity sequences. It is made easier by the increasing use of automatic recording devices.

With the increasing development of more highly automatized (e.g., computerized) systems—as, for example, in process control industries like nuclear power plant facilities—the role of the operator changes and consequently the nature of the error also changes (see Meister, 1995, in press). In automated systems, where formerly the operator "ran" the equipment the computer now does so and the human becomes primarily a monitor of information provided by the system (e.g., via meters, warning lights, and a detector of deviant system values, i.e., those exceeding parametric limits). Assuming that a malfunction results (which appears to be inevitable over time), the operator becomes a diagnostician and troubleshooter.

In such problem-solving situations, the definition of an error as a deviation from a required procedure no longer applies, because there are few formal troubleshooting procedures. During the diagnostic process the operator's major role is to acquire information, to develop and test hypotheses about what went wrong, and to make decisions—all highly cognitive processes, and all of which can be considered erroneous until there is a satisfactory resolution of the problem. The question is whether an action taken to secure information or to test a hypothesis should be considered an error if it does not immediately resolve the problem.

The activity notion of strategy and changes in strategy is peculiarly appropriate to these new situations, which will become increasingly the norm. The system malfunction we have described is essentially an accident in process and consequently the importance of investigating accident phenomena becomes even more important.

Because the accident is a problem-solving situation (how to avoid it, how to contain it if it occurs, how to avoid its most dangerous consequences), it is necessary to distinguish between the meaning and sense of the dangerous situation. If meaning determines the operator's understanding of a situation, then sense determines the situation's significance. Russian ergonomists distinguish positive significance from negative significance, so it is important to discover the relationship between them in dangerous accident situations.

For example, in a potentially dangerous situation an operator may disregard the situation in order to pursue a goal with higher significance for him or her. This was shown markedly in the Chernobyl tragedy, which occurred during the performance of a test program. Nuclear safety during a test requires the functioning of various emergency reactor safety systems. For example, only after activating the emergency power reduction system, can a test program be performed. For this purpose, the operator must press a special button (Medvedev, 1991). However, no one from the personnel staff did this.

Another safety system, the emergency core cooling system (ECCS), was turned also off. Normally, the emergency power reduction system switches itself on when certain parameters of the nuclear power station reach a particular value. It was discovered later that all five protection systems were turned off because these systems complicated the test program. From the self-regulation point of view, the significance of the test program for personnel was very high. Failure to perform the test program could result in severe punishment, even termination. The position of nuclear operator was extremely prestigious with a high salary. Nothing had ever happened during a test program before, which is why the significance of the test program was very high, and that of the safety program was low. The main desire during task performance was to finish the test as quickly as possible; consequently safety requirements did not have a strong sense for the personnel.

In this situation, the operator did not formulate adequate goals of his actions and tasks as a whole.

Kotik (1987) studied the relationship between the significance connected with anxiety, the chance of dangerous events, and the evaluation of consequences. He demonstrated that the more a person negatively evaluated the consequences of his or her own errors, the lower the chance a dangerous event would cause anxiety. This means that the negative significance of events is associated with a higher probability of danger.

This suggests that not only cognitive but also emotional-motivational components of activity can influence the accuracy and reliability of an operator's performance. El'kin (1962) demonstrated that when people estimate the duration of time intervals, if they were punished for errors by receiving an electric shock, the accuracy of the estimates increased. Experimental feedback information also increases accuracy of task performance. Klimov (1969) studied the accuracy and speed with which subjects reacted to stimuli. The subjects received information about their errors but not about their reaction time. Accuracy increased and the number of errors decreased by five times. Reaction time also decreased but not so significantly. In reversed saturations, when subjects received information about reaction time but not about errors, Klimov secured the opposite results.

The accuracy with which one reads an aviation instrument often depends more on the significance (importance) of this instrument than on the visual features of the instrument. For example, an aircraft's attitude indicator has a rough scale; the distance between scale elements is about 5° and the distance between the numerals is about 15°. This instrument is very significant for pilots, because they learn to read the horizon of the aircraft with an accuracy of about ±1.3°, which is much higher than their precision in reading other displays, which have a more detailed scale (Kotik, 1978). In the same way if task significance changes, the precision of task performance also changes.

The speed with which information is perceived varies with the strategy selected. For example, if an automobile driver is on a street whose speed limit is 25 mph (miles per hour), if he or she is in a hurry but is afraid of getting a ticket, he or she selects a speed between 25 and 30 mph, knowing from experience that this minor violation will not result in a ticket. This means that the speedometer can read between 25 and 30. The driver will quickly glance at the speedometer and note that, if the position of its pointer is between 25 and 30, he or she is driving at a safe speed. If the pointer is to the left of this area, he or she can increase speed, and if it is to the right, he or she should decrease speed. This strategy eliminates the requirements for a precise quantitative evaluation of the pointer's position and increases the speed of perceiving information.

Kotik (1978) found that pilots in flight revealed that they spent very little time (.3–.5 seconds) observing individual displays. To understand why, he used a tachistoscope under laboratory conditions to present pilots with photographs of individual aviation instruments. The pilot had to read the instruments as quickly and precisely as possible. Before the study began, subjects were told the kind of instruments to expect. Readings were considered erroneous if errors were more than one interval on the scale. Results indicated that, for required accuracy, a pilot should spend 1.2 seconds reading each instrument. The pilot's experience and skill level did not increase the speed of the readings.

Kotik (1978) hypothesized that, in actual flight, a pilot perceived not isolated displays but logically interconnected items of display information. It is therefore not as important for a pilot that an individual scale pointer is not in an exact position, but how far it deviates from the specified position, in other words, how far it exceeds tolerance limits. The pilot compared the required scale value with the real one. Hence if the pilot possesses the image of the situation, it helps him or her to perceive information from instruments more efficiently.

Other studies by Kotik (1978) indicated that the pilot organizes the information from his or her individual instruments into an interrelated, dynamic whole—a gestalt as it were. This conclusion would not be unfamiliar to any pilot or student of aviation psychology. What is significant in the Russian explanation of the phenomenon is the emphasis on image as the organizing mechanism.

The pilot's image of flight is dynamic and changes as the situation changes and his or her evaluation of the consequences of actions changed. Russian ergonomists see this as an example of the importance of the self-regulation process. For example, the image of flight depends on the function blocks "goal," "formation of task," "psychological representation of relevant task conditions," "difficulty of task," and "task significance." Results of the pilot's actions through feedback can affect these function blocks. As a result the image of flight can be changed.

The Russian concept of error emphasizes what in the West would be called "error tolerance." An action can produce an output that varies on some sort of scale. For example, a pipe must be machined to a tolerance of ±.05 mm. Any pipe machined from +.05 mm to −.05 mm is acceptable; in such a situation zero tolerance would be desirable, but an error of machining is acceptable as long as it is within parametric limits. If the helmsman's instructions are to steer the ship south (180°) but he or she actually steers between 175° and 185°, the variation (although technically an error) is within tolerance because the ship is certainly steaming south.

In the example of machine tooling, when the deviation from zero approaches the outer limits of the allowable parametric values, something called "reserve of precision" is reduced. The tolerance ±.05 mm can be considered a reserve; this is a tolerance within which the worker can function and only when she or he exceeds ±.05 mm is the output in error. The Russians therefore distinguish between precision, which is the output tolerance, and error, which means exceeding that tolerance.

The worker always has some particular reserve of precision, and the less it is, the more requirements for precision there are. According to the model of self-regulation (Fig. 2.6) the reserve of precison can have both objective and subjective meanings. Depending on the subjectively accepted "reserve of precision," the operator can use different strategies to attain task precision. In Russian ergonomics, the concept of reserve precision has been further detailed in the form:

$$\delta_{jres} = D_j - \delta_{jmin} \tag{4.7}$$

where δ_{jres} is the reserve of precision, D_j is the admissible deviation of parameter j, and δ_{jmin} is the minimum amount of deviation (Kotik, 1978).

The minimum amount of deviation is related to specific parameters of the system and can be defined in the following way:

$$\delta_{jmin} = \delta_{1j} + \delta_{2jmin} + \delta_{3jmin} + \delta_{4j} \tag{4.8}$$

where δ_{1j} is the amount of deviation according to the parameter j, δ_{2jmin} is the minimal amount of deviation while perceiving and processing information, δ_{3jmin} is the minimal amount of deviation during control action execution, and δ_{4j} is the amount of deviation in transition from the operator's control action to instrumentation.

From the preceding it is apparent that Russian theorists assume that each stage in the operator's processing of information and response activation has a "reserve of precision." Although the concept of reserve of precision should be acceptable to the West (even though the terminology is strange), the extremely molecular manner in which it is presented may be questioned.

The precision of the perceptual process is increased by matching input information with a "system of expectation" and by creating an image of the goal and the situation. The strategy of the attentional process continuously changes according to the image of the goal, image of the situation, and consequences of the operator's actions.

Kotik (1978) introduced the notion of reserve of precision in the same way he used reserve of time. Reserve of precision can also be objectively given as a tolerance value and subjectively accepted by an operator. Sometimes the two will not match and might even contradict each other. Therefore, the precision of task performance depends on the strategy of work activity that is derived from the self-regulation process (see Fig. 2.6).

Reserve of time reflects all situations as a whole and can limit the change in time of all parameters. Reserve of precision must be determined in relation to specific physical parameters. In any task, an operator has one reserve of time but different reserves of precision.

The complexity of a task can be increased by limitations in time and precision. According to self-regulation, increased precision requirements influence the following self-regulation function blocks: "subjective parameters of success," "subjective parameters of admissible deviations," "assessment of sense of task," and from these all other function blocks. As a result, the strategy of activity can significantly change. In simple tasks, where achieving required precision is not difficult for an operation, the significance of the task may be reduced. Hence, motivational intensity may also be reduced. If a task has a high precision requirement, it becomes more significant for an operator and motivation can be increased. If a task for an operator is very difficult, increasing precision can reduce significance of task for an operator, and as a result, the operator may decrease his or her subjective criteria of precision. Kotik (1978) demonstrated the relationship between reserve of precision and significance of task to be an exponential one. The less reserve of precision the more significance of tasks:

$$S(\delta_{res}) = e^{-\eta \delta_{res}} \tag{4.9}$$

where $S(\delta_{res})$ is the significance of task as a function of reserve of precision, and η is the mean arrival rate of event.

$$\eta = 1/\bar{\delta}_{res} \tag{4.10}$$

where $\bar{\delta}_{res}$ is the average reserve of precision for a group of operators. The equation was proved on the basis of a study in which pilots were presented with an actual flight task and specified precision requirements. The subjects were asked to evaluate the probability of their capability to perform the required task.

All the subjects were given a paper with the precise description of individual tasks, involving level of admissible deviations of the control parameter (D) according to the precision requirement. Subjects wrote the probability of failure to achieve the task's goal based on the required level of precision and the subject's opinion about his or her own ability to achieve these requirements. Probability of failure to achieve the task goal was considered an index of its significance.

For any individual task, the minimum admissible value of error (δ_{min}) with which the task could be performed was determined by experts. For any permissible deviation of control parameter (D) was calculated the reserve of precision (δ_{res}). Because the subjects evaluated the probability of success for different levels of (D), the experimenter was able to compare the experimental significance (S^*_{res}) with various values of δ_{res} (see Fig. 4.6).

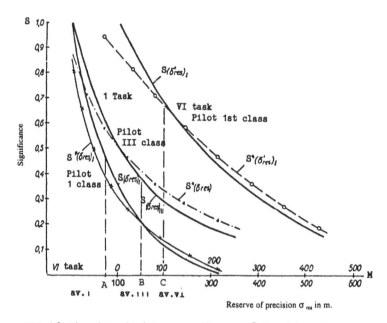

FIG. 4.6. The relationship between significance $S^*(\delta_{res})$ and the existence of reserve of precision (experimental curve). Theoretical curves $S(\delta_{res})$ depicted by a solid bold line. From Kotik (1978). A—δ_{res}, for I class pilots performing task I; B—$\bar{\delta}_{res}$ for III class pilots performing task I; C—$\bar{\delta}_{res}$ for I class pilots performing task VI.

$S^*(\delta_{res})$ described experimental data that demonstrate the relationship between significance (S) and the existence of reserve of precision (δ_{res}). Experimental data also demonstrate that the less reserve of precision the more significance of task. The theoretical curves $S(\delta_{res})$, $S(\delta_{res})$III, and $S(\delta_{res})$ are represented by a solid bold line.

Figure 4.6 presents the data for a simple task I and more complicated task VI. For simple task I, data about the pilots' first and third levels of expertise are presented. On the horizontal axis the reserve of precision is given in meters (δ_{res}), and on the ordinate axis the significance (S) is presented. A, B, and C are the average reserves of precision for the first and sixth task.

Experimental and theoretical data fit each other on a statistically significant level. This means that on the basis of theoretical calculation $S(\delta_{res})$, it is possible to predict the significance of the task for the operator. From these results, one can conclude that the activity of an operator in achieving required precision is one of the basic mechanisms of achieving the goal. Reserve of precision can also be used for an approximate evaluation of the task's significance.

Failure to achieve a goal is only one important factor that generates significant emotional reactions to situations; achieving a goal is another. However, achieving a goal usual generates much less emotion because failure to achieve a goal is a relatively infrequent event for the operator. As a result, failure is much more significant for an operator. Consequently the intensity of the desire to avoid failures and their consequences can be used as a major criterion for evaluating the significance to the operator of his or her task performance.

In the previous study, the significance of the task depended only on the reserve of precision. The significance of the task can also depend on the significance of a goal without any relationship to reserve of precision or reserve of time. For example, the significance of a goal for an operator can be very low. Increasing task precision criteria in such situations may be ignored by the operator, who may develop his or her own subjective criteria of precision, which can be lower than any objective requirements for precision, thus lowering quality of task performance (increasing rate of error). One can influence the acceptance of precision requirements by increasing the significance of the goal, for example, increased rewards for the achievement of required precision and punishment for a decrease in quality. The relationship between precision and speed of performance, which is termed in American literature as the speed-accuracy trade-off, is also important. A person tends to make more errors if the speed of work is significantly increased.

Instructions concerning precision also have physiological effects. For example, Rutman (1974) studied how instructions influenced the bioelectrical activity of the brain during the performance of different reactions. He discovered (based on electroencephalograph measures) that instructions "act[ed] with maximal precision" activated visual systems centers in the brain. These centers are connected with evaluations of stimulus in different complex situations. As a result, precision of action increased but the readiness to react decreased.

Sergeev and Romanenko (1965) studied the relationship between precision and stability of task performance. They discovered that when the requirement for precision of task performance was very high, the stability of the task performance, according to precision and time criteria, was reduced.

Accuracy and speed criteria are also influenced by stimulus intensity. For example, when subjects received weak signals almost at threshold level, their accuracy was a more sensitive indicator of performance than reaction time. Bardin (1969) formulated a principle, according to which reaction time can be used as a criterion of performance when optimal intensive signals are presented (this is called approach to operative threshold.) When one uses weak signals (called approach to absolute threshold), accuracy is a more appropriate criterion.

The interaction between quality and speed of work in a laboratory situation was studied by Chebisheva (1969). Subjects had to separate variously colored wooden sticks to the beat of a metronome. In the beginning the metronome pace was slow but soon increased, which meant that the subjects had to separate the wooden sticks more quickly. The slower pace caused the subjects to perform the task with a high error rate. Subjectively, this pace was evaluated as inconvenient. Increasing the pace to a subjectively optimal level increased the quality of the task performance. However, further increase of the pace to a level evaluated as "stressful" and as "very difficult to achieve" resulted in a decreased quality of task performance. It was also discovered that training at a stressful pace (higher than optimal) positively influenced the performance at the slower paces. As a result, the subjectively evaluated optimal level of task performance became higher. This means that training of speed-related tasks should be performed at different paces, sometimes higher than the actual required pace.

Shadrikov (1979) also studied the relationship between an operator's precision and speed, in a task in which subjects had to place pins in a specific order on a pin board. Shadrikov used three types of instructions: Act as quickly and precisely as possible, act as quickly as possible, and act as precisely as possible. The first instruction was the least efficient, which suggests that at the initial stage of skill acquisition students should pay attention only to precision and only when some skill has been acquired should they pay greater attention to speed performance.

The study of an operator's activity precision presented in this section and the experimental data about the precision of positioning actions described in section 2.3 support the concept that a person may continually change his or her strategies based on self-regulation. As a result, the precision of activity may vary widely, depending on the individual's evaluation of task conditions, task difficulty and significance, subjective criteria of success, and feedback. Precision in performance considered as an aspect of self-regulation requires consideration of all the factors involved in that self-regulation.

4.5 THE RELIABILITY OF AN OPERATOR'S WORK ACTIVITY

To increase the reliability of an operator's work activity (in the West this is called "human reliability") is a major ergonomics problem. In Section 4.3 a taxonomy of error was presented. One of the assumptions underlying that taxonomy was a distinction between reversible and irreversible error. The latter, when it has extreme effects on the system, was termed *failure*, and differentiated from errors that had no such extreme effects.

In the former Soviet Union, two separate approaches to evaluating the reliability of human performance were used: systemic-technical and anthropocentric, the latter based on the concept of activity. Representatives of the systemic-technical approach attempted to develop analytical methods of evaluation an operator's performance reliability. One of the major approaches is the method suggested by Gubinsky and Efgrafov (1977).

The essence of this method is that an operator's activity is divided into individual hierarchial levels, each of which is represented as a specific structure. For smaller units of activity analysis error rates are determined. Error rates for smaller units are combined mathematically to arrive at a numerical value for total performance reliability.

Readers of Western human reliability literature (Meister, 1995) will recognize the approach as similar to that of THERP and one that presents many of the same difficulties that Western human reliability methods possess. There is, first of all, the problem of collecting the elemental error rate data, a problem that has not been solved in the West, much less in the former Soviet Union. The method, although very formal, is flawed because it makes certain unproved assumptions. Among these assumptions are: independent units of activity, singular effects of task performance, failure to recognize variations in operator strategy, and the independence of failure effects (when in fact irreversible errors may have a "cascade" effect on the system). The method of combining error rates assumes complete independence of activity elements, which is as incorrect for Gubinsky and Efgrafov (1977) as it is for THERP.

For evaluation and prediction of an operator's performance reliability, Russian theories often make use of the exponential law of reliability. When the failure rate of task performance is considered as a constant, the law of reliability is described by the following formula (Dhillon & Singh, 1981; Lomov, 1982):

$$F(t) = e^{-\lambda t} \tag{4.11}$$

where F(t) is reliability during the required time of performance and λ is the failure rate.

This law can be used when errors do not influence each other, a condition that is unlikely to be found very often. The failure rate for repeated operations can be determined from:

$$\lambda_j = n_j/N_j T_j \tag{4.12}$$

where λ_j is failure rate of the "jth" type, N_j is the number of performance operations of the "jth" type, n_j is the number of failures, and T_j is the average time of performance for operation t_j.

Another way of evaluating an operator's performance reliability is whether she or he performs within required time limits. Failure to do so is an error, and because this is an irreversible error, can be considered a failure. One wonders, however, how one would conceptualize the performance of an individual who performs within time limits, but does so incorrectly or inadequately.

The probability that the task can be performed within the required limits can be defined from the following formula (Shibanov, 1983):

$$P = P\{\tau < t_1\} = \int f(\tau)d\tau \qquad (4.13)$$

where τ is the time performance of the task, t_1 is the limit of time beyond which performance can be considered a failure, and $f(\tau)$ is the distribution of time performance of the task by operator.

Note that t_1 can be either constant or random. If t_1 is random, determining P becomes more complicated. If τ and t_1 are distributed according to the normal distribution, the calculation of P becomes simpler. In Fig. 4.7, the distribution of the time performance of the task is shown, as well as the distribution of reserve time with the parameters τ, σ_τ, and and t_1, σ_τ.

If we know the law of distribution, the parameters, and the time limit, we can calculate the probability of performing the task in the required time according to the theory of probability. The dashed area in Fig. 4.7 demonstrates the probability of performing the task in the time that exceeds the permitted requirements. However, the authors of this method did not describe how they calculated the meaning of different parameters or how they got these data for calculation.

This example demonstrates the problems Russian scientists encounter when they try to use mathematics for ergonomic studies. The main problem here is that there are insufficient input data to be used for a quantitative approach.

The relationship between the human and the machine is a complex one. In accordance with system theory the human is considered as a component of the total system. At the same time the system is subordinate to his or her commands. This human control of the system is being progressively eroded,

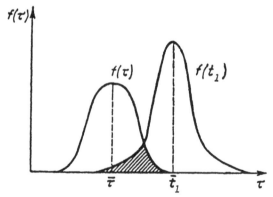

FIG. 4.7. The distribution of the time performance of the task and distribution of reserve time with the parameters τ, σ_τ, and t_{res}, σ_t.

however, by extensive computerization. Previously, one thing that differentiated human and machine was the former's ability to sense an error he or she had made and then to correct that error (where possible). If the machine failed it could not signal any causal mechanism; it simply stopped functioning.

With increasing computerization and particular the inclusion of sensor components as part of automatization, the system can determine on its own when it is failing or in process of failing and alert the operator to the need for diagnostic and corrective action. This is in small part a form of consciousness, although it would be premature to assume that the machine has a sense of individuality.

Mention was made earlier that the human's role vis-à-vis the machine had changed from a controller who "ran" the machine to an overseer of automatic processes, a diagnostician, and a troubleshooter. Because this role is preeminently one of information processing, proper ergonomic design of the system should center around determining the nature and amount of information to be presented by the system to the operator to optimize his or her cognitive functioning. This places great priority on analysis of the errors that may be made by the operator during the diagnostic/troubleshooting process. Because error reduction must now be designed into the system from very early stages of design, it is necessary to predict the kinds of machine malfunctions that may occur, the symptomology these malfunctions may manifest, and how the symptomology can best be presented to the operator. Although operator errors may still be deviations from prescribed procedures (where these still exist), we now have a new error concept: the error of interpreting symptomology. Such an error is far more complex and difficult to anticipate than a procedural error, although Meister (1992) suggested a procedure for doing so in connection with advanced nuclear power plant design.

The reliability of an operator's performance is, of course, not stable and may change over time. Increasing skill and experience tend to make performance more stable, but daily changes in motivation tend to destabilize performance; reliability studies must therefore be more than momentary.

Any study of operator reliability must begin with a determination of the major factors that impact his or her performance. Next, it is necessary to determine how these factors influence his or her performance reliability. This may involve an investigation oriented around the error taxonomy described in the previous section, or some variant of task analysis. Any empirical data bearing on the operator's performance, such as error rates for elemental acts, should then be applied in the manner characteristic of making a human reliability prediction. If empirical data are lacking, it is permissible to use expert judgment under controlled conditions.

One method of determining the impact of the operator's performance on system reliability is to measure the reliability of the system performance with and without inclusion of operator contributions to that performance. In effect one subtracts the operator from system performance. Because operator error always has a negative effect on system performance, the amount of that effect can be determined mathematically. This procedure was employed by Meister and his colleagues (1970) in analyzing the reliability of several aircraft systems after failures resulting from human errors were described from the reliability record. System reliability increased significantly. However, such a procedure does not take into account the operator's positive influence on system reliability. We consider briefly some data that follow from the systemic-technical approach. The reliability of operator from activity point of view is discussed next.

A study bearing on the question of operator influence on system performance was conducted by Kotik (1978). This involved the pilot's detection discrepancies in flight attitude from two instruments. When the discrepancy between two instruments was more than 5°, a red indicator was illuminated to signal the failure of one of the two instruments. Which instrument was faulty was determined by the pilot's examining other instrument data. Even though the signal did not show which instrument had failed, the probability of discovering the failed instrument with the warning signal was $p = .999$. When this indicator was absent, $p = .928$. This means that the failure warning increased the probability of more effective information processing.

Mileryan (1974) studied pursuit tracking under severe operating conditions (vibration, noise, temperature, fatigue). The vertical movement of the tracking stimuli on the screen was controlled by moving a lever forward and backward. Horizontal movement was accomplished by pushing two foot pedals. Performance adequacy was measured by proportion of time when controlled element coincide with the target. The equation is

$$r = t/T \tag{18}$$

where t is the total time when performance coincides with the target and T is the total time spent tracking.

It has been pointed out in previous discussions of the mechanisms of self-regulation that the operator creates a subjective model of the objective reality. Where the objective reality is somewhat ambiguous, the operator's subjective model determines his or her strategies, which in turn influence the reliability of task performance. Management of a system requires the operator to regulate not only the system, but also him or herself as a part of a system. This is especially the case when the individual's role is to monitor, supervise, and troubleshoot an automated system.

Task significance influences performance strategy, particularly in cognitive tasks and some sensorimotor tasks. The chess player's thinking processes have been used as a model; Tikhomirov (1984) developed a special methodology for this purpose. Experienced chess players were asked to solve chess problems that varied in complexity, by discussing the solution process while an assistant moved the chess pieces. The verbalization was recorded on tape. The galvanic skin response of the left palm was recorded. A decrease reaction in skin conductance was considered as an emotional reaction to the chess problem. If the emotional reaction appeared 1.5 seconds after the problem was presented, the problem was considered to be the cause of the emotional reaction. If the emotional reaction appeared earlier, it was considered as a general state of arousal.

Later, the galvanic skin response was compared with the tape-recorded solution of the problem. Subjects were also debriefed following the experiment.

It appeared that the psychogalvanic responses coincided with the process of solving the problem. These responses usually appeared before the subjects' verbal formulation of the principles used in solving the problem. It was therefore hypothesized that conscious verbal actions follow the "unconscious emotional solving" of the problem. The subjects sense how to solve the problem but cannot yet verbalize its solution. Psychophysiological indices are defined as reflecting unconscious emotions that also reflect the significance of a task for subjects.

In one experiment, the subjects were required to verbalize their solution processes while consciously repressing any emotion. If the psychogalvanic response occurred, the experiment was interrupted. Under these conditions, the response was much reduced and subjects solved only the simplest tasks successfully. Tikhomirov (1984) concluded that the emotional regulation of activity while solving problems significantly increased the efficiency of task performance. It is possible that emotional reactions can be used as the criteria for the detecting unconscious problem solving, which at a subsequent stage becomes conscious problem solving.

Another experiment was conducted by Kotik (1978). An operator sometimes performs a complex task well but fails at a simple task. From the self-regulation point of view, this means that there is an inappropriate relationship between the different components of the self-regulation process, with the operator selecting an inadequate strategy of task performance. In the process of selecting a strategy, the relationship between task difficulty and its significance plays a special role. The subject's feeling that a task is very simple and has a low significance can produce a lower level of motivation.

Kotik (1978) conducted a number of studies directed at discovering how task significance influences the operator's performance. One experiment included a number of sequential actions with varying levels of complexity. Twelve windows were located on a vertical panel, through which red or white signals were presented in a certain sequence. The subjects had to move a corresponding lever in a horizontal panel that turned the light off. The complexity of the actions was based on the time restriction.

Before conducting the experiment subjects had the opportunity to perform the various actions under different time restrictions. This was done to allow the subject to evaluate subjectively the difficulty of the various actions. According to the model of self-regulation described in chapter 2, increased task difficulty indirectly results in increasing the significance of the task. Time for the various tasks varied from .8 seconds to 3.0 seconds. The shortest time permitted (.8–.9 seconds) required a maximum mobilization of subject effort, which supposedly increased the significance of the action. There were two types of tasks: significant (emergency) and insignificant (routine). The time permitted for insignificant actions was 3.0 seconds. Subjects perceived the experimental procedure as testing their ability and therefore had a high motivational level. Before any trial, the experimenter provided information about the consequences of each action.

If subjects exceeded allowable time for routing actions, a bell rang. If the subjects exceeded allowable time for emergency actions, they received an electrical shock. Both time and errors were recorded. Pulse, breathing rate, and skin electroconductance were measured both before and during testing.

Before they performed significant (emergency) actions, subjects overtly manifested stress. They seemed uncomfortable, shifted their bodies repeatedly, and their skin developed blemishes. These phenomena were not observed before subjects performed routine actions. Obviously, time constraints, the nature of the task (i.e., its significance), and anticipated punishment sharply altered psychophysiological indices. The difference in these indices between significant and insignificant actions were statistically significant ($p < .05$). However, pulse rate was chosen as the basic criterion.

The observation of the external behavior of the subject as well as the changes in physiological indices allow us to conclude that in the performance of significant actions the activation of the organism increases. The physiological indices of significant actions that are performed in time were compared with significant actions performed beyond the time limit. When subjects performed emergency actions within the time limits, the average pulse rate was 93 beats per minute; when they exceeded the time limits the pulse rate averaged 88 beats per minute. To determine how preliminary anticipations influenced the performance of emergency actions false instructions were given (information about normal actions were given instead of information about emergency actions). In these cases, subjects significantly exceeded emergency time limits and physiological measures suggested that the activation level for emergency situations was low. All of the aforementioned is evidence for the fact that the activation of organisms essentially influenced the correct performance of emergency actions.

Activation level for Russian engineering psychologists represents an intensification of the individual's physiological processes, something akin to an increase in adrenaline level. This presumably creates a greater readiness to perform, especially in an emergency ("fighting") situation.

Reserve time (t_{res}) was calculated for all actions. Based on t_{res}, the significance of task performance (S_{res}) was determined according to the formula given in section 4.3. The study's results demonstrated that the task becomes significant for the operator when its time constraint was 1.0 second or less. Figure 4.8 suggests the relationship between the reliability of performance (P), significant actions (S), and the level of activation (A).

The curve demonstrates how changes in the significance of action influences performance reliability P(S) and level of activation A(S). Probability P and activation A are considered as functions of significance. Until the significance level increases beyond S = .3, subjects perform actions with the same high level of reliability. When task significance increases beyond S = .3, performance sharply decreases [solid curve P(S) in Fig. 4.8].

If the significant actions appeared unexpectedly and subjects reacted to them without any preliminary preparation [the dotted curve P(S)A = 0 in Fig. 4.8], performance decreased sharply whereas task significance increased.

When subjects knew about the appearance of significant actions because of preliminary warnings, reserve time was increased in comparison to when

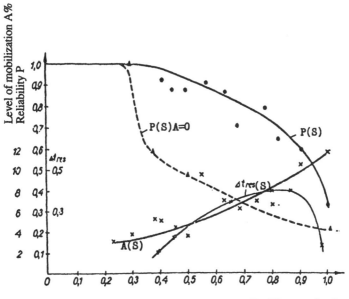

Significance of action S

FIG. 4.8. Relationship between significance, activation, and reliability. From Kotik (1978). P(S)—dependence of reliability of performance P from significance of action with preliminary activation. P(S)A = 0—dependence of reliability of performance P from significance of actions without preliminary activation. A(S)—dependence of the level of activation from the significance of actions. Δt_{res}(S)—dependence of increment of reserve time from significance.

subjects did not know about the significant actions. It is possible that warning increases the activation level and consequently the speed of performance. Differences exist between reserve time with warning conditions [curve P(S)] and reserve time without warning conditions [curve P(S)A = O]. In Fig. 4.8 curve $\Delta t_{res}(S)$ demonstrates how reserve time changes depending on the significance of an action and the existing preliminary activation.

When actions are relatively simple with a low significance (S = .3–.4), only slight activation is necessary to provide a high level of performance reliability. With an increasing significance level up to S = .9, the level of physiological activation increases, increasing speed of performance and reducing the decreasing performance reliability of complex and significant actions.

In this study, the influence of task significance upon performance was experimentally demonstrated. Moreover, if significance achieves a certain level, it sharply increases the subjects' physiological activation. Zarakovsky and Medvedev (1971) proved that the stability of activity possesses sharp discrete features, thus allowing the sudden transition from one level of performance stability to another. This mean that the subject changes his or her task performance strategy.

From a practical standpoint it follows that if in an emergency situation, special devices can keep the system's parameters within a set of required limits for a short time so that the operator has additional time to respond, this can sharply increase the reliability of an operator's performance.

The concept of "tuning" strategies to specific task situations appears to be important to Russian theorists. To them tuning is an important mechanism for increasing the reliability of task performance. This process is for them a complicated reconstruction of not only cognitive, but also the emotional-motivational components of activity that influence each other. This tuning occurs based on the self-regulation mechanisms described in chapter 2. Based on this mechanism the operator can concentrate his or her psychic and physical efforts according to task requirements.

One important mechanism of self-regulation is self-control. Russian scientists define self-control as the individual checking the correctness of his or her actions. This checking process, which is inherent in any motivated operator, is intensified by adequate feedback (Nikiforov, 1977; Popov, 1971). Nikiforov, for instance, studied different degrees of feedback by presenting subjects with tachistoscopic displays of various instrument readings.

In the control condition feedback was not provided; in one experimental condition only the words *correct* and *incorrect* were supplied. In a third condition subjects received information about their performance. In Fig. 4.9, performance is shown to vary significantly as a function of increasing feedback.

Russian psychologists feel that it is not enough to consider error probability in theorizing about operator reliability; the nature of the error, its

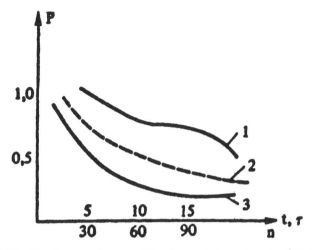

FIG. 4.9. The change in the probability of error during training in relation to the character of self-control. From Nikiforof (1977). 1—without self-control; 2—partial self-control; 3—full self-control; t—time in min; τ—number of trials.

consequence, and its significance to the operator need to be considered also. If the task has high significance for the individual, with great error consequences for the system, he or she will feel the error more than if the task has less significance and the error consequences are minimal. Implicit in this is the concept of self-regulation.

4.6 CONCLUDING COMMENTS

What Russian ergonomists do in their practice involving system developments is very much the same as that done by Western ergonomists. How much "hands-on" design consultation with engineers is performed is not clear, because Russian theoretical concepts presented in this chapter emphasize analysis of the design problem to a greater extent than do Western concepts. This may be misleading because this book is written from a theoretical and research standpoint in which analysis plays a much greater role than it does in actual design activities. In this chapter attention is paid to the engineering psychological aspects of design. The difficulty of translating engineering psychological concepts into concrete analytical design guidance bedevils Western human factors people also.

The same conceptual problems face Russian specialists as Western scientists. An example of this is the concept of human reliability, which Russians treat sometimes as accuracy (precision). One differences between Russian and Western human reliability specialists is that the former emphasize time constraints as a factor affecting performance much more than do Westerners.

Other important distinguishing features in the study of precision and the reliability features of operator performances is that the Russian specialist tries to study this problem from an activity point of view. Because of this, particular attention is paid to self-regulation and strategies of task performance. In this chapter it was shown that the operator possesses substantial functional and compensatory abilities. Because of this the reliability of the operator's performance depends to great extent on the adequacy of selected strategies of activity. An important factor of reliability is the significance of the task for the operator. Humans very often perform complicated tasks more reliably than simpler tasks. This can be explained by the fact that the more complicated the task is for the operator, the more significant it is for him or her. This in turn influences more efficient strategies to be used.

For example, we can demonstrate by referring to experienced pilots who successfully perform complicated flight tasks and then in simpler ground situations fail to perform adequately, which results in a crash of the aircraft. Strategies of activity depend on the specificity of task, conditions of task performance, individual orientation of the operator, and his or her past experiences. All of the preceding data are evidence of significant differences in laws that determine the reliability of humans and technical devices.

Mathematical methods of the theory of reliability, including those that are presented in this chapter, give only an approximate result in the evaluation of reliability of human performance. This can be explained by the fact that people use different strategies of activity in different situations. As a result the reliability of performance can greatly vary. Reliability of human performance can change in a gradual fixed manner, as was described by exponential functions, or change abruptly. The reliability of the operator's performance can change gradually while using the same strategies, and when using different strategies the change can occur abruptly for the same operator. In evaluating the reliability of performance it is necessary to take into consideration the adaptive abilities of the operator.

Attention should also be drawn to the Russian heavy reliance on physiological concepts as a means for explaining changes in performance. This is essentially the Western "arousal theory" that was popularized by Hebb (1955) and was discussed in greater detail in Schroder, Driver, and Streufert (1967), but that seems to have been ignored in Western ergonomics in the last few years. Physiological and psychological factors are considered as dynamic components of activity that are tightly interconnected. These in turn are determined by the interconnection of the energetic and information aspect of self-regulation.

Activity theory also emphasizes much more than does Western psychology the fact that what the individual responds to is not objective reality but rather a subjective model of that reality. An example of this is the attention paid to subjective critera for success, the role of feedback, and so on, which are considered important components of reliability. It may appear that although the distinction between objective and subjective reality is clearly warranted, in the great majority of work situations (i.e., in which personnel work to prescribe highly proceduralized sequences of instructions), the distinction between these realities is less important than in ambiguous work situations or those in which the operator has a choice of performance options.

In conclusion, this chapter shows that for achievement of required goals and the reliability of performance, the operator not only influences the technical components of the system, but consciously or unconsciously regulates his or her own behavior by using different strategies for the purpose of utilizing his or her energies and abilities more efficiently.

Examples of Russian Ergonomics Studies

5.1 OVERVIEW AND INTRODUCTION

The purpose of this chapter is to illustrate Russian ergonomics methodology. Much like the Western version, it turns out to be mostly experimental research, occasionally with a specific design orientation.

5.1.1 Aviation

Russian ergonomics methodology follows principles based on work activity theory. A specific example of how the principles are applied is in the field of aviation. Among those principles is that of test under both routine and emergency conditions. Multiple criteria and measurements—including flight parameters, eye fixation, physiological processes, and interviews—are employed. The inflight characteristics of the pilot's activity must be preserved in the simulator, because Russian scientists are aware that often results achieved in the laboratory do not accord with inflight results. There is an attempt to preserve the internal psychological content of the pilot's activity, but it is not clear how this can be done in an artificial laboratory environment.

Emphasis is also placed on reproducing operator characteristics. Another area of specific emphasis in Russian aviation research is the difference between cues received from the aircraft instruments and those received from the aircraft itself, for example, noise, vibration, engine thrust. Russian psychologists emphasize that the two types of cues may provide contradictory information, but that noninstrumented cues may precede those of an instrument nature.

As was pointed out in preceding chapters, Russian scientists pay much attention to temporal study results. They are also concerned about the difference between objective and subjective performance criteria. They specify that every aviation study should include physical flight parameters, time and error measures, the pilot's information-processing strategies, his or her internal psychic processes, psychophysiological factors, and subjective responses obtained in interviews.

5.1.2 The Application of Graph Theory to Design

This method is based on network models defined by sets or collections of nodes connected by arcs. These models exhibit in a concise form a great deal of information about the real world. Thus, these models are especially valuable during initial stages of ergonomic design where operators use a substantial number of informational displays and controls. The benefit of these models is that at more advanced stages they may be translated into mathematical and quantitative models. These models are well known to operation researchers and other specialists. The distinguishing feature of Galaktionov's (1978) approach, for example, is the stipulation of the precise sequence of procedures that empowers transition from one graph model to another in order to describe and analyze the structure of an operator's activity and its alignment with the technical components of the systems. These procedures are based on a series of defined stages. First, the specialist describes a logical sequence of events. Next, she or he describes a logical sequence of information, leading finally to the delineation of the logical sequences of actions of the operator. Because Galaktiontov presented structure of activity in broad strokes, this method should be supplemented by other, more detailed and precise procedures.

5.1.3 Algorithmic Analysis

This is a concept that attempts to represent the information-processing and logical conditions of task performance in symbolic form.

An algorithm can be a general principle, usually stated in equation form, that contains variables whose values, when changed, allow the algorithm to be applied to a number of situations involving those variables. Russian algorithms are highly proceduralized methods (based on logical rules) that can be used to describe how system components interrelate. An algorithm can also be presented as a table, a symbolic description, or a graph.

Although Western behavioral algorithms are usually based on generalizations from human performance, some Russian algorithms derive from logic. Whether based on logic or human performance, Russian algorithms are called "human algorithms" because the major units of analysis are actions and operations. On the other hand, some Russian algorithms tend to summarize very complex phenomena like stress in a shorthand, and, hence, overly simplistic manner. Russian psychologists have used algorithms for descriptive, analytical, and evaluational purposes and as a design aid.

5.1.4 Practical Examples

Two design-related examples of Russian ergonomics work are presented, the first in aviation, the second in manufacturing. The aviation examples deal with the effects of automation on pilot performance. The principle of joint control as an intermediate process between complete manual and complete automatic control is introduced and demonstrated to be superior to its alternatives. Another series of research studies deals with the controversy between inside-out and outside-in attitude displays, and the aviation examples conclude with an experimental comparison of two alternative cockpit configurations.

The manufacturing example describes improvements in the design of a hydraulically operated copying lathe. Observations and measurements were made of the existing machine, deficiencies were noted in the design, mock-ups of an improved design were built, and measures of operator performance that demonstrated improvement in that performance were taken.

5.2 EXPERIMENTAL METHODS
FOR STUDYING WORK ACTIVITY IN AVIATION

Although aviation as a field of investigation is important in its own right and for safety reasons, it is described here primarily as an example of general Russian experimental methodology.

Ergonomic evaluation of the aircraft system and its equipment is based on the study of a pilot's work activity. Russian experiments follow principles developed by Dobrolensky et al. (1975). These are:

1. Aviation experiments should begin with a study of the dynamic features of flight.
2. The basic psychological characteristics of a pilot's activity must be preserved in the experiment.
3. Cognitive processes should be considered in the context of actual task performance by the pilot.
4. The experiment should be conducted not only under routine but also under emergency conditions.
5. Results should be analyzed using combined indices as well as by the comparison of individual measures.

One method of studying a pilot's activity is to analyze the deviation of actual parameters of flight from required ones at different stages of flight. Another method is the study of his or her attention process, or what is now called situational analysis (Dobrolensky et al., 1975). One way of doing this is by comparing single- with multiple-task performance, the latter under time-sharing conditions. Particular attention is also paid to the pilot's strategy of gathering information. Strategy is usually viewed as being a conscious, deliberate carrying out of a plan of action, but in the Russian context it can equally well be unconscious and so studied by recording eye movements toward various instruments, to describe frequency and duration of fixation, the path the eyes travel, and the time taken to shift the eyes from one instrument to another. Russian psychologists believe that this method, in combination with others, permits analysis of the thinking process, that eye movements reflect the pattern of attention, which is at least partially cognitive.

Experiments study how pilots with varying experience perform different versions of the task under various conditions, such as changing time and precision requirements, task significance, and environmental factors. The pilot's mental and motor responses are analyzed with regard to motor actions; specialists investigate spatial, temporal, and forcing characteristics of movement, including the micromotions described previously.

Debriefing the pilot is also important, as is the comparison of objective performance data with the subjective interpretation of it by the pilots. Concurrent verbal protocol analysis is also utilized. The verbal responses of the pilot in actual flight are also recorded. Pilot opinion is also examined by questionnaires.

Psychophysiological criteria are also widely used: pulse rate, breathing frequency, electromiograms, galvanic skin response, sugar and cholesterol content in the blood, amount of adrenaline in the urine, and the like. All physiological reactions of an organism are analyzed within the context of task activity. For example, it was discovered that when a pilot ejected from the seat in a laboratory study, his or her pulse rate sharply increased in response to the command. This shift in pulse rate was more significant than that during the ejection itself (Isakov, 1974). The increase in pulse rate was not only the result of physical influences, but also of the psychological conditions of task performance.

Physiological reactions are also used to analyze the subjective significance of the task. Physiological data are important while studying task performance under emergency conditions. For example, it was discovered that, depending on the complexity of flight (manual, automatic, different altitudes, etc.), pulse rate, breathing rate, and galvanic skin response all significantly change (Brusnichkina, 1975). The same changes were observed when the pilots performed tasks that had different personal importance for them. This means that the significance of the task also influenced physiological criteria. The study of the physiological reactions of test pilots, regular pilots, flight students, and instructors permitted an interesting conclusion (Beregovoy et al., 1978): that the same physiological reactions that underlay the fear that disrupted task performance also underlay efficient performance of a significant task. This means that one cannot analyze physiological data apart from the total activity. The same physiological reaction can be interpreted completely differently in context of activity.

From the preceding discussion, it is quite clear that Russian separate techniques for studying pilot performance do not differ in any substantive manner from those used by Western psychologists. However, the method of their integration and interpretation of experimental results are different.

The specificity of using these methods is their integration, and interpretation of data based on their comparison. Another important specificity of using different methods is that the same task can be presented to subjects under different conditions, when their components can be changed. This permits the discovery of the specificity of the strategies of task performance and the mechanisms of self-regulation. Scientists analyze how pilots attempt to achieve the same goal under different conditions and how they change their strategies depending on the significant of the task. The specialists also try to discover how the task is performed by test pilots, regular pilots, and flight students under routine and emergency conditions. It is understandable that the same methods can be used in different ways depending on the theoretical principles that underscore these methods.

The first principle, mentioned earlier, is the study of dynamic features of the aircraft. The major dynamic features of aircraft are controllability and stability. Controllability is defined as the aircraft's ability to perform any maneuver under the pilot's control. Stability is defined as the aircraft's ability to return to the programmed flight regime if some external disturbance, such as wind, effects the flight program, without requiring the pilot to perform particular control actions. Stability and controllability are closely linked. Overstability complicates control of the aircraft (performance of different maneuvers). At the same time a higher level of controllability makes it easier to perform different maneuvers; however, it decreases aircraft stability.

To evaluate the dynamic characteristics of an aircraft in the first stage of study, ergonomists use different kinds of perspectives, which allow the dynamic characteristics to be changed in a wide range. In the second stage of study, dynamic characteristics are evaluated under actual flight conditions. Quantitative measures of the dynamic characteristics of the aircraft are compared with expert evaluation by pilots. One of the more known expert procedures is the Cooper–Harper rating scale (Cooper & Harper, 1969).

Russian aviation psychologists believe that the Cooper–Harper scale has some weaknesses, in that it does not sufficiently consider flight duration, pilot fatigue, and requirements for attending to instruments. In relation to the last, the scale considers attention only under extreme conditions.

Russian aviation psychologists have made suggestions for improving this scale. Tuvarev suggested a 5-point scale (according to Dobrolensky et al., 1975). Evaluation would be performed in three stages (Fig. 5.1). Scale values would be interpolated in the third stage for a more precise evaluation. However, it is questionable whether or not experts can more precisely evaluate the features under consideration. At the same time, though, it has been suggested that this procedure is easier to use than the Cooper–Harper scale.

Ratings by expert pilots are first made on the ground (in simulators) and then during flight. Sensor devices, which allow one to register automatically the parameters of aircraft flight such as attitude or glide slope, can then be compared with pilot ratings.

The second principle recommended preserving in the study the actual inflight characteristics of pilot activity. For example, it was discovered in a laboratory study with individual displays that the pilot's perception of information from these displays very often did not correspond to his or her perception of information from the same displays during actual flight (Kotik, 1987; Zavalova & Ponomarenko, 1980).

It was pointed out in chapter 4 that the pilot did not received individual items of data from individual instruments, but rather processed this information in a global, holistic manner. That manner was determined by the particular task being performed and strategies of task performance selected by pilots. In accordance with work activity theory, the pilot's goals, motives, and strategies must be considered. In any experimental procedure it is necessary to preserve the internal psychological content of the pilot's activity, including such fundamental notions as goal, task, conceptual model, and operative image, or, more broadly, relative task conditions and significance of task.

For example, according to Dobrolensky and colleagues (1975), when the pilot under laboratory conditions received information about a fault in the aircraft's engine, he or she processed the information in 3 seconds. During actual flight, the pilot processed the same information in 280 seconds. The reason for this tremendous discrepancy is that in the laboratory information was presented only about the engine fault, whereas in actual flight, the pilot must be aware of other flight information that competes with the fault-specific information. Laboratory tasks must replicate actual flight conditions or the resultant data will be irrelevant, if not actually erroneous.

The laboratory study must be conducted not only with test pilots but also with ordinary pilots, because the level of qualifications for these two types of pilots is different and their tasks are very often different too. To stress the aircraft into revealing design inadequacies, the test pilot often must create non-normal flight conditions, such as diving the craft at a particular attitude or climbing to an extreme altitude at full power.

The distinction between test pilots and ordinary ones is exemplified by a situation in which, during development of a single-seat aircraft, the test

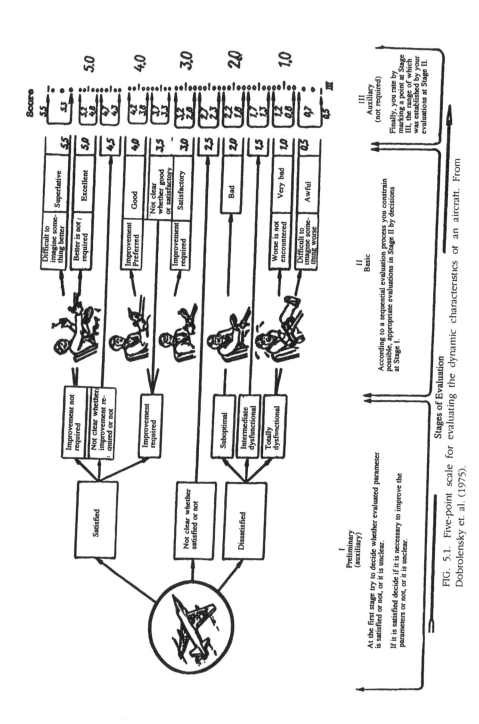

FIG. 5.1. Five-point scale for evaluating the dynamic characteristics of an aircraft. From Dobrolensky et. al. (1975).

194

pilot had to introduce a fault in the autopilot, observe the outcome for 5 seconds, and recover. The test pilot was able to do this with ease and the aircraft went into production. Later, however, when flown by an ordinary pilot, the autopilot failed and the pilot almost lost his life in consequence.

After aviation psychologists were asked to study the accident, they found that when the autopilot failed, the problem lay not in remedying the consequences of the fault, but in detecting it. Apparently, test pilots and ordinary pilots have different goals when performing this task (Zavalova et al., 1986). The test pilot's goal is to observe the behavior of the aircraft with faults she or he has her or himself inserted. If the same faults appear, the ordinary pilot's goal is focused on his or her comprehension of the situation and the determination of its causes. Only after this does the pilot formulate the goals of how to eliminate this fault and perform the necessary responses. Consequently, the task of recovery for an ordinary pilot took longer than the maximum 5 seconds predicted in experiment.

The third principle states that cognitive processes should be considered in the context of actual activity. Western psychologists may consider that under ordinary circumstances the pilot's goal is well formulated and almost always accepted. If a commercial pilot is supposed to fly from San Diego to San Jose, she or he would not ordinarily deviate to Las Vegas. When an emergency condition arises, she or he is aware that it is necessary to remedy the situation and maintain the aircraft's safety. The overall goal is clear; however, the intermediate goals that emerge during the performance of a particular task can be interpreted in different ways, or the pilot may set his or her own intermediate goals, which can also vary. Hence, the specificity with which various goals are developed during the flight affects the performance.

It is here, with the pilot considering options, that the point the Russians make about formulating and accepting goals becomes important. For example, in an emergency, should the pilot return to base or continue, should he or she set down immediately, or what? If an unexpected weather condition crosses the scheduled flight path, should the pilot attempt to fly over or around it, or through the weather, or what? To the Westerner it may appear as if the overall (high-level) goal is quite clear, but that subordinate (means–ends) goals are much more ambiguous, and require the attention the Russians give them. In creative tasks (e.g., scientists, researcher) and in ambiguous conditions or in situations in which there may be competing goals, task formulation/acceptance is, of course, very important.

The importance of task significance and subject expectations (interpretation of the study purpose) was seen in another study by Dobrolensky et al. (1975). The purpose of this simulator study was to evaluate the adequacy of four instrument configurations (A, B, C, D) for landing aircraft in zero visibility. A balanced design was used.

The initial trial suggested that A was superior, because pilots could land with relatively high accuracy, whereas accuracy with the other configurations was lower. Subsequent trials, however, indicated no significant performance improvement with A. Performance with the other configurations improved significantly over repeated trials. Specialists made the preliminary conclusion that instrument configuration A is unacceptable. However, in subsequent debriefings, pilots reported that A was much easier to use than the other configurations.

This is why the pilots concluded that they must demonstrate their ability to work with more difficult tasks. They consider these tasks as more important and, as a result, they pay more attention to the instrument configurations B, C, and D than A. The investigators concluded that because A was easy to use, subject interest in A was lower than for the other configurations. Moreover, A was downgraded because subjects considered the experiment to be a test of their ability to perform a complex task, and A made the task too easy for them. Later, experimental study proved that in situations when pilots relate to different instrument configurations equally, they demonstrate better performance with configuration A.

The fundamental means of analyzing a pilot's activity is what can be termed the "operative conceptual model" of his or her activity. In forming this model function blocks such as "goal", "formation of task," and "representation of subjectively relevant task conditions" play an important role. The "operative image of task" is a particular instance of what Russian scientists call "relevant task conditions." These conditions have a broader meaning than purely perceptual functions; they include also conceptual components of activity with which they are closely associated. It is the dynamic mental representation of a task that becomes most important for an operator during task performance.

In other words, he or she actively selects information from the environment. Based on past experience, the pilot analyzes the situation and goes beyond merely receiving information. He or she creates this information, interprets it, and performs particular actions based on it.

As indicated previously, the concept of the operator in Russian psychology is that he or she is highly active, not merely reactive; the use of the term *creates information* exemplifies the emphasis on proactivity. Most Westerners should agree that to a degree information is the joint product of both external and internal sources. The major difference between Russian and Western concepts is that the former emphasize the proactive aspect much more than do the latter.

The following is a good example of this concept. The ability of the experienced pilot to perceive information from instruments 1.5–2 times as quickly as an inexperienced pilot cannot be explained by the speed of the perceptual process. Younger pilots actually perceive information more quickly than do middle-aged ones. However, because experienced pilots can better forecast the dynamic feature of the situation and task performance, they can perceive and process information more quickly (Beregovoy et al., 1978). Their data are similar to that obtained by Kotik (1978) mentioned in chapter 4.

This can also be explained in terms of the activity model of attention (see Fig. 3.14). The mechanisms of voluntary attention are tightly linked to the goal of an activity and the significance of a task. Because of the existing forward and backward interconnections between the various function blocks in this model, the attention process can easily be adapted to a specific task. Information processing, deviation of attention, and switching of attention among task components cannot be correctly studied out of the context of the individual's strategies for performance.

In effect, Russian psychologists connect strategies with goal attainment. The goal (as well as task conditions, of course) narrows the individual's range of options as to how she or he can perform a task; therefore, a strategy must be adopted. In highly skilled repetitive tasks like throwing a lever in a production line, the options are essentially nil; in a complex, cognitive task requiring decision making, the number of options may be large. Under the latter condition, the operator must adopt a strategy; even to fail consciously to do so is to adopt a strategy, although one that may be fatally flawed. Russian psychologists do not, however, equate a strategy simply with selecting one of a number of options.

That Western psychologists are aware of strategy-developing activity by the individual is shown by their awareness of such phenomena as the speed-accuracy trade-off and decision-making games like chess that represent conflict conditions. The Western scientist would explain strategy in the context of information processing, the Russian as an integral part of self-regulation. They are interested in describing strategy analytically, which means referring the strategy to the various self-regulation blocks. What difference the two interpretations would make in terms of objective performance is difficult to say.

Beregovoy et al. (1978) studied how expectations influenced strategy. In a laboratory study they introduced a failure into the autopilot and then measured the time required to detect the malfunction. However, when they compared detection times in the laboratory with detection times for the same failure in flight, the results were not comparable. In the laboratory pilots spent much more time at the same task than in the flight situation.

The Russian psychologist would explain this by the fact that in real flight conditions a failure emerges within the context of events that precede the failure. The operator develops a conceptual model of flight, which is adequate for a routine situation. The pilot can predict the likelihood of various events occurring and correctly evaluates their importance. The failure emerges within the context of these events, which means that the strategy of activity involving the selection of information depends on this context.

Russian psychologists consider that within the ergonomics schema "input-interpretation-decision-making-action," the psychological mechanisms, which precede input information, are often ignored. A person does not merely perceive information, but formulates a goal of observation according to which he or she perceives information. Specialists who conduct this particular study on the first step try to reproduce only the unexpectedness of the failure and are unable to correctly represent the system of operator expectation and the goal of a particular task.

The fourth principle is that an experiment should be performed not only under routine but also under emergency conditions. The pilot's strategy during the performance of the same task in both normal and emergency conditions can be very different. Under routine conditions, any design weaknesses may not be discovered because these weaknesses may be elicited only by emergency stresses.

Dobrolensky et al. (1975) studied the pilot's ability to detect deviations resulting from engine failure as manifested by two sets of instruments. The deviation was presented during the first 20 seconds of flight and, if the pilot did not report it, it was considered pilot error. The task was presented to the pilot under both single-task and multitask conditions. Table 5.1 shows that under single-task conditions there was no discernable difference between the two sets of instruments, but under multitask conditions the first set of instruments was much superior, as indicated by probability of failure detection.

The authors of this study, however, overlooked a very interesting result. One would expect that complicating the conditions of task performance by adding additional tasks, for example, should increase the duration of visual fixation on the instruments. However, the fixation data from Table 5.1 suggest that the duration of visual fixation is longer under single-task conditions. This difference was statistically significant. This can be explained using the concept of self-regulation of activity. The increasing difficulty of the task

TABLE 5.1
The Role of Multitask Performance to Compare Two Equipment Configurations

Flight Conditions	Version of Grouped Instruments	Probability of Detection of Failure	Duration of Visual Fixation on Grouped Instruments (in sec)
Single task	I	.74	3
	II	.75	5
Multiple task	I	.68	2
	II	.38	4

Note. From Dobrolensky et al. (1975).

made it more significant to the operator, thus increasing the level of motivation. As a result, under more complex conditions, operators process information more quickly but more unreliably as well.

The introduction of emergency conditions during experiments has special interest for the analysis of accidents and preaccident situations. Studies show that the recognition of potential accident situations during flight consumed two thirds of the pilot's time (Zavalova et al., 1986). The recognition of an accident situation is complicated by the fact that it is multivariate, and the same signals may have multiple meanings. It follows that a pilot's training for emergencies should be oriented not so much to his or her motor skills, but rather to cognitive skills and the ability to correctly select and perform required strategies.

It has been pointed out previously that stimuli indicating the occurrence of an accident situation during flight can be both instrumental (received from instruments) and noninstrumental (received not from instruments but as a result of changes in aircraft position, vibration, noise, etc.). These stimuli (even those that are instrumented) often present imprecise and incomplete information about events.

Stimuli may provide different types of information: complete and relevant, incomplete but relevant, incomplete and irrelevant, and contradictory. In any emergency or accident situation, it is rare that all information provided is both complete and relevant. Because of this, the pilot's efficiency depends on his or her ability to interpret the situation, formulate the task, evaluate relevant task conditions, and determine the difficulty and significance of the task. If the imaginative components in task performance are predominant, Russian psychologists use the term operative image of the task to indicate the major components of relevant task conditions. In an ambiguous situation, relevant task conditions, or the operative image of the task, can be correctly formed by the pilot only after some exploratory mental and motor actions. These are involved in the development of hypotheses about what is going wrong or has gone wrong.

In the development of hypotheses and their subsequent validation or invalidation, a major role is played by the following function blocks: "assessment of the meaning of input information," "goal," "formulation of the task," "subjectively relevant task conditions," "assessment of difficulty," "sense of task," and "motivation."

Zavalova et al. (1986) studied pilot performance during a scenario in real flight in which one out of four motors failed. Information about the state of the engine was presented by meters and four warning lights. The pilot also received noninstrument cues, such as aircraft movement, engine sound changes, change in engine thrust, and so on.

Based on exchange of radio messages with the pilot during the flight, it was discovered that the operative image of the situation was first formed as a hypothesis that regulated the strategy of searching for further information. In the first stage noninstrument cues played the predominant role. Responding to these, the pilots observed their instruments for somewhere between .3 and .8 seconds (based on eye movement data). However, because of the inherent slowness of the instrument response, these were unable to display failure-related information quickly enough. As a result, the pilot very often developed an incorrect hypothesis of the situation. Because the task and the operative image of the task were formed incorrectly, the pilot searched for information that could prove his or her incorrect hypothesis of the situation to be true.

It was discovered that the real strategies of searching for information about failure essentially differ from strategies prescribed in formal instructions (see Table 5.2). From this table, one can see that in experimental flight, with imitated failure of engines, pilots first pay attention to noninstrumental signals. Only on the last step do they pay attention to warning signals (warning lights). This significantly reduces the reliability of pilots' actions in accidental conditions. This can be explained by the fact that, based on noninstrumental information, the pilot develops hypothetical operative images of the flight, which determine strategies of gathering information. This essentially differs from those strategies given in the instructions. It was discovered that a discrepancy between real strategies and formally developed strategies caused incorrect instructions and inefficient design of instrumental panel. These resulted in incorrect formation of the task and relevant task conditions. The pilot may overlook the warning indicator to shut the engine off, because the warning conflicts with his or her mental representation of the situation. The pilot also might process information from his or her instruments without analyzing the context in which it appears. Obviously, in an emergency situation, the most important thing is to search for information to develop hypotheses about what is wrong, and then select one hypothesis upon which to act. This may be more difficult to do in a very fast reacting system like an aircraft, but the same process of selecting and analyzing information, developing and testing hypotheses, can be more readily seen in a slow-acting system like a nuclear power plant (Meister, 1995).

Where the pilot's initial hypothesis cannot be confirmed by stimuli, the pilot's activity becomes one of testing other hypotheses, and what the Russians term *auxiliary explorative actions* develop automatically.

Russian psychologists consider that these exploratory actions as described in chapter 2 are both mental and motoric. The process of developing and testing hypotheses is to them exploratory. Exploratory components of activity include those processes that are connected with evaluation by the pilot's external environment.

The process just described in self-regulatory terms (detecting/recognizing the problem, gathering information, developing, validating/disconfirming hypotheses) is very similar to what Western scientists have described in such

TABLE 5.2
Comparison of Formal Instructions With Actual Flight Performance During Engine Failure

Emphasized in Formal Instructions

Stage	1	2	3	
Suggested sequence of gathering information	Increasing turn and deviation of aircraft	Signal light	Reading of engine instruments	
Expected reactions of pilot	Detection of engine failure	Comprehension of the cause of failure	Specification of the number of the failed engine and performing required actions	

Found During Flight

Stage	1	2	3	4
Actual sequence of gathering information	Noninstrumental signals: sense of angle acceleration, change in effort on controls, etc.	Instrumental reading (during first second after failure)	Confirming instrument readings	Signal light (30% of cases)
Reaction of pilots	Suspect engine failure	In 70% of cases did not receive proof and turned attention to other possible causes; in 30% of cases, understood causes	Continued search for required information	Received proof and began to act

Note. From Zavalova et al. (1986)

Bedny considers TOTE to have self-regulation elements, but it is also possible to describe the troubleshooting process in solely information-processing terms, as exemplified by the model of troubleshooting developed by W. B. Rouse and S. H. Rouse (1983) and Meister (1995). The Western troubleshooting models are, of course, much simpler than that of self-regulation; but these Western models focus on troubleshooting only, whereas the self-regulation model attempts to describe all work activity.

It was discovered that instructions very often ignored the real strategies of pilot activity when noninstrumental signals play an important role. In other cases, explorative activity sharply increased during an emergency situation. This can result in an accident.

Very often, ignoring the specificity of the self-regulation of activity in an emergency situation is connected with a careful description of only what the pilot must do in any particular situation. However, instructions that help pilot evaluate this situation, correctly formulate the task's goal, and extract relevant task conditions are ignored. For example, analysis of the data in Table 5.2 suggests that most of the more important noninstrumented signals, which were perceived by pilots when the engine failed, were ignored in real instructions. In a real situation, after receiving noninstrument signals pilots immediately turned their attention to their instruments. Only at the last stage of information processing did the pilot, in 30% of all cases, look deliberately at the warning light. This means that pilots very often ignore warning signals.

Russian psychologists explain the preceding by hypothesizing that pilots developed an image template during training that did not correspond to the operative image of the actual emergency. The function block "relevant task conditions" is crucial in an emergency situation because it is responsible for developing the dynamic mental representation of the task.

It was suggested that the instructions and instrumental panel do not correspond to real strategy of task performance. After the study of real strategies of task performance, more efficient instructions as to how a pilot should act when the engine failed were developed. It was also suggested that the position of signal lights indicating engine failure should be changed. An illuminated indicator with the word *danger* on it was placed on the instrumental panel in front of the chief pilot. Another illuminated indicator with the words *engine off* and the engine number was also placed above the configuration of engine displays. Later, in a simulator, and then in actual flight, this design configuration was tested and pilots immediately attended to these warnings.

The last requirement for conducting an experimental study is that of the necessity of using multiple criteria and the analysis of the relationship among performance indices. If we return to chapter 2 and consider the model of self-regulation, which is as complex as it is because human activity is a multidimensional phenomenon, various components of activity are interconnected and influence each other. For example, according to Russian theorists, increasing the complexity of a task increases its significance and the operator's motivation, which in turn, influences task performance; the temporal parameters of activity can change under these conditions.

Lowering the significance of the task and increasing its complexity can change the operator's subjective criteria of success and decrease the precision of task performance. Russian psychologists, therefore, conclude that all activity parameters are closely linked, which should be taken into consideration when developing an experiment. Because the theory is so complex, pragmatically it is desirable sometimes to limit experimental variables to the major function blocks.

Dobrolensky et al. (1975) and Zavalova et al. (1986) emphasized that exclusive concentration on so-called "objective" measures to the neglect of "subjective" measures in an experiment may result in erroneous conclusions. Most Western experimenters would agree. For example, during the study of the pilot's interaction with failure warning signals, Dobrolensky et al. discovered that there was a significant delay in reaction to the signals. They concluded that this delay resulted because the failure warnings were inefficient. However,. eye movement data and reaction times, when correlated with the pilot's subjective opinion, suggested that this delay was a rational one. Pilots consciously delayed their reactions to these warnings in order to increase the precision and reliability of their actions.

The following are the measurements that Russian experimenters feel should be taken in every study: (a) physical parameters (e.g., flight parameters such as glide slope and attitude) and those parameters outside of which the system will fail, (b) time, precision, and reliability of operator responses, (c) information-processing strategies employed by the individual and fixation and eye movement data, (d) internal psychic processes (e.g., strategies; decision-making and memory processes), (e) psychophysiological factors such as workload, stress, and other physiological indices, and (f) subjective responses of the subject obtained during debriefing.

All of these are identical with those used in Western studies. A potentially confusing factor is that the Russian psychologist uses the Western measure/method in a different manner. An example is pulse rate measurement. Western scientists might use pulse rate to measure general arousal resulting from excitement, as, for example, when an astronaut's pulse rate increases significantly as he or she ascends into orbit and then decreases after orbit is achieved. The same pulse rate might be used by Russian scientists to measure "task significance," which is a very different concept from simple arousal or excitement, because it possesses cognitive and motivational elements.

One other difference between Russian and Western experimenters should be noted. The Russians appear to be less concerned about the number of subjects who are tested in any single study than Westerners. Russians make much more use of the same smaller number of subjects in a series of interrelated experiments, whereas Westerners usually concentrate on a single experiment (although with several conditions) with many more subjects. Which procedure is more desirable (if either) is a matter of speculation.

5.3 ALGORITHMIC ANALYSIS
OF TASK PERFORMANCEE

In the former Soviet Union, the algorithmic analysis of system control was first suggested by Liapunov and Shestopal (1957). Later, this method was transformed by Zarokovsky (1966) for describing and analyzing the activity of operators during task performance. This method was widely used in the former Soviet Union.

What is called an algorithm includes a number of information-processing operations and logical conditions; the latter determine the sequence of these operations. Logical conditions can be best described in terms of "if-then" rules. For example, if the red bulb turns on, then press the right button. Here, we can see that logical conditions specify the conditions under which a particular operation will occur. We can extract two operations, information reception and response performance. In describing task performance algorithmically, it is first necessary to determine the units of activity into which the task under consideration will be divided, that is, the units of information and elementary actions. Units of information include an image, notion, judgment, command, and so on, which becomes a single pattern used by the operator during task performance. Elementary actions involve the transformation of information and energy. Examples are the acts of perceiving, thinking, extracting information from the memory, motor actions, and the like. One of the problems in the algorithmic analysis of task performance is that it uses actions and operations as units of analysis. To do this one applies the criteria described in chapter 1. Examples of the level of detail are: Read a display, make a decision, move a lever, and push a button. This is a human algorithm, which must be distinguished from a nonhuman algorithm, because they each use different units of analysis of activity. This problem is discussed in more detail in chapter 8. Task elements should be designated by special symbols, which become elements of the algorithm, following one another in logical sequence.

An algorithm can be presented as a table, as a formula using symbolic description, and as a block scheme. During the initial stages of creating the algorithm one uses a table, then a symbolic block scheme.

Before creating the algorithm the analyst develops a verbal description of the task and its goal, stimuli for task performance, task conditions, and temporal relationships of subtasks.

The task description is then divided into elementary actions and logical conditions. Elementary actions are called *operators* and are designated by special symbols. For example, O^α, where O means "operator" and α means "perceiving information." All operators with the α symbol are designated as receiving information under the name *afferent operator*. Symbol O^ε is associated with the execution components of activity, such as moving a gear or rotating a wheel. All operators with the ε symbol are designated as *efferent operators*. Logical conditions are designated by using 1.

The logical conditions have two values, 0 or 1. The symbol 1 for a logical condition must include an associated arrow with a number on top, for example arrow $\overset{2}{\uparrow}$. An arrow with the same number but a reversed position must be presented before the other operator, for example $\overset{2}{\downarrow}$. The syntax of this system is based on a semantic denotation of a system of arrows and superscripted numbers. An upward-pointing arrow denotes a logical state (when $1 = 1$) that requires skipping the next appearance of the superscripted number with an arrow (e.g., $\overset{1}{\uparrow}\overset{1}{\downarrow}$). A downward-pointing arrow with a superscript implies that the next denoted operator, which follows the logical conditions, is to be executed.

Operators utilizing the memory use the symbol O^{μ}. There is also a special designation to distinguish continuous and discrete processes in the algorithm. If elementary operators perform uninterrupted and the duration is important, one uses parentheses under the symbol O^{α}, for example, $\underline{O}^{\alpha}$. If two elementary operators perform simultaneously, the same number is shown as $\overset{1}{O^{\alpha}}$; $\overset{1}{O^{\varepsilon}}$. At the bottom right, close to the symbol, is an ordinal number designating the number of the algorithm, for example, O^{α}_1, O^{ε}_2, 1_1.

In addition, false logical conditions are designated by the symbol ω. False logical conditions must also include an associated arrow after them (e.g., $\omega\uparrow$). These logical conditions always have a value of 1. This means that they always require skipping to the next corresponding arrow with a superscripted number. This symbol is used to designate the repetition of an algorithm process, according to the arrow that follows after false logical conditions. Thus, the false logical condition is not related to task performance; it is used only for making more convenient symbolical description of algorithm. All members of the algorithm are ordered from left to right. In this way the logical condition with the corresponding arrow denotes the transition from one member of the algorithm to another.

Based on these symbols one can distinguish different stages of task activity. Zarakovsky (1966) provided an example. A rudder performs the task of keeping the ship on a specific course. The helmsperson looks at the compass and turns the wheel if necessary. If the compass needle moves to the left, he or she turns the wheel to the right. If the needle moves to the right, he or she turns the wheel to the left. When the needle comes to the required position, the rudder keeps the wheel in that position. Based on this verbal description one can present the algorithm in tabular form for this particular task (see Table 5.3).

Next, the algorithm is described as a formula. This simplifies the analysis of transition from one member of the algorithm to the next and quantitative evaluation of the algorithm's complexity. The algorithm logical formula looks like this:

TABLE 5.3
Algorithmical Description of Rudder Task Performance

Member of Algorithm	Description of Member of Algorithm
O^ε_1	Keep wheel in required position.
O^α_2	Take reading from compass.
l_1	Is there a deviation of needle from required position? If yes ($l_1 = 1$), transfer to O^ε_1. If no ($l_1 = 0$), transfer to l_2.
l_2	If needle deviates right ($l_1 = 0$), perform O^ε_3; if needle deviates left ($l_1 = 1$), perform O^ε_4.
O^ε_3	Turn wheel left.
ω_3	Always false logical conditions.
O^ε_4	Turn wheel right.
O^α_5	Take reading from compass.
l_4	If needle does not return to required position ($l_4 = 0$) perform O^ε_6. If needle returns to required position ($l_4 = 1$), perform O^ε_7.
O^ε_6	Keep wheel in the same position.
ω_5	Always false logical conditions.
O^ε_7	Return wheel to required position.
ω_6	Always false logical conditions.

Note. From Zarakovsky (1966).

$$\overset{1,6}{\downarrow} O^\varepsilon_1 \, \underset{1}{O^\alpha_2} \, 1_1 \uparrow \overset{2}{l_2} \overset{2}{\uparrow} O^\varepsilon_3 \, \omega_3 \overset{3}{\uparrow} \overset{2}{\downarrow} \, O^\varepsilon_4 \, \overset{3}{\downarrow} \overset{5}{\downarrow} \underset{2}{O^\alpha_5} \, l_4 \overset{4}{\uparrow} \, \underset{2}{O^\varepsilon_6} \, \omega_5 \overset{5}{\uparrow} \overset{4}{\downarrow} O^\varepsilon_7 \, \omega_6 \overset{6}{\uparrow} \quad (5.1)$$

In this formula O^ε_1 designates keeping the wheel in the desired position. O^α_2 shows that the helmsperson has taken a reading from the compass. If there has been no deviation from the direction, $l_1 = 1$, then one would return to the elementary operator O^ε_1. If there is a deviation, $l_1 = 0$, one proceeds to the next logical condition. If the needle deviates to the right ($l_2 = 0$), perform O^ε_3. If the needle deviates to the left ($l_2 = 1$), perform O^ε_4 according to arrow $\overset{2}{\uparrow}$. The false logical conditions indicated by ω_3 denote that after O^ε_3, the following operator, O^ε_4, should be omitted. One will then interpret this formula for other logical conditions in the same way. Consider ω_5. After O^ε_6 according to ω_5 we should always move, according to $\overset{3}{\uparrow}$, to $\overset{3}{\downarrow}$ and the process will repeat again. ω_6 means that after finishing O^ε_7 the process must be repeated again according to $\overset{6}{\uparrow}$ and $\overset{6}{\downarrow}$.

This algorithmic formula shows in a symbolical way all of the actions required under all possible conditions and a precise description of the logical sequence of actions performed by human operators. In addition, it should be noted that an agorithmic formula is not always required. When the task involves a lot of actions performed by the operator, it becomes difficult to develop and comprehend this formula. In such cases, the algorithm can presented as a table, or as a block scheme. An algorithmic formula becomes especially useful in those cases when a specialist tries to evaluate a complexity of the algorithm at the later stage of analysis. In Fig. 5.2 the block scheme represents the same algorithm. The diagram is similar to the flow charts developed by computer programmers. In this diagram, elementary operators are designated by a rectangle and logical conditions by a diamond. The meaning of the rectangle and diamond can be indicated verbally, or by means of the symbols used earlier. Transition of one member of the algorithm to another is designated by an arrow. Any logical condition has two outputs depending on the condition's meaning. False logical conditions, ω, always have one output. Landa (1974) introduced the notion of the human algorithm, which differs from the nonhuman algorithm and particularly from the computer algorithm. Here, we consider one kind of human algorithm. Differences between human and nonhuman algorithms are discussed in more detail in chapter 8. The human algorithm allows one to describe a logical sequence not only of physical but also of mental actions.

Zarakovsky (1966) used a logical formula of an algorithm for quantitative evaluation of the complexity of task performance. For this purpose he suggested a number of measures and described procedures for obtaining them. As was mentioned before, all elements of the algorithm can be divided into three basic groups: afferent operators (O^α), efferent operators (O^ε), and logical conditions (1). The number of afferent operators can be summed according to the formula ΣN^α, the number of efferent operators according to ΣN^ε, and the number of logical conditions according to ΣN^l. Then the general number of algorithm elements (symbols) can be determined according to the following formula:

$$N = \Sigma N^\alpha + \Sigma N^\varepsilon + \Sigma N^l \tag{5.2}$$

Zarakovsky (1966) suggested that the equation indicated to some extent the required amount of knowledge and skills for a particular task. This

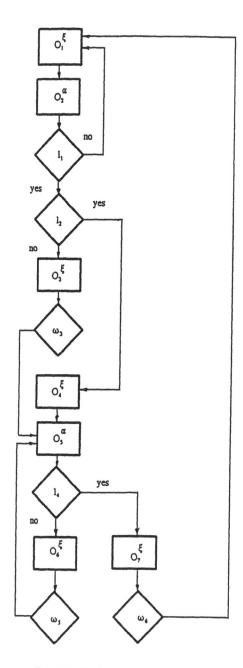

FIG. 5.2. Block diagram of algorithm.

implies that skills and knowledge depend on the number of operations in the task. To those familiar with the relevant literature this may seem like a rough approximation.

For the calculation of the predominant type of activity in a particular task one should determine the relative numbers of the different elements of the algorithm. For this purpose the following formulas are used:

$$N_{\alpha} = \Sigma N^{(\alpha)}/N, \; N_{\epsilon} = \Sigma N^{\epsilon}/N, \; N_{\iota} = \Sigma N^{\iota}/N, \tag{5.3}$$

where N is the total number of elements in the algorithm.

In the following stage the number of variants of the algorithm is determined. Because logical conditions have only two outputs, the number of different versions of algorithm is:

$$N_{w} = 2^{N} \tag{5.4}$$

The formula exaggerates the number of variants in actual usage. In an actual situation the number of different versions of the algorithm is less. To specify the actual number one must use experimental data.

The entropy (H_{ω}) of the variant of realization of the algorithm is:

$$H_{\omega} = -\Sigma \; P_{\omega} \; log_2 \; P_{\omega} \tag{5.5}$$

where P_{ω} is the probability of the variant of the algorithm ω. Equations 5.4 and 5.5 indicate the likelihood of a different version of the algorithm.

Another element (V) describes the measure according to the following formula:

$$V = N/\tau \tag{5.6}$$

where V is the dynamic intensity of task, N is the number of member of algorithm performance, and τ is the average time of a task performance. The more elements of the algorithm and the less time there is for their performance, the more stress occurs.

Western psychologists will recognize this as stress determined by the availability of time to perform required actions. Although time availability is a major component of stress, other factors, such as task difficulty, enter in. Time availability as the sole determinant of stress would be appropriate only for very simple tasks. What we see here is an attempt to mathematicize very complex phenomena, which results in oversimplification.

In conclusion, two additional measures are used to evaluate the stereotypic and logical complexity of task performance. Task performance is stereotypic if it has only a single consequence. If some part of the task is stereo-typed, elementary operators in the algorithm will follow without logical conditions between them. For example, in considering the algorithm dis-cussed previously, the symbols O^ε and O^α do not have logical conditions between them. As a result, after performance of O^ε, the operator always performs O^α.

If any algorithm has an uninterrupted sequence of symbols, these can be used to calculate the stereotype of task performance by calculating the number and length of these sequences. This is the mathematical mean of the number of operators in the algorithm that are presented in an uninter-rupted sequence:

$$Z = \Sigma\, P_n\, X_n \qquad (5.7)$$

where X_n is the number of operators in particular groups with lengths 1, 2, 3 . . . n numbers and P_n is the probability of this group.

The opposite characteristic is the dynamic characteristic of task perform-ance or the flexibility of actions in different situations. In order to evaluate this characteristic Zarakovsky (1966) introduced the concept of logical com-plexity, which measures the irregularity of information processing. The more groups there are in the algorithm with logical conditions and the longer these groups are, the greater the logical complexities. The logical complexity is defined by the formula:

$$L = \Sigma\, P_m\, X_m \qquad (5.8)$$

where X_m is the number of logical conditions in groups with lengths 1, 2, 3 . . . m numbers and P_m is the probability of this group. Let us calculate Z during the algorithmic analysis of a single task involved in maneuvering a ship. The elementary operators, O^ε and O^α, were extracted with lengths 1, 2, 3, 4, and 5 in succession, the frequency of which were .59, .33, .04, .02, and .02, respectively. According to Equation 5.7:

$$Z = 1 * .59 + 2 * .33 + 3 * .04 + 4 * 02 + 5 * .02 = 1.55 \qquad (5.9)$$

When the value of Z increases the stereotype of task performance increases also. The algorithmic method is widely used in the former Soviet Union, but the limitations of this approach for evaluation of complexity of task performance have become obvious, such as the use of incommensurable units of measurement.

5.4 THE APPLICATION OF GRAPH THEORY
TO CONTROL-ROOM DESIGN

Automated systems often consist of a large number of components and hundreds of control displays, which makes it difficult to determine what should be displayed. As was pointed out previously, the operator's responsibility in such systems is to monitor the system's instruments to determine when parametric limits are about to be exceeded, troubleshoot the system if a malfunction occurs, and in general maintain production. Successful performance of these tasks requires an accurate mental representation (situation awareness; see Endsley, 1995) of the current and future state of the system by the operator. In order for the operator to work efficiently in this context, it is important to develop efficient control room designs. Because information is the common denominator in these designs, it is necessary to determine what kind of and how much information must be presented to the operator. And, because the important information in automated systems relates not to control of the system (since this is performed by computers) but to impending or actual malfunctions, one must first perform a malfunction analysis before proceeding to design the displays that present the malfunction indicators.

Galaktionov (1978) used graph theory to develop an analytical method to determine the information to be displayed in the control room, and what controls are required for operator. Graph theory studies the interrelationships among objects. Objects may be represented abstractly as a set of points, and the relationship between the various objects may be represented by lines connecting them. Such a representation is possible for problems in a wide range of fields, including ergonomics. The graph can be considered as a geometrical figure consisting of points (*nodes*) and lines (*arcs*) that connect these points. Arcs demonstrate the frequency of interconnections. Based on this description, one can evaluate the probabilistic characteristics of activity structure. In some cases one can use a matrix transition from one node to another. Galaktionov developed a sequence of procedures for this purpose. The following description is highly abbreviated because the procedures are quite complex. They have much in common with Western procedures that describe the cause(s) and effect(s) of system malfunctions, such as "fishbone" or cause-and-effect diagrams.

The first step in the methodology is to divide the proposed system into the smallest number of functionally related subsystems. After that, ergonomists specify what kind of tasks are connected with a specific subsystem. At the next stage of analysis, possible malfunctions of a particular subsystem are listed. These malfunctions are called *basic events*. From them, models are built that emphasize the underlying exogenous causes of the malfunctions. Each individual model is called the *primary causes graph event* and is shown in Fig. 5.3, which depicts hypothetical malfunction A (basic event A). The next level

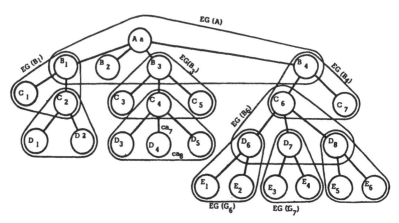

FIG. 5.3. Primary causes graph event for hypothetical malfunction A. From Galaktionov (1978).

of this graph represents immediate (auxiliary) causes of the malfunction (B_1, B_2, B_3, B_4). Each cause or event is a node of the graph and is connected with the top nodes of the graph (basic event A) by arcs. This connection represents one level of the graph and is called the *elementary graph*. Elementary graph is denoted by a rounded triangle and labeled EG (A), elementary graph for basic event A.

Malfunction events B_1, B_3, and B_4 have their own primary causes: C_1, C_2, C_3, C_4, C_5, C_6, and C_7. These events, together with events B_1, B_3, and B_4, form the elementary graphs of the second level of a hierarchy. They are related to EG (B_1), EG (B_3), and EG (B_4). An elementary graph for event B_2 does not exist. Event B_1 is major in relation to events C_1 and C_2. B_3 is a major event for events C_3, C_4, and C_5, and event B_4 is a major event for C_6 and C_7. This decomposition of causes can be conducted until no further causes can be imagined. Auxiliary events correspond to specific control actions (small ca_1, ca_2, etc.). The primary causes graph event is completed when all its branches are completed by a specific primary cause and all are connected with controls. Figure 5.3 is hypothetical and exemplary because not all analyses proceed to this level of detail.

The next stage involves constructing a new graph for determining the checkpoints. Each checkpoint is indicated by a node (event) on the graph, which should be provided with the display that presents the required information to the operator. Not all nodes require displays in order to present information.

At this stage, the primary causes graph event transforms into a checkpoint graph (see Fig. 5.4). At the following stage of analysis, ergonomists point out which nodes should be provided with individual controls, where the event requires a control action. These nodes (designated by ca) are known as points

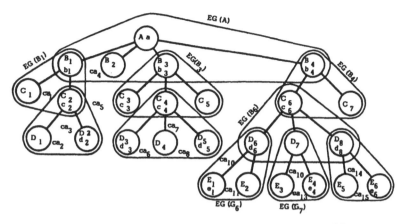

FIG. 5.4. Checkpoint graph for hypothetical malfunction A. From Galaktionov (1978).

of control. Figure 5.4 does not indicate the type of display or control needed. Rather, it only demonstrates which event requires an indicator or controls. To protect the operator from being overloaded by unnecessary information, it is necessary to determine which events (nodes) cannot be examined by the operator. These points or events are called *bypassed checkpoints*.

Let us consider a hypothetical example in more detail. Figure 5.4 shows the elementary graph EG(A). It presents basic event A and four nodes, B_1–B_4, which are the causes of this basic event. However, the required and sufficient number of checkpoints equals three (B_1; B_3; B_4). B_2 is a bypassed checkpoint. Elementary graph EG(A) has four checkpoints: a, b_1, b_3, and b_4. Any event can be determined as the conjunction of the following parameters: B_1: a $\wedge$ b_1; B_3: a $\wedge$ b_3; B_4: a $\wedge$ b_4 (where $\wedge$ is the conjunction).

Event B_2 can be determined by eliminating events B_1, B_3, and B_4 as the causes of event A. In reality, if event A occurs and the operator checks B_1, B_3, and B_4 and discovers that they are not the causes of event A, this means that the cause of event A is B_2.

The basic features of the checkpoint graph can be described by the following statements:

1. The number of check nodes U_{cn} on the checkpoint graph is equal to the number of branches on this Graph V.

$$U_{cn} \approx V \qquad (5.10)$$

2. In any elementary graph, one branch is bypassed in conditions where the top of this graph can be checked or determined by using the method of elimination.

3. Any checked event related to specific nodes should be defined by one parameter alone.

4. The required number of checked parameters for a diagnosis of all graph events is equal to a number of graph branches minus the number of duplicated parameters.

$$N = V - \Sigma (d_i - 1) \tag{5.11}$$

The method for constructing and evaluating the checkpoint graph involves the following steps:

1. The significance of primary events that are the cause of basic events (malfunctions) must be determined. Significance levels can be determined by the following rules: An event (a) directly results in accidents and distortions of system processes, (b) leads to sharp decreases of power, productivity, and quality products, and (c) leads to minimally worsened production quality, decreasing productivity.

2. The required number of control nodes for diagnosis of graph events must be determined according to Equation 5.10.

3. The primary cause graph event is divided into elementary graphs.

4. Starting from the top, for any elementary graph, the required checkpoints are determined according to the second statement the standard graph.

5. The algorithm for determining basic events is developed. For situations when there are initial events with the checked primary cause, the algorithm formula has only two members: basic or major parameter (event) and the parameters that define the primary cause of the major event.

For events with unexamined primary causes, the algorithm includes parameters in the top corresponding elementary graph (EG) and all other inverted events that belong to this elementary graph. For example, events B_2 as a cause of event A in elementary graph EG(A) can be determined according to the following algorithm: B2: $a \wedge \overline{b}_1 \wedge \overline{b}_3 \wedge \overline{b}_4$. In other words, all bypassed (unchecked) nodes can be examined by using the method of eliminating other events. All parameters included in the formula for determining the primary cause events are connected by using the conjunction $\wedge$ symbol.

At the following stage we can compare the presented (developed) checkpoint graph with standard (the best) checkpoint graph. Here the primary causes graph event is transferred into a standard checkpoint graph. For these purposes, one bypass node should be determined in any elementary graph. These nodes should be evenly spread between the branches of the checkpoint graph. After that we calculate the number of check nodes in the real (developed) graph and compare this with the number of checked nodes in the standard (the best) graph. This allows us to determine the exceeded number of insufficient number or check nodes.

At the final stage, the *control graphs* are developed. In the primary causes graph event, for any checked parameters we determine control influences (actions performed by an operator with different controls). These data are introduced in the primary causes graph event. (Refer to control points ca_1–ca_{15} in Fig. 5.4). For any control actions, instructions are developed. At the first stage, the analyst develops instructions on how to start and stop a subsystem, switch to a different regime of work, and so on. Based on these instructions, one should develop algorithms for control of a subsystem.

Galaktionov (1978) also developed specific symbols: α - turns on, open; β - turns off, stop, close; γ (m) - switch or install in position m; α (N) - switch or install in a nominal position; δ (m) - check the position control (or meaning of parameters) that must be in position m; $\overline{\alpha}$ - open slightly, totally closed valve, bolt, slide, and so on, until a signal about a required position of control disappears; $\overline{\beta}$ - slightly closed, totally opened valve or other controls until a signal about its final position disappears. ε is a logical condition "if": $\Sigma \overset{\blacktriangle}{P_i} \ \Sigma P_i \ \Sigma \overset{\triangle}{P_i} \ \Sigma P_i$ (i.e., all parameters must be in a higher, lower, or not higher, not lower position. C is the connection or message with other persons; t is the delay time for an operation. In brackets, control actions and number of parameters are shown. Outside the brackets on the right side, the control action symbol and required final position are designated. On the left of the brackets, the initial meaning of the parameters is shown. For example, [β 19] $\underset{\blacktriangledown}{H}_{19}$: This means that an operator should perform a "closed" action. $\underset{\blacktriangledown}{H}_{19}$ represents the end of the operation and the parameters must be in a lower position (totally closed). Galaktionov's application of his method is discussed later.

The example presented in this section is the task of developing malfunction displays and controls for the control room of a cement manufacturing facility. The basic task is to find the answers to three questions: What can malfunction? What causes the malfunction? What can be done to correct the malfunction? Once the answers to these questions are know, the task of developing appropriate displays to alert the control-room crew to impending and actual failures is relatively simple. In modern automated systems the behavioral design problem is one of controlling information, particularly about impending failures, because the operator's tasks in such systems is to act upon information. The hardware (human engineering) form in which the information will be presented to the crew is a fairly mundane subsequent problem.

As the first step, the functional flow diagram of the cement malfunction process is described (see Fig. 5.5). For this purpose all processes are divided into functional-technological subsystems that have different output events. These output events are called basic events, Basic events $\overset{\blacktriangle}{B}_{37}$ designates

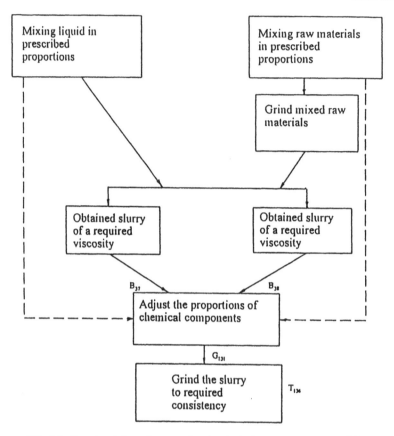

FIG. 5.5. Functional flow diagram of technological processes. From Galaktionov (1978).

increasing slurry viscosity. In Fig. 5.6, the primary cause graph event for basic event $\overset{\blacktriangle}{B}_{37}$ is shown. The nodes of the graph can be considered checkpoints where display or controls are required. The primary cause is those nodes that have specific control actions and can be performed with the existing control mechanisms. This figure demonstrates one of the end products of the Galaktionov (1978) procedure.

Event $\overset{\blacktriangle}{B}_{37}$, shown in Fig. 5.6, together with events a, b, and c create an elementary graph. A primary cause does not exist between events a, b, and c because they have their own causes from their elementary graphs. However, the primary causes are only in graph a. This graph has three primary causes: (a) shutdown of electromotor 42 (designated by symbol $\underset{\blacktriangledown}{H}_{42}$), (b) disconnection of coupling 44 (designated by symbol $\underset{\blacktriangledown}{H}_{44}$), and (c) breakdown of the reduction gear or breakdown or disconnection of the crankshaft. Event 2 is not an examined event (nonchecked node). The operator can get information about this event by eliminating $\underset{\blacktriangledown}{H}_{42}$ and $\underset{\blacktriangledown}{H}_{44}$.

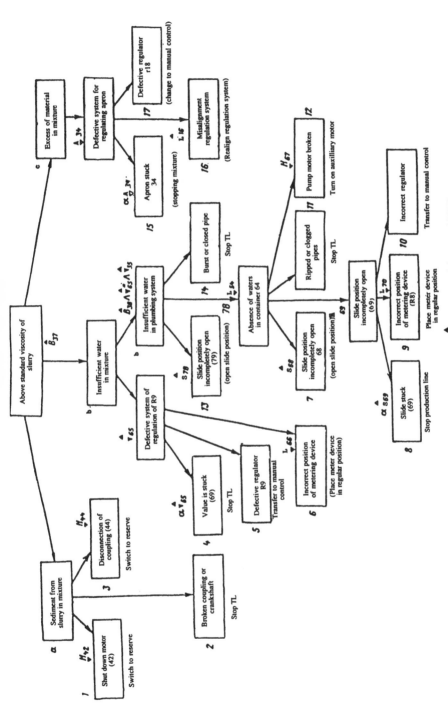

FIG. 5.6. Primary causes graph event for event $\hat{B}_{37}$ From Galaktionov (1978).

217

The subsequent stage involves the selection of checkpoints and the transformation of the primary causes graph event into the checkpoint graph. Events a, b, and c are not examined and therefore are eliminated from the checkpoint graph. Thus, an elevated hierarchical level of checkpoints includes the following events from the lower level of the hierarchy: events 1, 2, and 3 in the elementary graph with primary cause a, a wrong system of regulation of (A_{34}), a wrong system of regulation R_9 (V_{65}), and insufficient amount of water in a water pipe $(B_{38} \wedge, V^*_{65} \wedge, V_{65})$. From six possible events on this hierarchical level, one event should be an unexamined (bypassed point) event. In our example, event 2 was selected as the bypassed point. Therefore, for any six possible events for the second hierarchical level, the checkpoint graph has five examined nodes, and together with event B_{37} there are six checkpoints. We can use this same method to analyze other points on the graph.

The next stage includes the development of an algorithm for examining the primary events for basic events B_{37}. For these purposes, we use the checkpoint graph and the logical algebraic procedures previously described. We developed a special table with three columns: The first column is the number of primary causes for malfunctions, the second column describes the malfunctions, and the third consists of the algorithm for discovering the malfunctions. The number of malfunctions (wrong event) in this table corresponds to the number of check nodes on the primary cause graph event (Fig. 5.6). As an example, fragments of this table are presented (Table 5.4). This table includes only five malfunctions.

TABLE 5.4
Algorithms for Discovering the Malfunctions B_{37}

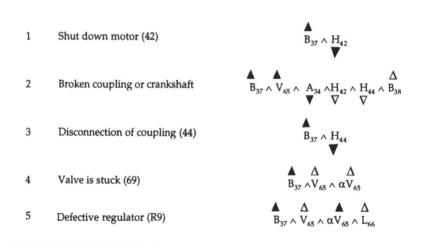

1	Shut down motor (42)	$B_{37} \wedge H_{42}$
2	Broken coupling or crankshaft	$B_{37} \wedge V_{65} \wedge A_{34} \wedge H_{42} \wedge H_{44} \wedge B_{38}$
3	Disconnection of coupling (44)	$B_{37} \wedge H_{44}$
4	Valve is stuck (69)	$B_{37} \wedge V_{65} \wedge \alpha V_{65}$
5	Defective regulator (R9)	$B_{37} \wedge V_{65} \wedge \alpha V_{65} \wedge L_{66}$

Note. From Galaktionav (1978).

Let us analyze one fragment of Table 5.4 in which five examined algorithms for a basic event $\hat{B}_{37}$ are presented. For this purpose, again consider the primary causes graph event in Fig. 5.6. All checkpoints in this graph are designated by symbols, which are, in turn, placed on the top of the nodes. For example, node "Shutdown Motor" has symbol $\underset{\blacktriangledown}{H}_{42}$ labeled on the top. In this case, a dark triangle with its top upside down designates this node. All other symbols operate the same way for designated checkpoints. Events a, b, c, do not possess their own check parameters and their corresponding control influences. This is the reason why, when we construct examined algorithms, these points are eliminated. From the checkpoint graph in each examined algorithm is included the nodes from the corresponding branch of the graph. Uninspected, or bypassed, checkpoints require using methods of inversion on the last step, the logical branch of the graph.

For example, if we have event 1, "Shutdown Motor", we should check only two events: namely, event B_{37} and primary cause H_{42}. Event a should be eliminated. If event 3, "Disconnection of Coupling 44," exists as our primary cause, parameters $\hat{B}_{37}$ and $\underset{\blacktriangledown}{H}44$ should be checked.

Using another example, let us analyze how we can construct the examined algorithm for the bypassed primary cause under 2, "Broken Coupling Crankshaft." Bypassed points can be check by using the method of elimination. On this hierarchical level, six causes exist: cause 1, 2, and 3, as well as V_{65}, B_{38}, and A_{34}. Bypassed nodes can be examined by using the method of elimination as shown in Table 5.4. $\underset{\blacktriangledown}{V}_{15}$, $\underset{\blacktriangledown}{A}_{34}$, $\underset{\triangledown}{H}_{42}$, $\underset{\triangledown}{H}_{44}$, and $\hat{B}_{38}$ represent the inversion of corresponding events.

The following example is the primary cause event 4, labeled "Valve Is Stuck." In this case, we need to check the following nodes: $\hat{B}_{37}$, $\hat{V}_{65}$, $\alpha\hat{V}_{65}$. Event 5, "Defective Regulator R9," is not examined. In this case, the examined algorithm includes the nodes $\hat{B}_{37}$ and $\hat{V}_{65}$ as well as the inversion of events $\alpha\hat{V}_{65}$ and L. In the last stage, the designer constructs the control graph based on the symbols described earlier.

On the basis of developed procedures it is possible to decide that displays describing slurry viscosity, engine malfunctions, couplings, amount of water, apron regulation, and so forth, and controls should be provided.

Procedure is a formalization of what human factor specialists in the West have been doing less formally for many years (developing a failure modes and effects analysis). They use a less structured methodology, which emphasizes verbal analysis based on answers to the questions with which this section began: What can malfunction? What are the possible causes of the malfunction? What can be done to correct the malfunction? All three of the answers to these questions can be seen in Fig. 5.5. Graph theory and failure modes and effects analysis can supply the same answers, although failure modes and effects analysis is less mathematically rigorous than graph theory. However Galaktionov's (1978) method can be used as analytical procedures for ergonomical design.

5.5 EXAMPLES FROM PRACTICE

This section presents some examples from practice that are connected with the development and improvement of various kinds of equipment. In the former Soviet Union ergonomics recommendations are made mostly for astronauts, the military, manufacturing, and industry. The first area considered here is aviation. In this section, as distinguished from section 5.1, design problems are emphasized. The first question addressed is function allocation between the pilot and the aircraft.

5.5.1 Flight Function Allocation

Three kinds of aircraft control are possible: manual, semiautomatic, and automatic. Manual control is used in the more complicated stages of flight such as takeoff and landing and during system failure. Manual control of the aircraft is required when the pilot must process large amounts of information in a short time. In flight at altitude, semiautomatic and automatic (e.g., autopilot) subsystems are switched on. However, these subsystems contain weaknesses that may affect the pilot. Studies (e.g., Zavalova & Ponomarenko, 1968) demonstrate that after flying with automatic control for some time, the pilot cannot shift efficiently to manual control. Experiments and accident analyses (see Zavolova & Ponomarenko, 1970) demonstrate that, in cases of equipment failure, auxiliary explorative eye movement reflects the pilot's search for necessary information from both instrumented and noninstrumented sources.

According to Dobrolensky et al. (1975), during the transition from automatic to manual control in a malfunction situation, auxiliary actions can be observed in 67% of cases. The pilot begins to intervene in the control of the aircraft, and can perform improper actions that worsen the situation. From a self-regulation point of view, these actions stem from the development of a hypothesis of what is wrong, the creation of a plan of action, the performance of actions, and evaluation of the result. Based on these actions, the original hypothesis of what is wrong is either accepted or rejected; if the latter, the process is repeated with a different hypothesis. From this follows that specificity of receiving new information and method of its interpretation depends on the pilot's actions. Description of pilot's activity is a consequence of stages that involved receiving information, processing information, and performance responses becoming insufficient. The Western theorist will find it in the troubleshooting model of W. B. Rouse and S. H. Rouse (1983).

The more unfamiliar the situation, the more these explorative actions emerge. Because of delays in the presentation of instrumented information, it is possible for the pilot to discover failures based on noninstrumented cues, even before instruments respond. Noninstrumented cues signaling failure immediately alert the pilot and result in preliminary explorative acts. If instrumented cues do not back up noninstrumented ones, a conflict arises in the pilot.

These contradictions are less likely to appear with manual control (Beregovoy et al., 1978; Dobrolensky et al., 1975). This can be explained by the fact that with manual control, the pilot has greater knowledge about what influences she or he exerts on the aircraft. Because of this, she or he can more easily predict changes in the flight pattern. From this, it follows that for quick recognition and correct interpretation of an unexpected situation motor actions are important. These motor actions also have a cognitive function.

When a task is highly time limited (e.g., an emergency landing), an increase in such exploratory actions may be undesirable because each action consumes time. Because the pilot is more personally engaged in manual than in automatic flight, she or he is probably more alert and hence more responsive to minor cues.

Based on the advantages of manual control, Beregovoy et al. (1978) and Zavalova et al. (1986) introduced the principle of joint control by the pilot and automatic system. This system, which is shown in Fig. 5.7 (Dobrolensky et al., 1975), permits effective involvement by the pilot when a system of automatic control exists (S.A.C.). When the pilot assumes control, switch K_1 in Fig. 5.7 turns on and switch K_2 turns off, immediately transferring control to the pilot. In the automatic control system, only those components are shut down that would interfere with manual performance.

Joint control of flight decreases the pilot's workload, reduces fatigue, and permits urgent intervention by the pilot (Beregovoy et al., 1978; Dobrolensky

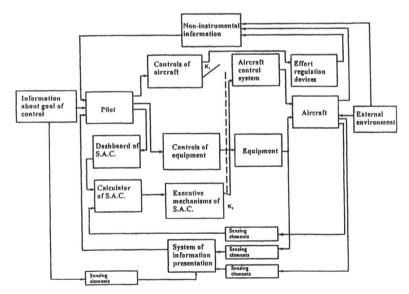

FIG. 5.7. Joint control of flight. From Dobrolensky et. al. (1975).

et al., 1975). It was discovered that under joint flight conditions, the pilot spent more time looking outside the cabin. For example, in one study the duration of eye fixation outside the cabin under manual control varied from 2.7 seconds to 14 seconds, and in joint flight control from 11.1 seconds to 30 seconds. In spite of increasing time breaks between gathering information from the instrument panel during joint flight, flight quality was not decreased.

Dobrolensky et al. (1975) demonstrated that under joint flight conditions, flight parameters were more stable than under automatic flight. Joint flight conditions increased the reliability of the pilot's actions when he or she transferred from automatic to manual control. For example, in one experiment, the degree of readiness for manual control after 1 hour of automatic flight was evaluated. Deviations from required parameters of flight during the transition from joint flight and automatic flight to manual flight are presented in Fig. 5.8. If the transition from joint to manual control occurred after 1 hour of joint flight, there was little or no deterioration of manual flight efficiency. After 1 hour of automatic flight, when the pilot transferred to manual control, some deterioration of manual flight parameters was found when compared with the performance of the same task after 1 hour of manual control only. These data were statistically significant. This study showed that periodic correction of the automatic control system by the pilot improves flight stability.

Comparison of quality of flight and motor activity discovered that increasing precision of flight during the joint control can be achieved by the pilot with fewer motor actions. This can be seen in Table 5.5 where the number of movements by the pilot while manipulating two controls during joint and manual control are contrasted.

During automatic flight the pilot paid more attention to stimuli outside the cabin than during manual flight (Zavalova & Ponamarenko, 1968). This may cause some loss of information from cabin instruments. A number of studies (Dobrolensky et al., 1975; Ponomarenko & Rudney, 1973; Ponomarenko & Zavalova, 1970) discovered, based on eye movement data, that the selection

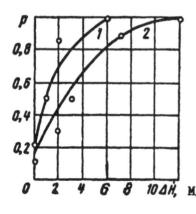

FIG. 5.8. Deviations from required parameters of flight. From Dobrolensky et. al. (1975). 1—after joint control; 2—after automatic control.

TABLE 5.5
Comparison of Motor Activity During Manual and Joint Control of Flight

Controls	Number of Movements per Minute	
	Joint Control	*Manual Control*
Ailerons	Average 3.1	Average 6.3
	Range from 0 to 7	Range from 0 to 14
Altitude control	Average 1.3	Average 8.0
	Range from 0-4	Range from 3 to 16

Note. From Dobrolensky et al. (1975).

of information during joint flight was quite similar to that of manual flight. When a pilot performs control actions, she or he expects to obtain a particular result, which she or he then tries to examine on the appropriate instrument displays. The pilot's attention attracted to the required instruments almost automatically. In automatic flight, however, the pilot had to make a conscious effort to direct his or her attention to cabin instruments.

In other experiments, conducted by the same authors, the time required for detecting different faults in automatic and joint flight was compared. After 30 minutes of actual flight, failures were introduced on the speedometer. Flight speed errors varied from 100 km/hr–200 km/hr. In automatic flight, detection time sometimes approached 160 seconds, whereas it did not exceed 5 seconds during joint flight. In Fig. 5.9, the changes in probability and time of failure detection during joint and automatic flight are presented.

In automatic flight, it appears when evaluated psychophysiologically, that the interaction between the visual and motor systems of the pilot was dis-

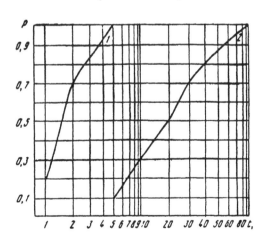

FIG. 5.9. The change in probability and time of failure detection during joint and automatic flight. From Beregovoy et. al. (1978). 1—joint flight; 2—automatic flight.

TABLE 5.6
Average Duration of Visual Fixation in Different Kinds of Flight

Kinds of Flight	Duration of Visual Fixation in Seconds					
	Attitude		Variometer		Altitude	
	Mean	Standard Deviation	Mean	Standard Deviation	Mean	Standard Deviation
Manual flight	.68	.19	.60	.23	.58	.21
Joint flight	.65	.25	.56	.20	.60	.20

Note. From Dobrolensky et al. (1975).

turbed. The opposite result was found in joint flight. It is hypothesized that, in joint control, motor activity preserved the coordination between the pilot's visual and motor systems in the same way as during manual flight. As a result, the gathering of visual information in joint and manual flights is similar. For example, in Table 5.6 the duration of visual fixation by the pilot on various instruments during manual and joint flight is similar.

An hour of automatic flight produced changes in fixation duration after transition to manual flight; one does not find such changes after the pilot transitions from joint flight (Table 5.7).

In automatic flight, the bioelectrical activity of the muscles of the working arm decreased. In joint flight conditions, the electrical activity of the muscles was the same as during manual flight. It was also discovered that the impulses of the electromyogram emerged before the motor actions began, thus in-

TABLE 5.7
Duration of Visual Fixation in Manual Flight (in Percentages)

Kinds of Flight	Duration of Visual Fixation			
	Up to 1 second	Up to 2 seconds	Up to 3 seconds	More than 3 seconds
Manual flight	73.9	16.7	5.0	4.4
Manual flighat after 1 hour of automatic flight	59.0	34.8	3.5	2.7
Manual flight after 1 hour of joint flight	70.	16.	7.6	6.4

Note. From Dobrolensky et al. (1975).

creasing the performance speed of those actions. This muscle readiness was higher in joint flight control. Based on this and other similar studies, Zavalova et al. (1971) introduced the principle of the active operator. The essence of this principle lies in the fact that in automated systems the operator must be actively engaged in a goal directed activity. The operator can not act simply as a reserved component which is engaged in task performance only in case of technological malfunctioning.

Attitude Displays

Another question considered by Russian scientists is the well-known problem of aircraft attitude display. At the present time, two versions of attitude display exist: moving aircraft whereby the earth or horizon is fixed with the aircraft moving relative to it (outside-in display); moving horizon whereby the aircraft is fixed with the horizon moving relative to it (inside-out display); see Fig. 5.10.

There are advantages and disadvantages to both displays. Information from each display requires a mental transformation in presenting information and is not completely accurate in determining aircraft position. In this situation an important question arises: Which kind of indicator is better? In the former Soviet Union the outside-in display predominates, whereas in the West the reverse is true (Ponomarenko & Zavalova, 1981). Presently opinion differs over which to use.

Hasbrook and Rasmussen (1973) conducted a comparative study of these two indicators in which two groups of subjects took part. One group included less experienced pilots (400 flight hours), the second more experienced pilots (more than 1,000 flight hours). Errors and reaction time were measured. Results showed that with the outside-in display the number of errors in the

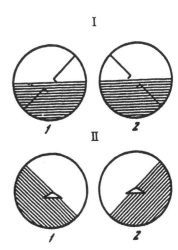

FIG. 5.10. Two versions of aircraft attitude design. From Zavalova (1986). 1—outside-in display; 2—inside-out display.

two groups was similar. With the inside-out display the less experienced pilots made more errors. The authors recommended the outside-in display. Similar results were found by Beringer, Williges, and Roscoe (1975).

The uncertainty in study results caused Zavalova et al. (1986) to conduct a further study. After interviewing pilots, they recommended that when the view of the actual horizon is blocked, it is better for the pilot to use the outside-in display.

In another study by the same authors, an additional attitude display (inside-out) was installed on the windshield, along with an outside-in display on the instrument panel. Following flight, interviews with the pilots revealed that differences between the two displays were not noticed when the pilot could see the actual horizon. Most pilots thought that the inside-out display on the windshield was also an outside-in display.

Pilots oriented the aircraft based on the horizon when it is viewable and not on the displayed attitude. When, however, the horizon was difficult to see, they immediately noticed that they had an inside-out display on the windshield. In visual flight, when the pilot can see earth, the indicator did not matter because there was no doubt about the aircraft's position. Based on this study, the suggestion that the attitude display on the windshield should possess a moving horizon line that matches the natural one was refuted.

Under complex flight conditions, there may be a possible contradiction between sensoriperceptual and conceptual levels of activity regulation. Under these conditions, the inside-out display may be less desirable because it requires additional mental operations to perceive the degree of roll of the aircraft. These mental actions include coordinating the mental presentation of the aircraft with the geocentric reality. The essential concept of the geocentric system of coordination is that the pilot's image of flight is that his or her aircraft is moving with relation to earth.

The image of flight that emerges for people who are not pilots is "aircraft centered" (the earth is moving and the aircraft is stable). Zavalova and Ponomarenko (1984) studied pilot and nonpilot perceptions in a simulated and actual flight landing of an aircraft. The inside-out attitude was more suitable for the aircraft-centered system of coordination, and thus less suitable for pilots.

Similar results were obtained when the gymnast's orientation in space was studied (Gaverdovsky, 1985). This study demonstrated that when an experienced gymnast performed complicated rotations, he or she perceived the earth to be stable. However, when an inexperienced gymnast performed the same maneuvers under a trainer's supervision, he or she usually perceived the earth to be rotating.

Dobrolensky et al. (1975) also compared two alternative cockpit configurations experimentally in a simulator. What is significant about this study is that it reveals the predilection of Russian ergonomists to utilize all possible criteria in their work, reflection perhaps of the comprehensiveness of their work activity theory. Their criteria included flight parameters, eye movement measures, a comparison of flight with and without secondary tasks, physiological measures, and subject interviews. Apparently only the use of secondary tasks suggested that one of the two configurations was preferable.

5.5.3 Design of a Hydraulically Operated Coping Lathe

Another design-related example of Russian ergonomics studies is the revised design of a hydraulically operated copying lathe with programming control. This example is one of the best achievements by Russian ergonomists (Munipov, 1983). The study involved the analysis of various prototype designs.

The studies began with observation of worker performance with existing equipment. Measures were taken of time to perform, frequency of control display use, and worker posture.

The most significant deficiency of the existing equipment was the separation of the area of visual control from the motor operations of the lathe operator, which required excessive physical effort. The design of the protection shield was faulty because it interfered with the worker's perception of this equipment.

Based on preliminary data, ergonomists developed a special mockup that replicated major elements of the lathe (jaw chuck, tail stock, etc.). Using this, worker postures and motor actions were analyzed and the bioelectrical activity of the subjects' muscles in different work postures was measured (see Fig. 5.11).

A physical model of the equipment was then built. Again, the postures and movements of the workers were analyzed. Figure 5.12 demonstrates the posture of the worker with both the old and new versions of the protective shield.

The weight and size of the protective shield was decreased and the glass area increased. An additional control panel was placed on the tail stock, which greatly simplified task performance, especially when cutting big machine parts. The width of the lathe was reduced and a space for the lathe operator's foot was created, which improved the worker's posture.

During the study of motor space on the lathe, the "vectorial-coordinate" method of analysis and design of workplace was used (Zefeld, Munipov, & Tchernyshova, 1983). *Vectorial* means the direction of activity, and *coordinate* refers to the system of coordination in which an activity is performed.

The comprehensiveness of the work activity theory, with its emphasis on motives, goals, and internal states of the operator, leads to a somewhat optimistic philosophy of what is required for a valid experiment. Beregovoy et al. (1978) stated that the experiment must reproduce not only the physical and behavioral aspects of an activity but also the motives, goals, and internal

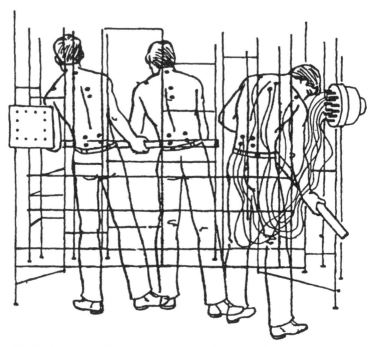

FIG. 5.11. Lathe mockup used for measuring bioelectrical activity of subjects' muscles. From *Ergonomics, Principles and Recommendations* (1981).

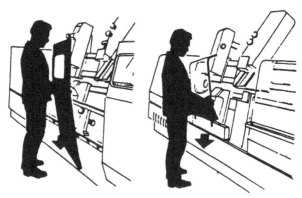

FIG. 5.12. Old and new versions of the protected shield. From *Ergonomics, Principles and Recommendations* (1981).

structure of that activity. They suggested a preliminary psychophysiological analysis of the actual task to determine the content of mental and motor actions during task performance. For them, every experiment must reproduce the psychological aspects of interaction between the operators and their machines, including inner psychic processes and the emotional state of the operator.

As a philosophy, the statements these authors made are admirable but one may question whether any experiment, which is an inherent artificial situation, can reproduce the internal state of the operators, even where subjects are the same operators whose behavior is being studied. Studies have shown repeatedly that subjects recognize immediately that they are in a nonoperational situation, and this undoubtedly affects their perception of the experimental task and its performance.

Beregovoy et al. (1978) also stated that the experiment can model the operator's actual activity. However, this is untrue. In an experiment one models (simulates) the physical characteristics of the environment and the equipment; such simulations can be highly realistic. However, the behavioral activity performed in this simulation is only imitative. What subjects perform during an experiment is a real activity that only imitates that observed in the operational environment; the latter is a sort of template against which the experimental activity can be compared. Experimental activity cannot be identical to operational activity, because the experimental situation is inherently artificial. This point was also made by Meister (1989).

5.6 CONCLUDING COMMENTS

In this chapter, major attention is paid to experimental methods that can be used in the design process. Analysis of the presented material demonstrates that the activity approach does not use behavioristic principles when human behavior is presented as a sequence of reactions to a sequence of stimuli. Russian methodology much more coincides with cognitive psychology, particularly that related to human information processing. Activity is considered from the point of view of the requirements that are presented to perception, memory, thinking, and attention processes. Differences are embedded in a manner in which activity is the described task performance. According to the activity approach, activity can be described and analyzed from the standpoint of motive that impels the activity and from the standpoint of actions and operations that are considered as a process subordinated to a conscious goal. These types of units correspond to the level of detail to be provided in the task description. These units are elements that characterize the structure of activity, which depends not only on these components but also on the specificity of their interrelationships. This is why in analyzing experimental results, heavy emphasis is placed on holistic concepts, because of the role these play in activity theory.

Explanations are often given in terms of the dynamic features of the situation and the development of operator strategies. The operator is presumed to influence the objective features of the experimental situation by developing what Russian psychologists call subjectively relevant task conditions. This suggests that even in very molecular situations the subject not only selects from the objective features of the situation but also interprets those features in a idiosyncratic way. Western researchers would agree that subject perceptions of the experimental situation do make a difference in performance, but would not emphasize this maxim as much as do Russian scientists.

In the theory of activity, psychic processes can be considered as subjective reflections of the objective world. The psychic itself is included in activity and is the result of a particular system of mental actions. This system of actions should correspond to the goal and motives of activity. For example, during the study of the method of perception of information by pilots, specialists should pay attention to the importance of the operative image, conceptual models, goal-directed strategies of gathering information, and so forth. The process of self-regulation provides a dynamic reflection of a situation that must correspond to the performed task.

In pilot activity, the mechanisms of self-regulation continually provided transformation of operative images of flight. Usually, transformation of images are performed in accordance to the logic of activity. In accidental conditions, logical transformations of images can be destroyed. The requirement for urgent formation of an operative image that can be adequate for an accidental situation becomes very important. Very often, the operator must explore the situation before he or she is able to create adequate images of this situation. In accidental situations, explorative activity very often emerges automatically. The operative image of accidental situations influences strategies of selecting information, its interpretation, and the program of performance actions. This is why attractive features of warning signals become insufficient, whereas the significance of signals and their position in the structure of activity become very important.

Other aspects that appear in emergency conditions are connected with the understanding of how the operator can redirect his or her attention to the dynamic components of activity that proceed accidental situations. In other words, this is the ability of the operator to reinstate the interrupted image of task. This situation can also cause a new accidental situation. As can be seen, in the framework of the concept of self-regulation, new approaches to the evaluation of the operator's activity in accidental situations are developed.

The concept of self-regulation of activity and, connected with it, the dynamic aspects of activity are compared to those studies in American literature known as "situational awareness." It is obvious that the concept of activity can be effectively used in developing that concept. In this case, situational awareness that is studied in aviational ergonomics reveals a general psychological basis that permits a deeper understanding of the relationship between conscious and unconscious components of work activity.

In work activity, much attention is devoted to the comparison of objective experimental data with the results of observation by a specialist and self-observation by subjects. Experimental procedures emphasize multiple criteria of measures and greater concentration on physiological indices.

In this chapter, not only is an experimental procedure presented but also procedures that can be used in an analytical study. Algorithmical methods developed by Zarakovsky and the method based on graph theory developed by Galaktionov can be applied for these purposes. As a descriptive analytical tool like the functional flow and operational system diagram used in Western practice, the Russian procedures also have value. The major purpose of algorthmical method is to determine the logical sequence of performed actions. There are two ways of developing an algorithm: logically (a particular relation between a design configuration and anticipated human behavior is acceptable because it appears logical), or on the basis of human performance results derived empirically (personal performance shows particular characteristics that have assumed a relationship to physical variables). Because logic only partly describes the behavior of the operator, the logical algorithm does not completely describe human performance. However, in combination with other procedures this method can be valuable. Galaktionov's method can be efficiently used in the beginning of the design process for determining information on controls that is required for the operator. This method can be particularly efficient when a lot of different instruments and controls exist. However, high complexity of methods requires more time and resources than developers (at least those in the West) would be willing to allocate. Methods developed by Zarakovsky and Galaktionov, like any other behavioral conclusion, must be validated empirically if they are to be taken seriously. An investigative tool in experimentation or system development may have merit, but it should not be used for actual design work unless its validity has been previously established.

Systemic-Structural Analysis and Design of Work Activity

6.1 INTRODUCTION

Activity should not treated as an aggregation of responses to diverse stimuli. Rather, activity is an organized system with a discrete structure. Thus, systemic-structural analysis constitutes a basic paradigm for approaching the study of activity. The system structural approach enables the delineation of an activity as a coherent system of distinct units with specific relationships among these units. Integrity, complexity, organization, and logical sequence constitute the most important integrative features of such systems. These integrative features define the nature of the system's behavior. Integrative features of activity are largely determined by the activity goals and the functional structure of self-regulation. The internal differentiation and the specificity of the relationships among the elements determine the complexity of the system. Among the fundamental notions for system structural analysis are "system," "subsystem," "elements," "structure," "environment," "hierarchical organization," and "intralevel and extralevel" relationships.

Systemic-structural analysis enables the development of new conceptual tools for describing the structure of activity. However, the systemic-structural approach must be seen as merely a general heuristic based on broad principles of analysis and description of complicated systems. This approach does not, by itself, provide practical solutions. It is a theoretical and philosophic framework for approaching practical problems. Thus, practical application of the approach can be quite daunting. Incorporation of these principles into practical design remains something of an art form.

In this chapter a new anthropocentric theoretical concept of design based on systemic-structural principles is described (Bedny, 1979; Bedny, 1987). The following basic principles are subsumed under this theoretical concept of design:

1. The stages and levels of analysis, as well as languages for description of the system, are determined. Four stages of description and analysis of activity during task performance are inferred. Each stage of analysis consists of separate levels, and each of these levels utilizes a distinct language of description and unit of analysis. Each involves a series of qualitative and quantitative procedures.

2. There is a loop structure to the design of activity. The essentials include a sequence of nested dependencies in which each step requires continuous examination of the unfolding results of the design process, as well as continuous implementation of corrective steps. Under this principle the operator's activity and physical configuration of equipment continually improved through this process. The loop structure principle suggests feedback influences at different stages and levels of the design process.

3. The principle of optimization is a step-wise procedure leading to the optimization of an operator's activity. Equipment acceptability may be evaluated in terms of its capacity to optimize performance.

The goal of the design process discussed in this chapter is the creation of models of operator activity and the use of these models to design equipment, as well as the design of efficient methods of performance. Models of activity reflect the structure of an operator's behavior, which is comprised of various actions and their logical interactions, captured by the appropriate language of description.

The initial qualitative stage of description and analysis employed those methods that emphasize a very detailed task analysis and include such familiar techniques as operational sequence diagrams, decision action diagrams, functional flow diagrams, and the like. Because American readers are familiar with these qualitative techniques, described in such books as Meister (1965, 1971, 1985), they are not described in this chapter. Emphasis is placed on the unfamiliar procedures and particularly those that involve quantitative analysis, which is performed to optimize design alternative. Optimizing design here means selection of the optimal concept that satisfies performance requirements according to particular criteria, rather than the selection of the best version of design.

From the Western perspective, Russian design analysis, which is quite different from that of the West although they both start from the same premises, represents a potentially powerful tool for the ergonomist. In particular, it reminds us that complexity, which is responsible for a major part of task difficulty, is something that should be noticed by Western ergonomists, who might find that research on this attribute would be well worth their while.

In summary, behavioral design analysis proceeds in four stages conceptualized as a series of hierarchial models: qualitative description analysis (essentially a highly detailed task analysis); algorithmic analysis, involving consideration of logical conditions; temporal analysis, involving deconstructing the task into its time elements; and quantitative analysis of complexity on the basis of various criteria, such as the amount of attention demanded by the various activity components, the time that these elements require, the opportunity to perform these elements sequentially or simultaneously, stress, and so on.

6.2 OVERVIEW OF SYSTEMIC-STRUCTURAL ANALYSIS AND DESIGN OF WORK ACTIVITY

Russian ergonomists conceptualize design as a description of not-as-yet existent objects and the specification of this description for the purpose of its subsequent transformation into material form.

From the position of the theory of activity, one of the important principles of design is that of the active operator (Zavalova, Lomov, & Ponomarenko, 1971). According to this principle, the operator does not simply react to a stimulus, but actively implements responses directed toward achieving a conscious goal. According to this principle, the operator should be actively involved in task performance.

This may appear to be a tautology, because all operator performance must involve some sort of activity. However, the point is that there are usually several ways of accomplishing a task goal, and some of these involve more activity than others. For example, as was shown in chapter 5, total automatization of a flight by means of computers is undesirable because this causes a reduction in the pilot's reliability during an emergency situation. This is why the principle of joint control during flight was developed (Zavalova et al., 1977), involving allocation of functions between the pilot (manual control) and his or her automatic systems. It was discovered that when a pilot is involved in manual control, he or she has a great knowledge about what influences he or she has on the aircraft and what the current state of flight is. These influences provide the formation and correct understanding of the goal of performance and the correct selection of relevant task conditions. Based on the principle of the active operator, motor activity will always be an important component of an operator's performance.

This principle was also demonstrated by a recent American scientific study (M. J. Adams, Tenney, & Pew, 1995). The introduction of the so-called "glass cockpit" in newer commercial aircraft (where cathode-ray tubes replace hundreds of gauges and instruments) permits one to eliminate many functions previously assigned to the crew. When a malfunction occurs, the computer either automatically corrects it or presents the required information about the problem to the crew. However, with the new glass cockpits, pilots lose their awareness of what is happening, because the cockpit generates a psychological distance between the pilots and the aircraft and its environment. Thus, complex automatization may have a negative influence on operators. An analysis of the relevant literature suggests that automatization is not a reduction of the operator's active involvement in performance but a change in the functions between human and machine.

Another important principle of design from the standpoint of activity is the anthropocentric approach to design that was discussed in section 4.1. According to the Russian anthropocentric approach, the system being designed is a structure of activity that must be compared with the physical characteristics of the equipment that produces (in probable meaning) this structure. Meister (1976) described this as the relationship between task and equipment characteristic. This is why Russian ergonomists make extensive use of what the West calls task analysis. But they believe that existng methods of task analysis cannot completely describe design requirements, because, in part, the language of task description is vague and the description of overt processes is not distinguished from that of covert or inner processes such as the appreciation of goal significance, motivational state, and strategies. They would probably consider that these methods ignore these essential aspects of activity.

They also feel that even if inner states can only be inferred (because they cannot be observed), they must be formally described. In the design process, it is important to distinguish between the description and analysis of the task (Meister, 1985). Only after describing the task does the specialist have the opportunity to analyze it. From the activity point of view, it is necessary to describe the structure of activity during task performance before analyzing it. Activity is not a sum of stimuli and reactions, but rather is a system that consists of actions and operations that have complicated interconnections. This is why systemic-structural analysis becomes very important for the theory of activity. This approach makes it possible not only to study separate units of activity but also to define functional relationships among elements and their properties in the structure of activity. During the design process, the language of description and procedures used must be standardized. This permits specialists involved in the design process to interpret the design correctly. Design is, of course, based on the system concept, one of whose principles is loop structure (feed forward and feedback). Design is, therefore, not strictly or even usually a sequential process, but highly iterative, with subsequent operations providing information that may cause the designer to modify earlier concepts.

Because activity is multidimensional, it cannot be described by a single model, but requires a hierarchy of models that describe activity related to design with different methods and various levels of detail. The systemic-structural principle of design activity suggests four basic stages of design analysis (see Fig. 6.1), in which design analyses are ordered according to the loop structure principle. The latter steps sometimes require reconsidering the preliminary steps of analysis. The coordination of qualitative and quantitative methods of description is very important. At any stage of design analysis activity can be described with different levels of detail, beginning with gross description followed by a more detailed one.

The design analysis begins with the selection of the most important tasks to be described and analyzed. At the initial qualitative stage the ergonomist, based on expert opinion and analysis of prototypes, specifies the goal of an activity and its significance. Accuracy requirements, time constraints, safety, conditions of task performance, and task interactions are taken into consideration. The analyst then arrives at conclusions about the possible emotional-motivational state of the operator during task performance. In this connection, factors to be considered are the repetitiveness of the operation and the potentials for variability in the method of performance. The next step in the qualitative analysis is the analysis of the cognitive aspects of the task. The analysis of the task into stimulus perception, information processing, and decision making was termed the *informational basis of activity* (IBA) by Shadrikov (1982). IBA differentiates between information used in performing tasks (feed forward) and feedback information based on the consequence of task performance.

IBA bears such striking resemblance to Western information-processing theory that it is difficult to discern significant differences between them, but these include an emphasis on stimuli characteristics, memory requirements

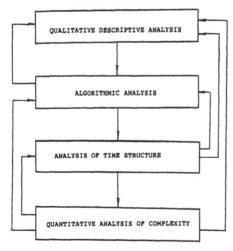

FIG. 6.1. General scheme of systemic-structural analysis and design of work activity. From Bedny (1987).

(information that must be kept in the working memory), feedback information, and the operator's motivational state. All of this produces a preliminary verbal description of the task.

Overall task activity is then divided into individual actions. This requires determining the subgoals that must be achieved during the performance of any particular action because such subgoals are the major criteria for separating one action from another. The analysis should point out the kind of action to be performed (sensory, perceptual, cognitive), the psychological specificity of an action, potential difficulties in performing the action, and the consequence of failure to achieve subgoals. The action is linked with equipment and tools during performance.

The more significant actions are extracted and described based on criteria of accuracy and speed, time constraints, and consequence of unattended goals. Analysis of cognitive processes is unfortunately insufficient to describe the structure of activity during task performance. The same psychological processes may be involved in different tasks, however the structure of activity can be different. It depends on the content of actions involved in task performance and the strategy of task performance.

If different strategies are employed, different actions to perform the same task may result, or the same actions can be performed in different ways, which is why actions should be divided into operations. The theoretical concepts described in the previous chapters of this book may suggest alternative ways of task analysis. Analysis of significance of goal and evaluation of the method of achievement of the goal permit evaluation of motivational aspects of activities during design process.

Such units of analysis as task, action, and operation have a hierarchical organization and permit one to describe the hierarchical structure of an activity. Any changes in the separate components of an activity immediately influence other components of the activity. Because of this, the systemic method of describing an activity is necessary.

The qualitative analysis includes also consideration of potential errors, their relationship to individual actions, their causes, and their possible influence on achievement of the goal (Nebylitsyn, 1965). Rubinshtein (1946) suggested a method of comparing correct and incorrect actions during task performance, particularly those errors that could result in safety hazards. Errors can arise as a result of lack of required information (Hacker, 1980).

Error analysis depends heavily on error data recorded from prototypes of earlier systems or from experimental testing in earlier design phases. Lacking such data, the analysis must depend on analytical procedures that are described in Bedny (1987) and Kotik (1978).

The verbal description of the task can then be translated into a symbolic one, using such techniques as link analysis and decision-action diagrams. Symbolic models are used to describe tasks with increasing levels of detail, so that, for example, if a process chart is developed, the same symbol can be used to describe a single action, a series of actions, or an entire task. Supposedly, this enables the analyst to develop a number of models with hierarchical structures. Verbal and symbolic description is not a strictly sequential process; one can move from a verbal description to a symbolic one, and vice versa.

When the analysis cannot gather any more information about the task, he or she should proceed to the second analytical stage known as algorithmic analysis (this method was described in chapter 5). This analysis is performed because Russian ergonomists feel that many important components of activity cannot be described based only on the preceding methods. In algorithmic analysis, the task is divided not only into actions, but also into algorithmic elements, operators, and logical conditions.

The operators in the algorithm represent actions (motor and mental) as performed by the human. This can be a single action or a group of interconnected actions integrated into a subsystem by a more general subgoal of the task. The logical condition is a decision-making process that determines which actions are selected; complex logical conditions include simple ones, all of which are integrated by the general subgoal. The algorithm determines the sequence of performing various actions. The level of detail of algorithmic analysis often cannot be specified in advance, and may have to be subsequently modified.

The algorithm can be probabilistic as well as deterministic (Bedny, 1987). Deterministic algorithms have logical conditions with only two outputs, 0 and 1. Where the operator's activity is multivariate (determined by logical analysis and expert judgment) probabilisitc algorithms, in which logical conditions can have any value from 0 to 1 and can transfer from one algorithmic element to another according to their probability, are employed. In the process of developing this type of algorithm a matrix of interactions of algorithmic elements can be used.

The third stage of behavioral analysis involves task time structure development. All the activity elements are translated into temporal indices that indicate their duration and distribution over task time. This permits one to determine which activity elements can be performed simultaneously and sequentially. Changes in equipment characteristics can lead to changes in task time structure, and hence, Russian ergonomists feel it is possible to analyze equipment design on the basis of that time structure. Time analysis can also be applied to manufacturing operations to increase their efficiency. Time structure also permits quantitative evaluation of task complexity, the fourth stage of the suggested concept of design. This stage also includes an analysis of probabilistic characteristics of task performance, because the measures of complexity can be developed if we know the probabilistic features of task performance. The quantitative evaluation of task complexity permits an evaluation of equipment configuration, forecasts the speed of skill acquisition, and develops one of the most efficient and safe work methods.

According to the activity approach, activity has a structured organization, in which all components can influence each other. Because of this, the activity approach describes activity holistically. The level of detail of description corresponds to units of analysis in activity theory. The more detailed units of analysis involve motor and mental operations. These units correspond to those used by American ergonomists. For example, such units of behavior as "detect," "receive," "calculate," "identify," "move," "press," given as a sample of classification of task behavior by Meister (1971), correspond to those that, in the theory of activity, are called operations. However, whereas American specialists simply list a sequence of actions or describe task performance in terms of human information processing, Russian scientists, in addition, attempt to describe the logical structure of holistic activity by using additional units of analysis, such as actions, operations, and function blocks. As result of this design process, specialists developed a hierarchical structure of models. These models include qualitative, algorithmic, temporal, and quantitative descriptions of optimal methods of task performance, permitting the specialist to supplement one method of description with another. The existence of interconnections between stages and levels of design analysis according to the loop structure principle allows the specialist to specify and correct developed models of activity. Sometimes, depending on the goal design process and specified task, particular stages of the design process can be eliminated.

As an illustration let us consider a situation in which only the qualitative and algorithmical stages of analysis will be applied. The third stage, which involves temporal analysis, can be omitted, and the quantitative stage can be restricted to only a probabilistic analysis of task performance. The behavioral design task is to lay out a control panel consisting of three displays and nine controls (see Fig. 6.2). In this example we try to demonstrate that using a system of logical interconnected stages and procedures (not the independent technique currently used) permits us to solve design problems in an analytical way without resorting to experimental procedures.

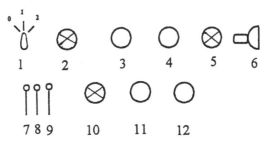

FIG. 6.2. Visual displaces and controls are used in task.

TABLE 6.1
Algorithmical Description of Task

Member of Algorithm	Description of Member of Algorithm
O^{ε}_1	Move knob 1 from position O into position 1.
O^{α}_2	Read information from digital indicator S2.
1_1	If the digit is even, press button 3 and then button 4. If the digit is uneven, perform O^{α}_7.
O^{ε}_3	Press the corresponding buttons.
O^{α}_4	Read information from the digital indicator S2 and S5.
1_2	If S2 > S5, move switch 6 up. If S2 < S5, move the switch down.
O^{ε}_5	Move switch 6 into the required position.
O^{ε}_6	Move levers 7, 8, and 9 into the required position.
O^{α}_7	Read information from S2 and S10.
1_3	If the sum is even, press buttons 3, 4, 11, and 12. If the sum is uneven, press buttons 11 and 12.
O^{ε}_8	Press corresponding buttons.
O^{α}_9	Read information from indicator S2.
1_4	If, after pressing the last button, indicator S2 reads an uneven number, move knob 1 into position 2. If indicator S2 displays digit 1 or 3, move knob 1 into position O. If S2 displays numbers 5, 7, or 9, press buttons 3, 11, and 12.
O^{ε}_{10}	Press corresponding buttons or move knob 1 into the required position.

In the first stage the specialist presents a preliminary verbal description of task (this stage is not represented). The verbal description is then transformed into an algorithmic description (Table 6.1), the second stage according to suggested design concept. After that, the specialist returns to the first stage of design and develops a symbolic description of the task (Fig. 6.3).

This description is based on the operational sequence diagram (OSD) that was developed in the West. OSD is a technique for plotting the sequential flow of information, decisions, and actions through the performance of a system or

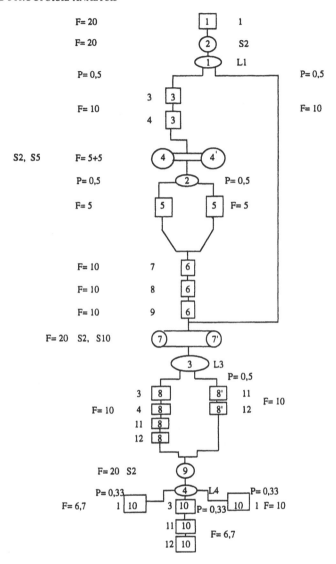

FIG. 6.3. Symbolic model of task performance.

subsystem missions. This flow is portrayed symbolically using standardized symbols to represent the reception of information, decision making, and control responses. This technique was modified in the following ways.

A square represents control manipulation, a circle represents receiving information, and an oval represents the decision-making process (see Fig. 6.3). The beginning of a task is shown at the top of the diagram and the end is shown at the bottom. Their interconnection between symbols is depicted by solid lines. During the development of symbolic models, we employed the following rules.

Inside each symbol the number of an algorithm element is placed. For example, O^ε_1 is designated by a square with a number 1 in it, which is placed on top of the model. O^α_2 is designated by a circle with the number 2 inside.

Logical conditions have separate numbers placed inside an oval. For example, 1_1 is designated by an oval with the number 1 inside of it. Because some algorithm elements include several indicators or controls, the symbols can be designated by the same numbers. However, these numbers are recognizable by dashes. For example, operator O^α_4 suggests perceiving information from indicators S2 and S5. Because perception is a continuous process it is depicted by two circles connected by two horizontal lines. The circles each have the number 4 inside, yet the one farthest to the right is designated by a dash.

To the left or right of the symbol are numbers and symbols associated with each display or control. For example, on the top right (see Fig. 6.3), the number 1 outside of the square designates the number of the control shown under number 1 in Fig. 6.2.

In the next stage, the specialist evaluates the probabilistic features of task performance. (From the second stage of design analysis we transfer to the fourth stage.) Following this the frequency of task performance, the frequency of using separate controls and indicators, and the frequency of different outputs from logical conditions are determined. Suppose the task is performed 20 times and any logical conditions have equal output probabilities ($p = .5$). This permits calculation of the frequency of use of various algorithm elements. For example, in Fig. 6.3, the frequency of use of Switch 1 and Indicator S2 equals 20. This requires the operator to make a decision 1_1 (oval with a number 1 inside) the same number of times. From this logical condition there are two branches, the probability of each being $p = .5$. From this it follows that if an operator performs this task 20 times, the frequency of use of control 3 is related to the algorithm element O^ε_3 is 10. All frequencies can be seen to the right and left in Fig. 6.3 and designated by the letter F.

The interaction of controls and indicators is determined by using mathematical set theory. This is done by grouping all controls and indicators into a subset depending on their relationship. If an algorithm element uses only one control or indicator, this element is considered a unite set designated as a square or a rectangle, inside of which is a number or symbol that corresponds to the number of controls or indicators. For example, there are four sets O^ε_1, O^α_2, O^ε_3, and O^α_4. (Logical conditions 1_1 are connected with O^α_2 and O^ε_3 and do not have their own set.)

| 1 | S2 | 3; 4 | S2; S5 |

Because the same controls and indicators with different algorithm elements can be combined in various ways, the set should be drawn according to their overlapping characteristics. For example, the second algorithm member O^α_2 includes a unit set, S2, and the fourth algorithm member O^α_4 consists of two elements, S2 and S5. Algorithm member O^α_7 also includes two elements, S2 and S10. Algorithm member O^α_9 has one element, S2. This means that unit set S2 can be included in a two-element set S2 and S5, as well as in a two-element set S2 and S10. Therefore it can be drawn in the following way:

Both of these sets overlap because of the same unit set S2. As a result it can be drawn the following way:

Based on a probabilistic analysis of the symbolic model (see Fig. 6.3), a matrix is created that indicates the interrelationships between controls and indicators (see Fig. 6.4). The top and left side of the matrix represent the control and indicator numbers. Inside any matrix cell the frequency of associations between controls and indicators is shown. The bottom of the matrix sums the total number of associations. For example, to determine the number of interconnections of S2 with control 1 (numbers outside symbols with the letter *F*) in Fig. 6.3, one scans from top to bottom for a total value of 20 + 6.7 + 6.7. This is then entered in cell S2-control 1 in Fig. 6.4. The number 20 is the frequency of interconnections between symbols 1 and 2. Number 6.7 is encountered twice and reflects the interconnection between symbols 9 and 10 for the left and right branch in the bottom of Fig. 6.3.

One can also determine significance (based on expert evaluation) of any interrelationship by applying a 3-point scale, with 3 being most valuable, and 1 being least valuable. Multiply the frequency of an interrelationship by its value and this indicates its importance. In the case of the example, the significance of different indicators and controls is the same. As a result, the weight of the interrelationship equals its frequency.

In the last step the specialist transfers activity analysis into design. The relationship between all sets of control and indicators is graphically represented in Fig. 6.5 by lines with numbers that designate the frequency of association. This drawing is modified by the frequency of connections between the elements of a set and their importance. Separate sets for controls and indicators should be drawn. (The largest set should be completed first,

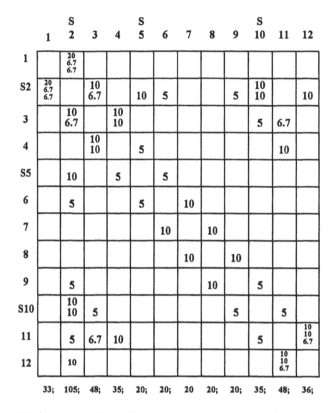

	1	S 2	3	4	S 5	6	7	8	9	S 10	11	12
1		20 6.7 6.7										
S2	20 6.7 6.7		10 6.7		10	5			5	10 10		10
3		10 6.7		10 10						5	6.7	
4			10 10		5						10	
S5		10		5		5						
6		5			5		10					
7						10		10				
8							10		10			
9		5						10		5		
S10		10 10	5						5		5	
11		5	6.7	10						5		10 10 6.7
12		10									10 10 6.7	
	33;	105;	48;	35;	20;	20;	20	20;	20;	35;	48;	36;

FIG. 6.4. Matrix of interrelationships between controls and indicators.

then the smaller one included in it, after that the overlapping sets, and finally the unit set). Determine the weight of interrelationships between set elements, then the weight of the interconnection between including sets, and after that the interconnection of overlapping sets. If all interrelationships have equal value, priority should be given for interconnections inside sets, then for included sets, and after that for overlapping sets.

Based on gathered data the layout of controls and displays can be developed (see Fig. 6.6).

The analytic methods described in this section correspond to the first and second stages of analysis of the suggested concept of design described earlier in this chapter. These analytic methods also involve probabilistic analysis of the structure of activity during task performance, which corresponds to the beginning step of the last stage of the systemic-structure concept of design. To further understand this theoretical concept, the problem of creating a temporal structure of activity and an evaluation of the complexity of task performance should be discussed, and is outlined in the following section.

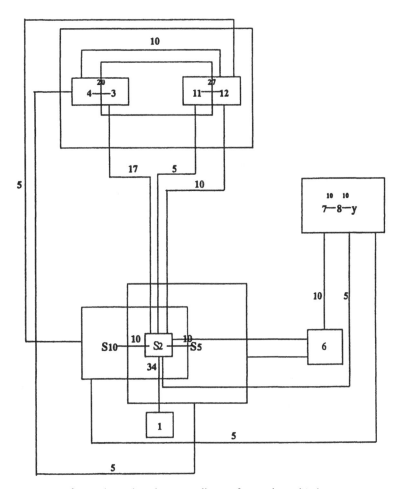

FIG. 6.5. Relationships between all sets of controls, and indicators.

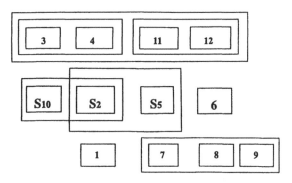

FIG. 6.6. Layout of indicators and controls.

6.3 TEMPORAL STRUCTURE OF WORK ACTIVITY

Although Western ergonomists make use of time (e.g., reaction time) as a performance measure, the concept of the "temporal structure of activity" is not, in the opinion of Russians, sufficiently recognized. Russian ergonomists see it as highly significant in design.

Sometimes, ergonomists confuse the timeline chart, which indicates the duration of tasks or their elements, with the temporal structure of activity during task performance. Hence, it appears that a clear understanding of activity time structure does not exist.

According to *A Guide to Task Analysis* (Kirwan & Ainsworth, Eds., 1992) the American National Standards Institute defines timeline analysis as "an analytical technique for the derivation of human performance requirements which attend to both the functional and temporal loading for any given combination of tasks" (p. 135). The basic units of analysis in this procedure are tasks or task elements.

The temporal structure of activity consists of the logical sequence of activity elements, their duration, and the possibility of their being performed simultaneously or sequentially. As has been indicated previously, Russians distinguish between task elements and activity elements. Timeline analysis and time structure analysis of activity are not the same. Russian psychologists (Bedny, 1987; Zarakovsky & Pavlov, 1987) differentiate between psychological (activity elements) and technological (task elements) units.

The same task elements can be performed by different activity elements, which means that the same task can be performed in a different way. For example, "move lever," "press a button," "take a reading from the instrument," are all typical elements of a task. Depending on the conditions of performance, the same task can involve different activity elements. For example, the task of moving a lever can be performed both consciously or automatically. These two conditions require totally different elements of activity. Examples of activity elements are move arm, detect, decision making, and the like, which can be performed in different conditions in different ways. Ignoring the differences between these units can result in inadequate task analysis.

One can use both timeline and time structure analysis, depending on the particular needs of describing analysis. Initially, specialists should work with task elements and user timeline analysis; later they translate the timeline results into time structure of activity. When specialists attempt to describe the structure of activity, they employ only activity elements. Most Russian ergonomists feel that it is important to utilize standardized units of activity to determine their duration. For example, in the *Handbook of Engineering Psychology* (Lomov, Ed., 1982), five tables are presented that list typical elements of activity and typical elements of task—for example, hand and arm motions, body movements, reactions to visual and acoustical signals, decision making of the "if-then" type—with the duration required for their performance. The weakness in this listing is that each element does not have a precise description, and the data elements are unrelated to each other.

Another approach for determining temporal data is an experimental one. Donders' (1862) method, widely used in cognitive psychology, is called the method of subtraction. A person performs two mental tasks, X and Y, where $Y = X + K$. Psychologists measure the time of performing X and Y, and then subtract time T_x from T_y to derive and get T_k. This procedure permits one to determine the duration of a mental process even if this process cannot be directly observed.

Another approach involves using predetermined time systems (see, e.g., Karger & Bayha, 1977; Van Santen & Philips, 1970), the most efficient of which for determining the duration of motor activity is MTM-1. MTM-1 also includes some time data for simple recognition and decision making. For this purpose, MTM-1 uses eye focus or fixation (EF) time as the time required to focus the eye on an object and look at it long enough to recognize readily distinguishable characteristics. EF can be used to determine the duration of a simple decision-making process (e.g., pushing a red button if the stimulus is red, pushing the green button if the stimulus is green). Unfortunately, no data are available in these time systems for more complex cognitive activities. Moreover, the interaction of activity elements is vital. For example, Bedny (1975) conducted a study in which subjects performed a sequence of motor, perceptual, and decision-making actions. The duration of each action and the entire task was measured. These data were then compared with MTM-1 estimates. It was discovered that the duration of particular actions depends not only on their specific characteristics but also on how they influence each other and the strategies subjects used during task performance.

From this study, it was concluded that predetermined time systems are used incorrectly. Such systems start by breaking down the task into individual elements. As a result, the strategies used by performers are ignored. Use of predetermined time systems must begin with a qualitative analysis that involves consideration of the task goal, the significance of the task, and possible strategies for goal attainment.

The advantage of experimentally derived systems, such as MTM-1, in comparison with experimentally developed data in engineering psychology is that in these systems all elements have a precise detailed description and classification. In predetermined time systems, all data are integrated. For example, after performing "reach" one can perform "grasp"; later "move," "release," and so on. However, in engineering psychology the elements are merely named; they do not have precise description and are not integrated into a holistic system, which makes difficult to use these data to describe complex behaviors.

In consequence, where ergonomists cannot secure data describing the duration of cognitive actions, the last two stages of the activity design analysis (temporal structure analysis and complexity analysis) cannot be performed. If, however, cognitive data can be derived from other sources, such as experiments, predetermined time systems, and so forth, the full analysis can be performed.

Hacker (1980) attempted to explain the time structure of activity during skill acquisition by assuming that, in the initial stage of skill acquisition, a student cannot perform actions simultaneously or one action right after another. Because of this, pauses emerge between the motor and cognitive components. However, when skills are being acquired, pauses are reduced and actions are performed one immediately after another. At the next stage, students are able to perform different actions that overlap one another, or even simultaneously. Pauses that appeared in the beginning are considered empty spaces of time (see Fig. 6.7). However, the assumption cannot be supported. Even assuming that such pauses exist, the operator is still processing information during these pauses. For a novice, performing mental and motor actions simultaneously can be very complicated, which is why she or he interrupts the motor action. During skill acquisition, mental and motor actions become simpler and the

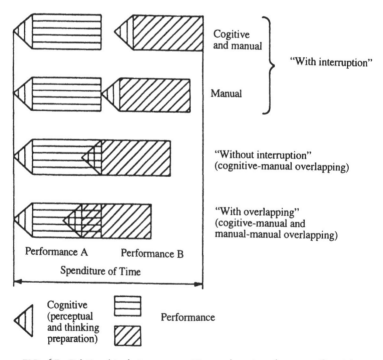

FIG. 6.7. Relationship between cognitive and motor elements of activity according to Hocker. From Hocker (1985).

person can perform them concurrently or can change strategy so that inter-ruption time between motor actions is reduced. Therefore, one can conclude that empty time during task performance does not exist.

The following are stages of developing a time structure of activity:

1. Determine the content of activity with the required level of decom-position for defining their elements.
2. Determine the duration of elements while considering their mutual influence on each other.
3. Define the distribution of activity elements over time, taking into ac-count their sequential and simultaneous performance.
4. Specify the preferable strategy of activity performance and its influence on the duration of separate elements and the total activity.
5. Determine the logic and probability of transition from one temporal substructure to another.
6. Calculate the duration and variability of activity during task performance.
7. Define how strategies of activity change during skill acquisition, and estimate what is intermediate and final about the time structure.

The determination of which activity elements can be performed simultane-ously, and which ones must be performed sequentially, is an integral part of performing a temporal analysis of activity. This determination depends on the goals, motives, significance of action and tasks, and, eventually, a strategy of task performance. For example, in a dangerous situation, when actions have a high level of significance, an operator performs them sequentially, even if they are simple. However, in a normal situation, where the consequences of error are not severe, the same simple actions will be performed simultane-ously. The strategies of activity also depend on the logical components of the work process and the complexity of separate elements of activity.

Time structure of activity cannot be developed until it is determined what elements of activity can be performed simultaneously, and what elements can be performed only sequentially.Gordeeva and Zinchenko (1982) extracted cognitive components of motor actions. They related these components to programming, control, and correction blocks in the model of self-regulation. The greater the duration of these function blocks and the more complex motor actions or motions are, the more concentration of attention they require. They found that if two motions are performed together, their simultaneous combi-nation depends on the phases of performing these actions. If the cognitive components of one motion coincide with the motor components of other motions, they can be performed simultaneously. If the motor phases of both motions coincide, they can also be performed simultaneously. However, the cognitive phase of two motions cannot be performed simultaneously.

MTM-1 provides certain rules that suggest when motions can be combined; if these conditions do not exist, actions must be sequential. These rules are based on the following assumption: The higher the level of controlled processing or level of attention is required during motion performance, the more difficult their simultaneous combination is. The control process can be categorized as low, average, and high. Simultaneous motions can be performed if they are of low or average complexity and do not require high level of attention. Two motions of a high level of complexity that require a high level of attention can be performed simultaneously only under specific conditions where the point at which the motion is terminated is in the normal field of view. If the point of motion termination is out of the normal field of view, its final steps must be performed sequentially. The more experienced a performer is, the greater his or her ability to simultaneously combine certain motions. This statement was proven experimentally (Karger & Bayha, 1977). It was also discovered that two motions performed in parallel required more time than one, particularly when these motions are complex and required more attention. If two motions are simple, the duration of their performance did not significantly increase. If two motions with low and average level of attention are performed at the same time, their performance durations are identical to that of two motions performed separately.

MTM-1 rules are limited because they consider only visual control. However actions require also motor, mental, and other types of control. For example, motor actions requiring a high level of examination, as by touch, can sometimes be performed simultaneously regardless of the field of view.

The MTM-1 system does not consider the ability to combine mental actions, and mental with motor actions. However, in this system there are data about the ability to combine EFs. These microelements are associated with the process of perceiving stimuli and simple decision making. According to MTM-1 these should be performed sequentially because they require a high level of attention. From this, it follows that mental operations involving recognition and decision making cannot be performed simultaneously.

This conclusion was also supported by studies in cognitive psychology. In Broadbent's (1958) study, when unskilled subjects attempted to listen to one message while answering another, accuracy deteriorate. According to Glezer and Nevskaya (1964), if two objects appear in the visual field, recognition of the second object starts after a .7 probability of recognition of the first. S. D. Smirnov (1985) discovered that the ability to perceive stimuli significantly deteriorates when a person simultaneously attempts to perform thinking operations. Smirnov also discovered that if stimuli always have the same meaning and are encountered in similar contexts, information processing becomes almost automatic. However, if the same signal is encountered in different situations and requires different interpretations, verbalization is involved and recognition requires a high level of attention and conscious information processing. According to Lindsay and Norman (1992), automatic attentional processes can proceed concurrently, because the conscious part of the attention process is restricted. The model of attention described in chapter 3 suggests that the more involved long-term memory and the regulative integrator were in information processing the more difficult simultaneous performance becomes. The function block "regulative integrator," which is responsible for coordinating energetic and informational components of attention, becomes overloaded as the task becomes more complex. This means that it is difficult for a person to allocate his or her attention between two goals of activity. In this situation, the human prefers to switch attention from one goal to another. Kahneman's (1973) model of attention similarly suggests that increased task complexity requires more attention.

The following sums up what is known about the possibility of combining mental or mental plus motor components of activity:

1. The simultaneous recognition of different stimuli is possible if they are well structured, the number of stimuli is no more than three or four (based on working memory capacity), and the stimuli are familiar. In other cases, input information should be received sequentially.

2. If an operator recognizes well-known stimuli in a familiar situation, mental actions can be simultaneously combined with motor actions whatever the level of attention required.

3. If an operator recognizes familiar stimuli in unfamiliar situations, motor actions that require only a lower and average level of concentration of attention can be performed simultaneously.

4. The decision-making process and motor actions that require a high level of attention should be performed sequentially.

5. Simple decision making (e.g., choice between) can be simultaneously performed with motor actions that require a lower or average level of attention.

6. Cognitive components should be performed sequentially.

7. Simultaneous performance of activity elements that might result in working memory overload (e.g., require simultaneous keeping in memory) different data (items) should be performed sequentially.

8. In stressful situations, or when personnel are not highly skilled, all activity elements requiring high levels of attention should be performed sequentially.

Another problem is determining the duration of mental actions for use in the design process. For this purpose it may be possible to use a predetermined time system like MTM-1. This requires additional rules about the microelement EF:

1. If stimulus characteristics are amorphous and cannot be easily distinguished (e.g., radar, sonar), additional time for analysis of the stimuli is required (two EFs instead of one).

2. In cases when signals are easily recognized, only one EF is required. Half of this EF time involves recognition and the other half decision making.

In the final stage of design when analytical models are tested experimentally, some corrections are possible. This is a common step not only for ergonomic but also for more precise engineering design.

6.4 DESIGN OF TIME STRUCTURE OF WORK ACTIVITY

The rationale for time structure analysis is that activity is a process embedded in time. The analysis cannot be performed, however, without taking into consideration the concept of work pace. For example, when the operator works under time-restricted conditions it is important to determine reserve time and its variation in time. One example of how human factor specialists ignore the work pace is the attempt to use Fitt's law to determine duration of task performance (Epps, 1986; McLeod & Sherwoods-Jones, 1992). This law suggests that motor components of a task are performed with maximum speed, and that the operator's actions are independent and therefore do not influence each other. Data presented in previous chapters refute these assumptions.

Unfortunately, a precise definition of work pace does not exist. Gal'sev (1973) defined work pace as the frequency of motions performed per time unit. However, this definition is unsatisfactory. For example, an operator can perform the same number of repetitive motions at the same time, but vary their amplitude, which affects their pace. The definition of work pace becomes more complicated when the activity involves not only motor, but also mental actions. This book defines pace as the speed of flow of different components of activity whose structure is organized in time. The operator's ability to sustain a specific pace during task performance is crucial.

Bedny (1975) defined three levels of pace: very high, high, and average. A very high activity pace is slightly slower than the operator's reaction time for various stimuli. This pace is possible only in those cases when an operator reacts to isolated signals, using discrete actions in highly predictable situations. For example, an operator has a high level of readiness to push a button or throw a switch when a particular signal appears.

A high pace is that in which an operator performs a sequence of mental and physical actions in response to the appearance of different signals. It is essentially the same as that reported in MTM-1. This condition lowers the degree of an operator's readiness to perform particular actions. An average pace is that in which an operator performs tasks at his or her own subject time scale (there are no time constraints).

The following is a practical example of developing a time structure of production operation and operator tasks. We begin with this because industrial ergonomics is extremely important in the former Soviet Union. Initially, we use the MTM-1 system to analyze tasks in which motor activity predominates. Following this, data from Lomov (1982, Ed.) are used to determine the duration of cognitive components.

In the experimental example, personnel had to fill a pin board with 30 pins according to specific rules. Optimally, subjects used both arms to fill in the center row with pins, starting with the hole closest to them, and then working upward and outward. However, 10 pins have a clearly visible flute. Therefore, the pins were put in the holes according to specific rules:

1. If the pins are regular (without a flute), they can be installed in any position.
2. If a fluted pin is picked up by a subject's left hand, it must be placed so the flute is below the hole.
3. If a fluted pin is picked up by a subject's right hand, it must be placed so the flute is above the hole.

If a subject erroneously installed a pin, he or she could correct it before finishing the task. This error increased performance time. A subject could make no more than one error per five installations. All the tasks were divided into 15 similar subtasks and were repeated 15 times. This means that for subtask, "taking and putting pins," the probability of error equals 1/75.

Because the sequence of actions in this operation depends on logical conditions, an algorithmic description should be performed first. This description is given in the following way: Afferent operator is designated by symbol O^α, logical conditions connected with decision making are designated by symbol L, and logical conditions connected with making decisions about correcting errors are designated as 1. Logical conditions for when fluted pins are absent in both the left and right hands can be designated as $L_1 = (l_l = 0$ and $l_r = 0)$. Logical conditions for when fluted pins are present in the left hand and absent in the right hand can be designated as $L_2 = (l_l = 1$ and $l_r = 0)$. Logical conditions for when fluted pins are present in the right hand and absent in the left can be designated as $L_3 = (l_l = 0$ and $l_r = 1)$. Logical conditions for when both the left and right hands possess fluted pins can be designated as $L_4 = (l_l = 1$ and $l_r = 1)$.

Because the number of pins is known, including those with a flute, the possible different pin combinations during task performance can also be predicted. Based on these data, we can calculate the probabilistic features of production operations (Table 6.2). An algorithm of task performance

TABLE 6.2
Probabilistic Features of Task Performance

Logical Conditions	Probability
$(l_1 = 0$ and $l_r = 0) = L_1$	4/9
$(l_1 = 1$ and $l_r = 1) = L_4$	1/9
$(l_1 = 0$ and $l_r = 1) = L_3$	2/9
$(l_1 = 1$ and $l_r = 0) = L_2$	2/9
$1_{1.er.} = 1$	2/75
$1_{r.er.} = 1$	1/75
$1_{1.er.} = 1$ and $1_{r.er.} = 1$	1/75

Note. From Bedny (1987).

allows only one error with a probability of 1/75. Based on the obtained data, we can describe the algorithm in table form (see Table 6.3). Based on this description we can develop a logical scheme for an algorithm:

$$\downarrow^{8} O^\varepsilon{}_1 O^\alpha{}_2 O^\alpha{}_3 L_1 \uparrow^{1} {}^{(1-3)} L_2 \uparrow^{2} {}^{(1-2)} \downarrow^{1} {}^{(2)} L_3 \uparrow^{3} \downarrow^{2} {}^{(2)} \downarrow^{1} {}^{(3)} L_4 \uparrow^{4} \downarrow^{1} {}^{(1)} O^\varepsilon{}_4 \omega_5 \uparrow^{5} \downarrow^{2} {}^{(1)} \tag{6.1}$$

$$O^\varepsilon{}_5 \omega_5 \uparrow^{5} \downarrow^{3} \acute{O}^\varepsilon{}_5 \omega_5 \uparrow^{5} \downarrow^{1} O^\varepsilon{}_6 \downarrow^{5} O^\alpha{}_7 l_6 \uparrow^{6} {}^{(1-2)} l_7 \uparrow^{7} {}^{(1-2)} \downarrow^{6} {}^{(1)} O^\varepsilon{}_8 \downarrow^{7} {}^{(1)} O^\varepsilon{}_9 \downarrow^{5} \downarrow^{6} {}^{(2)} \downarrow^{7} {}^{(2)} \omega_8 \uparrow^{8}$$

The scheme requires some explanation. The false logical condition ω_8 demonstrates that the algorithm should be repeated 15 times. The numbers in parentheses on top of the arrows demonstrate possible outputs from logical conditions. For example, L_1 has an arrow and above it the numbers 1–3 in parentheses. These numbers demonstrate possible outputs from logical conditions. For example, $\uparrow^{1}{}^{(1)}$ corresponds to situations when $l_1 = 0$ and $l_r = 0$.

If $L_1 = 0$, work is L_2. If $L_2 = 0$, the algorithm can work according $\uparrow^{1}{}^{(2)}$ (move to L_3). If this condition does not exist, the algorithm can work according to $\uparrow^{1}{}^{(3)}$. Regular logical conditions possess only two meanings, 0 and 1. In this example, logical conditions possess more than two outputs with different probabilities. In other words, the algorithm has a probabilistic character. In the algorithm, three false logical conditions always exist and are designated by symbol ω_5. This means that after putting pins in, the subjects transfer to $O^\alpha{}_7$ and l_6. According to algorithm rules, $O^\alpha{}_7$ and l_6 should be performed only after the pins are incorrectly put in holes.

Knowing an algorithm of task performance and the probabilistic features of task performance, one can describe the temporal structure of activity. In Fig. 6.8, a temporal model of activity during installation of pins is graphically depicted.

The horizontal segments index the duration of the individual elements of activity as described by the MTM-1 system. These elements are specified

TABLE 6.3
Algorithmical Description of the Operation "Filling a Pinboard With 30 Pins"

Members of Algorithm	Description of Member of Algorithm
O^ε_1	Move both hands to the pin box and grasp two pins.
O^α_2	Determine the type and position of the pin in the left hand.
O^α_3	Determine the type and position of the pin in the right hand.
L_1	If a flute in either the right or left pin is absent ($1_1 = 0$ and $1_r = 0$), perform O^ε_4. If this condition is not observed ($L_1 = 0$), transfer to L_1. If L_2 is also not noticed ($L_2 = 0$), transfer to L_3. If L_3 is not observed ($L_3 = 0$), transfer to L_4.
L_2	If the left pin is fluted and the right pin is not ($1_1 = 1$ and $1_r = 0$), a decision must be made to install the right pin in any position and to turn the left pin so the fluted side would be placed inside a hole, and perform O^ε_5. If the flute exists on the right pin instead of the left ($L_2 = 0$), transfer to L_3. If L_3 also equals 0, then transfer to L_4.
L_3	If the left pin is not fluted but the right pin is ($1_1 = 0$ and $1_r = 1$), a decision must be made to put the left pin in any position and to turn the right pin down so the fluted side is up when put inside a hole, then perform $\acute{O}^\varepsilon_5$. If L_3 is also not observed ($L_3 = 0$), transfer to L_4.
L_4	If both pins have a flute, ($1_1 = 1$ and $1_r = 1$), then a decision must be made to install the left pin according to L_2 and to install the right pin according to L_3, (perform O^ε_6).
O^ε_4	Put pins into a hole in any position.
ω_5	Always false logical conditions.
O^ε_5	An unfluted pin should be put in a hole in any position. A fluted pin must be installed according to Li_2.
$\acute{O}^\varepsilon_5$	An unfluted pin should be put in a hole in any position. A fluted pin must be installed according to L_3.
O^ε_6	Install a right pin according to L_3, and install the left according to L_2.
O^α_7	Notice that a pin was put in wrong.
1_6	If one pin is in an incorrect position, transfer to O^ε_8. If the pins are installed correctly, transfer to ω_8.
1_7	If both pins are installed incorrectly, transfer to O^ε_9. If the pins are installed correctly, transfer to ω_8.
O^ε_8	Remove one incorrectly installed pin from its hole, and install it again according to the proper instructions.

(Continued)

TABLE 6.3
(Continued)

Members of Algorithm	Description of Member of Algorithm
O^ε_9	Remove both incorrect pins from their holes and install again according to the proper instructions.
ω_8	False logical conditions always point out the repetition of the subtask 15 times.

Note. From Bedny (1979).

by symbols above the segments. At the left of each line are designated which physical or mental elements are involved: right hand (RH), left hand (LH), mental process (MP). In the MTM-1 system, microelement EF describes the perceiving of signals as simple decision making, such as "yes" or "no" decisions. EF designates the duration of mental actions. A dashed line indicates that another algorithm member is performed at the same time. For example, O^α_2 has a duration of ½ EF and is designated by a solid line, but if performed together with M22B, which belongs to O^ε_4 or O^ε_5 or O^ε_6, it is designated by a dashed line. The probability of occurrence of a particular element of activity in time structure is designated with the letter P.

Some symbols used in Fig. 6.8 require additional explanation. R32C means "move the hand a distance of 32 cm to a precise position" (requires a high level of concentration of attention); G1C1 means grasp a nearly cylindrical object with a diameter of more than 12 mm; M22B means move an object 22 mm to an approximate location (requires an average level of concentration of attention); mM10C means move an object 10 mm to an exact location (requires a high level concentration of attention). The symbol m before M means that the arm continues movement after the first movement (M22B) without interruption; T90S means turn the object with a small weight 90° (from 0–2 pounds); P2SE means position a symmetrical object (align, orient, and engage one object with another object), which requires light pressure and is easy to handle; RL1 means normal release performed by opening fingers. One can designate other elements in the same way according to the rules of the MTM-1 system.

There are three situations: picking up two pins (one in each hand) without flutes; picking up two pins, one with a flute; and picking up two pins, both with flutes. Let us consider O^ε_4, O^ε_5, and O^ε_6. O^ε describes a situation in which two pins are picked up without flutes, with both arms moving simultaneously. In the beginning, this movement was performed by M22B, and at the end, it was performed by mM10C. This is because the final step of moving the pins

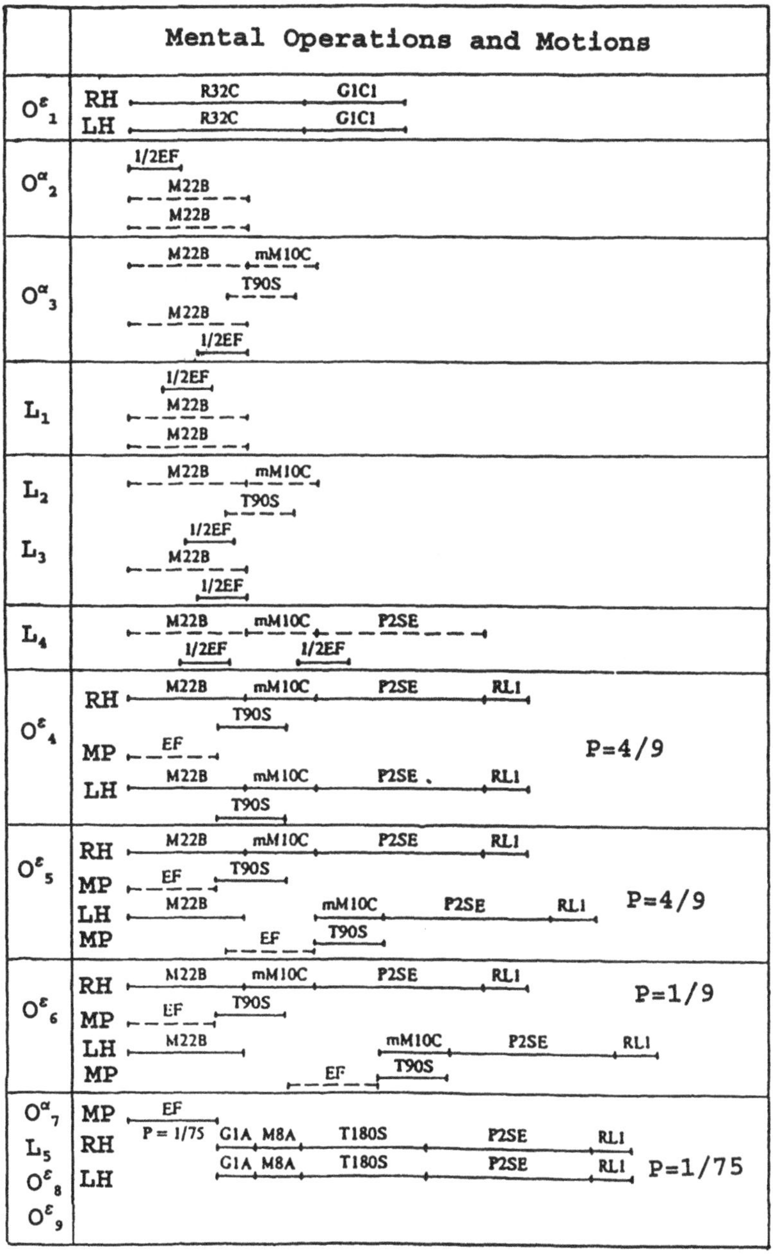

		Mental Operations and Motions
O^{ε}_1	RH	R32C GlCl
	LH	R32C GlCl
O^{α}_2		1/2EF
		M22B
		M22B
O^{α}_3		M22B mM10C
		T90S
		M22B
		1/2EF
L_1		1/2EF
		M22B
		M22B
L_2		M22B mM10C
		T90S
L_3		1/2EF
		M22B
		1/2EF
L_4		M22B mM10C P2SE
		1/2EF 1/2EF
O^{ε}_4	RH	M22B mM10C P2SE RLI
		T90S
	MP	EF P=4/9
	LH	M22B mM10C P2SE . RLI
		T90S
O^{ε}_5	RH	M22B mM10C P2SE RLI
	MP	EF T90S
	LH	M22B mM10C P2SE RLI P=4/9
	MP	EF T90S
O^{ε}_6	RH	M22B mM10C P2SE RLI P=1/9
	MP	EF T90S
	LH	M22B mM10C P2SE RLI
	MP	EF T90S
O^{α}_7	MP	EF
L_5	RH	P=1/75 G1A M8A T180S P2SE RLI
O^{ε}_8	LH	G1A M8A T180S P2SE RLI P=1/75
O^{ε}_9		

FIG. 6.8. Temporal model of activity during installation of pins. From Bedny (1987).

to the hole requires a high level of attention. At the start of a movement of the hands, the subject recognizes the pins and decides that he or she can orient them in any position. Recognition and decision making are noted as microelement EF, designated by a dashed line. Further, without any stopping motions by the arm, the subject turns both pins 90° to put them in the vertical position (T90S). Then follows the simultaneous installation of pins into a hole with the method P2SE. After that, the subject releases the pins.

When one pin is fluted, the strategy of performance changes. At the beginning of movement the subject recognizes that one pin is fluted and makes the decision (EF) to turn the other pin (without a flute) into a vertical position. At the same time, a turn is executed that moves a pin 90° (T90S). Then the subject installs the pins and releases (P2SE and RL1). The decision of how to orient the fluted pin in the left hand is performed immediately after deciding that the first pin can be installed in any position, which is why the second EF usually follows after the performance of the first EF. We are not going into more detailed description of task performance. Here, it should be pointed out only that the preceding temporal description of task performance can help predict strategies that can be used by an operator. It is of interest to compare the time structure of activity in Fig. 6.8 with a timeline chart of the same task in Fig. 6.9.

If we consider the temporal structure of an activity displayed in Fig. 6.8 and the timeline chart in Fig. 6.9, we can see that they are totally different methods

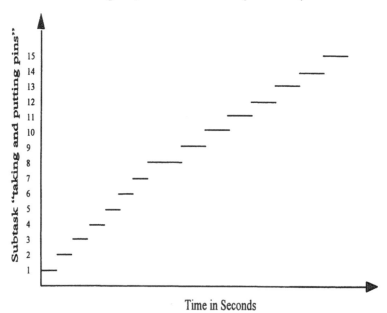

FIG. 6.9. Time line chart of the task "installation of pins." 1–7 flute absent; 8–flutes present in both pins; 9–15 flutes present in one pin.

of describing time performance. A timeline chart uses the duration of a task or subtask as a unit of analysis without any relation to what the operator did during this time. However, an operator can perform tasks of different complexities at the same time. Hence, timeline analysis gives only a rough evaluation of the operator's workload, particularly where mental activity dominates.

In another study (Bedny, 1979) seven subjects performed various sequences of actions depending on the information presented on a control panel. The operator performed two tasks sequentially. The first task involved a potentially hazardous situation; the operator had to make a decision to switch to an alternate work method that permitted the continuation of work without stopping the operation. However, the situation could become so dangerous that the operator had to stop the operation. The decision to do this was the operator's second task.

Performance time was recorded and then compared with the previous analytical method of calculating time performance. There was statistically significant agreement between the experimental data and analytical estimates.

Here we examine the two tasks in greater detail.

6.4.1 First Task

The operator hears an emergency signal that directs his or her attention to an indicator display that presents six symbols, each designating a particular equipment subsystem. If an emergency arises, one of these symbols begins to flash. The operator then determines which request button must be selected. According to the pointer position on the indicators, he or she determines the degree of danger involved. If that degree is level one, he or she eliminates the dangerous situation and continues. If the danger degree is two, the operator stops the operation.

6.4.2 Second Task

Depending on the information that is displayed on the five indicators, the operator can use all the controls or only some of them to stop the operation. The first indicator is a digital display that can present only number 1 or number 2. This information demonstrates which equipment should be shut down first. Based on this information the operator turns the switch to the appropriate position. He or she then uses a lever that can be moved into one of four positions. However, before this can occur, the operator must depress the button at the top of the lever. Depending on the position of the pointer on the display, the operator can use a 10-positional switch. A green or red indicator on the control panel may be illuminated. Depending on which indicator is lit the operator presses the green or the red button.

The algorithm description of the second task is shown in Table 6.4. The algorithm description of task two permits one to describe its temporal structure. Duration of perceptual and mental operations is based on data presented in Lomov (1982). This includes reading the pointer display −.4 seconds; time for eye travel time −.002 + .004α (where α is angle of eye movement); the recognition of the signal −.4 seconds; perceiving acoustical signal when the operator does not expect this −.35 seconds; decision making .29 seconds. The duration of motor actions can also be determined by data from Lomov. For example, using throwing the switch at .7 seconds, moving the arm to the individual control can be determined by the equation .16 + .006R (where R = distance of movement). However, for determining duration of motor actions we used system MTM-1, which is more precise.

The decision making involved in this operation is of the simplest type; it merely requires the operator to recognize the initiating stimuli and to remember the (usually) binary rule associated with it. Because of this, stimulus recognition is conflated with decision making for a duration of .3 seconds. In more complex decision making the operator checks his or her performance, which increases duration. The existence of repetitive checking actions was discovered during a study of astronaut tasks (Hrunov, Hrachaturaynz, & Popov, 1974). The temporal model of activity during performance of the second task is presented in Fig. 6.10.

Decomposition of activity during task performance to the molecular level as previously described is characteristic of Russian ergonomists (Gordeeva & V. P. Zinchenko, 1982; Lomov, 1982; Volkova et al., 1979; Zarakovsky & Pavlov, 1987), but is rarely employed by Western scientists, unless the tasks to be analyzed are inherently molecular, that is, highly simple, repetitive, and proceduralized, as in some industrialized assembly operations. Western ergonomists might suggest that Russian ergonomists would have difficulty applying temporal structural analysis to situations in which the stimulus for action is obscure, there are difficulties in recognizing the stimulus, and decisions must be made on a probabilisitc basis.

The Russian ergonomist would respond that the more obscure the stimulus, the longer the duration for its recognition, and hence the easier it is to measure its duration. The probabilistic features of the task must always be evaluated, as was shown earlier in this chapter. Russians feel that a time structure can be developed for any task that can be described in algorithmic form. As they see it, the problem Western ergonomists have in recognizing the temporal structure approach is that Russians use different units of analysis (action, operation, and function block) than do Westerners. Each of these elements has its own (some times very short) duration. Western scientists also in some cases measure temporal characteristics of behavior that have very short duration. For example, they can measure time needed for recognition of the signal, or determine the reaction time for certain signals. However, they make

TABLE 6.4
Algorithmical Description of Second Task

Members of Algorithm	Description of Member of Algorithm
O^α_1	Look at first digital indicator.
1_1	If the number 1 is lit, move the switch to the left; if the number 2 is lit, move the switch to the right.
O^ε_2	Move the switch to the corresponding position with the right hand.
O^α_3	Determine whether to turn on the digital indicator or the signal light.
L_1	If neither the digital indicator nor the signal is turned on ($L_1 = 0$), move right arm to the lever, grasp the handle, and press the button with the thumb; if any numbers from 1 through 9 on the second digital indicator are lit ($L_1 = 1$), perform O^ε_8; if 0 is lit, perform 1_4.
O^ε_4	Move right arm to lever, grasp the handle and press the button with the thumb.
$O^{\alpha w}_5$	Wait for 3 seconds.
O^α_5	Determine the pointer's position on the indicator.
L_2	If the pointer is positioned between the numbers 1 and 4, perform O^ε_6.
O^ε_6	Move the lever to the position that corresponds to the number between 1 and 4.
O^α_7	Determine whether the second digital indicator or the signal bulb should be turned on.
L_3	If the digital indicator displays a number between 1 and 10 ($L_3 = 10$), move the switch to 1 of the 10 corresponding positions; if the digital indicator displays 0 ($L_3 = 1$), perform L_4.
O^ε_8	Move arm to switch and put it in corresponding position.
O^α_9	Determine which one of the last two lights (red or green) to turn on.
1_4	If the red light is turned on ($1_4 = 0$), perform O^ε_{10}; If the green bulb is turned on ($1_4 = 1$), perform O^ε_{11}.
O^ε_{10}	Move the right arm to the red button and press it.
O^ε_{11}	Move the right arm to the green button and press it.

Note. From Bedny (1987).

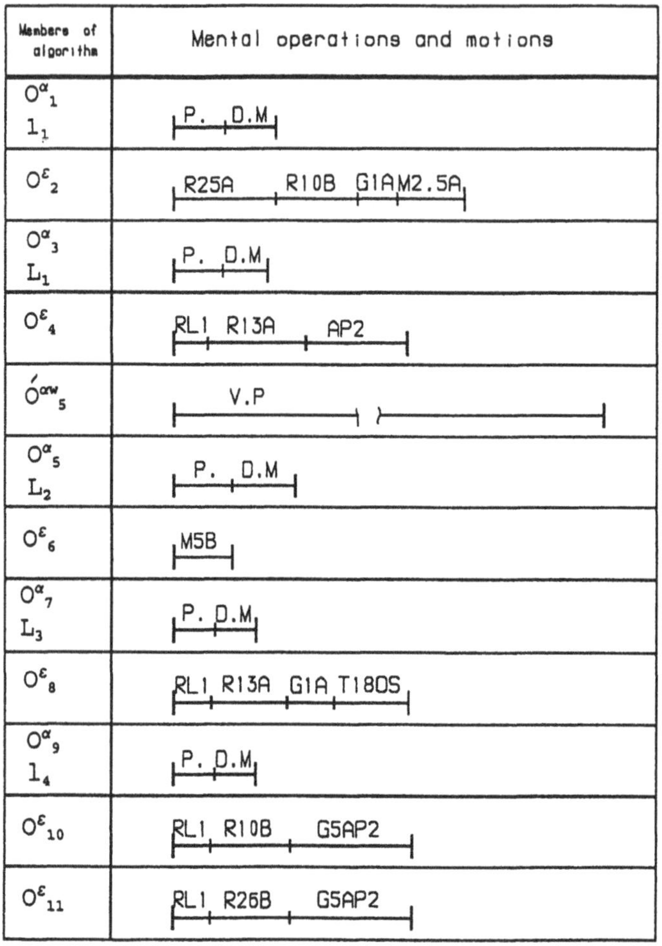

Members of algorithm	Mental operations and motions
O^{α}_1 1_1	⊢P.⊣D.M⊣
O^{ε}_2	⊢R25A⊣R10B⊣G1AM2.5A⊣
O^{α}_3 L_1	⊢P.⊣D.M⊣
O^{ε}_4	⊢RL1⊣R13A⊣AP2⊣
$\acute{O}^{\alpha w}_5$	⊢V.P⊣ ⊢————⊣
O^{α}_5 L_2	⊢P.⊣D.M⊣
O^{ε}_6	⊢M5B⊣
O^{α}_7 L_3	⊢P.⊣D.M⊣
O^{ε}_8	⊢RL1⊣R13A⊣G1A⊣T180S⊣
O^{α}_9 1_4	⊢P.⊣D.M⊣
O^{ε}_{10}	⊢RL1⊣R10B⊣G5AP2⊣
O^{ε}_{11}	⊢RL1⊣R26B⊣G5AP2⊣

FIG. 6.10. Temporal model of activity during performance of second task.

no attempts to describe the entire temporal system of behavior. Decomposition of activity into molecular units with the following description of holistic activity, which is composed of these units, is an important principle of systemic-structural analysis of activity.

6.5 THEORETICAL PROBLEMS OF EVALUATING TASK COMPLEXITY

The evaluation of task complexity is fundamental to ergonomics. If it can be done in advance of the actual equipment development one can evaluate the efficiency of the design to predict operator reliability and how long it will take to acquire skill. Two aspects of this problem should be distinguished: evaluating the complexity of existing tasks and translating complexity into design.

Task complexity evaluations are based on the assumption that the more complex a task is, the higher the probability that it will be difficult for a performer and will increase error. Currently in engineering psychology, two approaches to evaluating task complexity exist. One is based on selection of external criteria in relation to the internal structure of activity. This approach attempts to evaluate the complexity of a task by analyzing the characteristics of equipment. The second approach is based on using internal criteria of activity, such as the content of mental and motor actions, psychic operations, and verbal actions involved in performance. In the first approach, typical elements of the task are used as units of analysis, whereas in the second, typical elements of activity are used.

Typical elements of task cannot be used as units of measures for mathematically evaluating the complexity of task performance.

Some scientists (e.g., Mirabella & Wheaton, 1974; Venda, 1975) used a number of reactions (actions), controls, indicators, and so on, as units of analysis during an evaluation of task complexity. However, task complexity cannot be evaluated in this way. For example, in one task only one control is used, and in a second task three controls are used; however, it cannot be concluded that the second task is necessarily more complex. The first task requires a complex and precise movement of this control, and an error will result in an explosion. In the second task, an operator manipulates the three controls to particular stops, but there are no consequences for errors. Therefore, even though the first task involves only one control, it is more complex than the second task that uses three controls. The same operator can perform three similar but different decision-making actions of the "if-then" kind. For example, if the lighted green bulb lights, the operator should press the green button; if the red bulb lights, he or she should press the red button. This kind of decision-making action is performed sequentially. If the operator performs erroneous actions, he or she can correct those actions. But in a situation with, say, a fighter pilot, the pilot must make only one decision: when to attack the enemy. If the pilot makes an erroneous decision, he or she can be shot down. From these two examples, one can see that one action or decision-making process can be more complicated than three simple ones.

At present, when scientists attempt to evaluate task complexity they use noncommensurable units of measure (Mirabella & Wheaton, 1974; Venda, 1975; Zarakovsky, 1966). The quantitative method of evaluating task complexity suggests choosing units of measurement that permit a comparison of different elements of activity. In other words, it is necessary to transfer different elements of activity into one surface of measurement. This important problem has not yet been solved. As a result, an evaluation of task complexity is reduced to counting the number of actions, algorithmic symbols, the number of controls and indicators, and the like. However, from the mathematical point of view, this approach is incorrect because all these variables are noncommensurable units of measure. For example, we cannot tell which is more, 2 inches or 2 pounds. An analysis of the data in sections 6.1 and 6.2 allows us to formulate several theoretical statements that should be considered during an evaluation of task complexity:

1. Evaluating task complexity is only possible when one analyzes the structure of activity during task performance. Therefore, the typical activity elements rather than task elements should be used as units of analysis.

2. Because activity is multidimensional, several measures, instead of just a single one, should be used to evaluate task complexity. If only one measure is used, significant information about complexity can be lost.

3. Activity is a process. Therefore, a time structure of activity must be built before complexity is assessed. Intervals of time devoted to different elements of activity as units of measurement should be used to evaluate task complexity.

4. Any quantitative measure of complexity should reflect the possibility of simultaneous and sequential performances of activity elements and their probabilities of occurrence.

5. Quantitative measures should reflect the relationship between different components of activity and demonstrate how this relationship influences the complexity of task performance.

To develop activity-derived measures of complexity a preliminary classification system for typical duration elements of activity must be developed. The following criteria are used for classifying these units of measurement: substantial characteristics of activity during an interval of time, and the complexity of these activity elements during the interval.

According to the first criterion, one must distinguish the activity elements and then classify them by the behavioral processes performed during the time interval. Hence, one extracts typical elements of time devoted for motor and mental components of activity. The intervals of time devoted to mental activity are classified based on dominant psychic processes, for example, intervals of time for perceiving information, memorization, decision-making, and so on.

Intervals of time in which performance occurs must be distinguished from active waiting periods in which, for example, the operator waits for signals that inform him or her of the state of the work and what to do next. According to the second criterion intervals of time are evaluated according to their complexity. The complexity of time intervals depends on the level of concentration of attention during those intervals, the character of combining elements of activity, and existing emotional stress.

In the waiting period associated with securing information on the basis of which to continue operations, the operator attempts to predict what will occur. Konopkin (1980) labeled this the *intellectual waiting period.* He also distinguished the *physiological waiting period,* which is the general mobilization of an organism in response to an indeterminate situation. Both of these waiting periods are linked and it may or may not be possible to distinguish them. Underlying the waiting period is the state of the operator connected to neural center activations and level of wakefulness.

The levels of concentration of attention and stress in various time intervals influence task complexity. The level of concentration of attention can be considered an important criterion for evaluating task complexity. The more complex the task the more mental effort is required to the higher level of attention during task performance. Actions that require a higher level of concentration of attention require more time even if conditions are similar. For example, the MTM-1 system divides microelements into three separate levels of control (equivalent to levels of attention): low, average, and high. Obviously, performance is easier with low and average levels of control. Complexity of task correlates with time performance. However, sometimes more complex task can be performed in less time because the operator can mobilize his or her effort, combine different elements of task, and as a result increase speed of performance.

Bloch (1966) characterized attention as the intensity of neuropsychological energy, or the level of activation involved in performance. The higher the level of activation, the higher the level of wakefulness. Performance is a linear function of these two variables (see Fig. 6.11). In this scheme, attention is considered one of the levels of wakefulness. Therefore, it follows that the transition from sleep to wakefulness is, at the same time, a shift of attention; and the level of active wakefulness is, at the same time, a level of attention activation.

The next stage is connected to those levels of concentration of attention that convey emotions. In part A of Fig. 6.11, we can see that performance deteriorates after achieving a particular emotional state.

Task complexity can be linked to the activation level of the brain (see, e.g., Bloch, 1966; Lazareva, Svederskaya, & Khomskaya, 1979). The functional state of the brain can be described as either a specific or nonspecific level of activation. Global or general changes in the functional state of the brain are seen as simultaneous and regional changes in different subsystems (sensory, motor, and associative) and are considered nonspecific forms of activation. This kind of activation flow occurs on the basis of general, not specific stimulation and is tightly connected with the functioning of the reticular activating system. Specific forms of activation could relate with the content of activation and have qualitative differences. They are connected with different parts of the brain and produce different types of activation. Nonspecific forms of activation are connected with the difficulty of performing particular

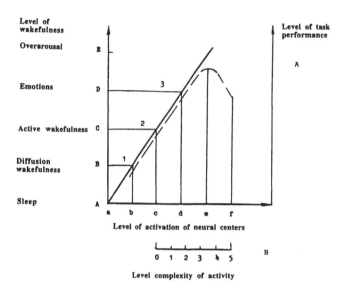

FIG. 6.11. Wakefulness, activation, complexity, and their relationship. From Bedny (1987). A—relationship between level of wakefulness and level of activation according to Bloch (1966), Level of wakefulness and attention. In P. Fraisse & J. Piaget (Eds.), *Experimental Psychology* (Vol. 3, pp. 97–146). Paris: University Press of France. B—level of complexity according to Bedny (1987).

activities; they are not directly connected to the content of activity (Alajanova, Slotinseva, & Khomskaya, 1979; Lazareva et al., 1979). The problem of task complexity associated with the difficulty of task performance is that specific activations may be neglected. Only particular ranges of wakefulness and nonspecific activation related to work activity are considered. Point "c" on the scale of activation (see part A of Fig. 6.11) corresponds to a minimal level of excitement for task performances. Point "f" corresponds to a maximum level of activation for task performance. After point "e," efficiency of activity decreases and after point "f" it starts to sharply deteriorate. Points "c" and "f" on the scale of activation correspond to points "C" and "E" on the scale of wakefulness. Point "C" corresponds to the level of wakefulness or attention when a human performs the simplest automatic actions. Point "D" corresponds to the situation when possible performance actions do not require a high level of concentration of attention and do not convey emotional attention. Starting form point "D," the level of attention sharply increases. This is a result of emotional tension. Point "E" is characterized by a high level of concentration of attention and emotional stress. After this point, efficient task performance is impossible and activity deteriorates.

The complexity of activity can also be viewed on a continuum. Actions that require a minimal level of wakefulness (attention) are the simplest, whereas actions requiring a maximum level of wakefulness are more complex. This continuum is illustrated as a straight line in part B of Fig. 6.11. Point 0 corresponds to the simplest level of activity and point 5 corresponds to the most complex. In the last case, activity cannot be performed with a high level of efficiency. If this 5-point scale is compared to the scales in part A of Fig. 6.11, it appears that the greater the concentration of attention that is needed, the higher the level of activation that is required. Task complexity or the difficulty of its performance can be compared to a level of wakefulness and activation. Zarakovsky and Magazannik (1981) also suggested that increasing complexity of the decision-making process is correlated with increasing emotional-motivational levels. They described two criteria for evaluating the complexity of the decision-making process. One is intentional and considers the probability of failure and strengths of motives that underlie individual activities; this criterion is connected with the significance of the decision-making process. The second is operational in nature and is associated with the specificity of processing information during decision-making operations.

The analysis of attention, wakefulness, and activation presented in this section corresponds to the model of self-regulation of activity in chapter 2 and the model of attention presented in chapter 3. The model of attention presented in chapter 3 suggests that an increase in the complexity of a task requires more resources of attention and, as a result, higher levels of motivation and activation of the neural center. The model of self-regulation in chapter 2 is also consonant with this hypothesis. The function blocks "significance" and "difficulty" in the model of self-regulation are intimately connected with the function block "motivation." From this it follows that the significance of the goal and the difficulty of its attainment are the basic factors that influence the motivation for an activity. From the self-regulation point of view, all of the aforementioned components are dynamic phenomena in the structure of activity. The data presented in literature (see, e.g., Alajanova, Slotinseva, & Khomskaya, 1979; Bloch, 1966) demonstrate that increasing task complexity incrementally increases the levels of concentration of attention and motivation. From the self-regulation point of view, the relationship between complexity and difficulty of the task and the interaction between the difficulty and significance of the task are also important. For example, as was shown, increasing the difficulty of task performance and decreasing the significance of the task can decrease the level of motivation.

The function blocks of self-regulation, which are connected with evaluating the difficulty and significance of the task, play a central role in integrating the cognitive and motivational aspects of the activity in the design process. Usually, the motivational aspects of an activity are ignored in the design process. However, the fundamental notions of the task's complexity, difficulty, and significance, and concentration of attention on the performed activity, permit the designer to take not only the cognitive and behavioral but also the motivational aspects of the activity into consideration in the design process.

268

The preceding theoretical data allow one to develop an ordered scale with different categories of complexity. Any category of complexity can be considered an interval. Within an interval, there is a continuum of complexity for different elements of activity related to the continuum of wakefulness. Any category of complexity includes complex and simple elements of activity. However, these differences inside the intervals can be neglected. The same principle used in technical science, which is "interchangeability," is employed in mass manufacturing processes. For example, the size differences in particular equipment parts can be neglected if these differences fall inside a particular range (tolerance).

An analysis of attention, wakefulness, activation, and self-regulation, as well as the MTM-1 system, enables one to precisely define five categories of complexity, as in part B of Fig. 6.11. For example, in the MTM-1 system there exists three groups of elements of activity depending on their complexity. The simplest group requires a low level of attention. For example, the element of activity "RA" (reach to object in fixed location) requires a minimal concentration of attention. More complicated is the element "RB" (reach to single object in a location that may vary slightly from cycle to cycle). This element according to the MTM-1 system requires an average level of concentration of attention. The most complicated is the element "RC" (reach to object jumbled with other objects in group so that search and select occur). This element requires yet a higher level of attention. We can scale these elements as the first, second, and third category of complexity.

In the MTM-1 system, an element of activity exists that is used when the operator should recognize the object and make an "if-then," "yes-no," kind of decision. This simplest cognitive element of work activity requires a high level of concentration of attention according to the MTM-1 system and can be related to the third level of complexity.

However, there are more complicated components of activity. For example, decision making when responses are unknown in advance is more complicated than when responses are known in advance. This type of decision making should be referred to the fourth category of complexity. Very often, the operator performs actions or makes decisions in an ambiguous situation. For example, the signal on the screen moves ahead, but the operator should move the lever back. This requires remembering instructions, a greater level of concentration of attention, and so on. Because of this, the mental actions connected with attention overload, the recognition actions involved with perceiving unclear signals, and the decision-making and performing actions in a contradicting situation can be placed in the fourth category of complexity. These actions require a greater concentration of attention than the simplest cognitive ones, which fall into the third category.

In some cases the operator performs a task in a stressful situation, such as during emergency conditions. Actions performed in a stressful situation

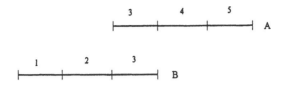

FIG. 6.12. Ordered scale with 5 categories of complexity. A—ordered scale
for cognitive activity; B—ordered scale for motor activity.

can be transferred to a higher level of complexity. This is why there is a
5-point scale for motor and cognitive activity (see Fig. 6.12). From this figure,
one can see that the motor and cognitive scales partly overlap each other.
The more complicated motor components of activity correspond to the more
simple cognitive components.

All of this allows one to develop a formalized system of rules and procedures that permit the translation of qualitative concepts into quantitative
indices. These rules and procedures enable one to identify a strictly monose-
mantically determined category of complexity of activity elements associated
with time intervals. The concept "complexity of time intervals" describes the
complexity of different elements of activity performed in a given time interval. The system of formalized rules, which is based on the theoretical
data described earlier, follows:

1. Time intervals for motions requiring a lower (A), average (B), or higher
(C) level of concentration of attention can be related to the first, second,
and third categories of complexity. This can be presented graphically as in
Fig. 6.13, part A, whereby three actions are designated by intervals A, B,
and C (low, average, and high levels of concentration). The complexity of
these intervals of activity is designated by the numbers 1, 2, and 3.

2. If two simultaneous actions requiring a high level of concentration of
attention (third category of complexity elements) are combined, this period
of time is related to the fourth category of complexity. This rule was concluded from experiments that showed that the combination of these actions
at the same time requires the highest level of resource mobilization, and
significantly increases their performance time (Hassan & Block, 1968). This
rule can be depicted graphically as follows. Suppose two arms perform two
motions that require a high level of concentration of attention (the third
category). Both actions are performed in the normal visual field. The motions
for the right (r) and left (l) hands begin and end at the same time. According
to this rule, the complexity of this time interval of activity equals four (see
Fig. 6.13, part B).

3. If two motions or motions and cognitive elements with different levels
of complexity are performed concurrently, the complexity category of the

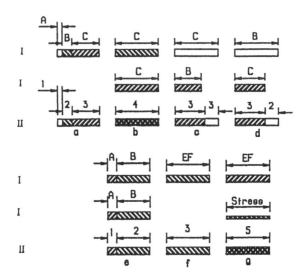

FIG. 6.13. Graphical interpretation of formalized system of rules for evaluation of complexity of time intervals for different elements of activity. From Bedny (1987). I—duration of activity elements; II—complexity of activity elements.

time interval for these simultaneous elements of activity is determined by the complexity of the more difficult element. The graphic interpretation of this rule is demonstrated on Fig. 6.13, part C. If a simpler action has more distance, and thus requires more time than a more complex action, the graphic evaluation of complexity is that presented in Fig. 6.13, part D.

4. If two simultaneous motions from the first or second categories of complexity are performed, any overlapping assumes the value of the first or second categories of complexity, respectively (the complexity of the over-lapping time does not change). See Fig. 6.13, part E. This rule was concluded from experiments that showed that performance of simultaneous motions that require a low or average level of attention (low and average difficulty) does not slow them down (Hassan & Block, 1968).

5. Time intervals related to simple decision making (an operator knows in advance how to react to a particular situation) can be related to the third category of complexity (see Fig. 6.13, part G). This rule follows from data obtained in the MTM-1 system. According to this system, the simplest deci-sion-making process requires a higher level of concentration of attention. According to Rule 1, it is third category of complexity. If an operator makes a more complex decision (does not know in advance how to react to varying situations), this interval of time can be related to the fourth category of complexity. Decision making in a contradictory situation, or in a condition where stereotypical decision-making processes are destroyed, is more com-

plicated than the alternative decision-making process. Therefore, decision making can be related to the fourth category of complexity.

6. Intervals of time devoted to activity where information presented to an operator interferes with other data, hindrance exists, feedback is destroyed, or performed actions contradict existing stereotypical behavior, should be transferred to a higher level of complexity. For example, if an arm moves a lever ahead and requires a low level of concentration of attention (first level of complexity) and then results in moving the control object back, the second level of complexity should be used.

7. If motor activity coincides with mental-perceptual and decision-making actions, this interval of time should be separately analyzed for cognitive and motor activity.

8. If an activity is performed in a stressful situation, the time intervals related to the third and fourth category of complexity should be transferred into five categories of complexity. Time intervals related to the first and second categories of complexity should be transferred into the third category. For instance, the operator performs a simple decision-making action (third level of complexity); however, this action is taking place in stress-related conditions. Such an action should be considered of the fifth level of complexity (see Fig. 6.13, part G).

This formalized system of rules and procedures permits the specialist, with an acceptable level of precision, to evaluate the complexity of different components of activity.

6.6 MATHEMATICAL FORMALIZATION
OF TASK COMPLEXITY

If one assumes, as we do, that task complexity is at the heart of human performance difficulties, then the quantitative determination of complexity becomes a *sine qua non* of ergonomics. The problem has not been solved; indeed, the very definition of complexity is at issue. It was pointed out previously that counting the number of equipment or human-machine interface components will not satisfy, because they are noncommensurable units of measure.

From the mathematical point of view, such data as the number of decisions the operator must make, the number of errors, the duration of task performance, and so on, also cannot be used as units of measure in analytical procedures. They can only be used as qualitative data that demonstrate the complexity of task performance. For example, as discussed earlier, sometimes a more complex task can require less time than an easier one, or an operator can perform one very complicated decision-making process that is more complicated than two simpler ones.

The most successful attempt to solve this problem was made by Zarakovsky (1966; see chapter 5 of this book). Zarakovsky's concept was to describe the task by developing an algorithmic formula, then using members (elements) of the algorithm as units of measurement. This method was very popular in Russia. However, we lack precise criteria for decomposition of the task. The same task can be described by algorithms that have different numbers of members. This can produce different measures of complexity for the same task, which is of course unacceptable. However, Zarakovsky's concept was interesting and influential. Russian psychologists feel that complexity measures should reflect the relationship between various activity components; evaluate workloads of memory, thinking, and perceptional processes; reflect difficulty of coordination of different motor actions, and so forth. Bedny (1981, 1987) suggested using intervals of time devoted to activity components as units of complexity measurement. These units of measure can be used only after developing the time structure of activity. The suggested method is based on the assumption that activity is a process; because of that we cannot use the typical elements of an activity (mental or motor operations) as units of measurement without indicating their duration. For example, measurement N_α, which was discussed in section 5.3, describes the relationship between afferent and other components of activity involved in task performance. This can be calculated not by using symbols of an algorithm, but by determining the proportion of time related to afferent components of activity in relation to the overall duration of algorithm performance.

Developing measures of task complexity involves several steps. As a preliminary step, the specialist should determine which elements of activity can be performed sequentially and which elements can be performed simultaneously. This step is required because elements of activity that perform simultaneously should be considered as more complex than similar elements that are performed sequentially. In the next step, the probalistic characteristics of activity during task performance should be described. Then, an evaluation of duration of different elements of activity is provided. After that the specialist calculates the mathematical mean for different components of work activity. Only after that can one calculate a proportion of time devoted to different components of activity during task performance.

Duration for an algorithm performance of task can be determined from the following formula:

$$T = \sum_{i=1}^{n} P_i \, t_i \qquad (6.2)$$

where P_i is the probability of the i-th member of the algorithm and t_i is the time of performing the i-th member of the algorithm.

Time for logical conditions, afferent operators, and executive (response) components of activity can be determined in the same way:

$$L_g = \sum_{i=1}^{k} P^1_i \, t^l_i; \quad T_\alpha = \sum_{R=1}^{q} P^\alpha_r \, t^\alpha_r; \quad T_{ex} = \sum_{j=1}^{m} P^o_j \, t^o_j \qquad (6.3)$$

where P^l_i, P^α_r, P^o_j is the probability of i-th logical conditions, r-th afferent, and j-th efferent operators; and t^l_i, t^α_r, t^o_j is the time performance of i-th logical conditions, r-th afferent, and j-th efferent operators.

Time related to recognizing and identifying suprathreshold task signals can be determined from the following formula:

$$T'_\alpha = \sum_{R}^{\alpha'} P^{\alpha'}_{rep} t^{\alpha'}_{rep} \qquad (6.4)$$

where $P^\alpha_{rep'}$ is the probability of emerging r'-th afferent operators appearing, characteristics of which approach to threshold value; and $t^\alpha_{rep'}$ is the time recognition or identification for the corresponding operator.

For the next step, one determines the relationship between time spent for logical conditions to time spent for response components of activity.

$$N_\alpha = L_g / T_{ex} \qquad (6.5)$$

where L_g is the time for logical conditions and T_{ex} is the time for response (executive) components of activity.

Activity can be stereotyped (repetitive) or changeable (variable). The performance of a stereotyped activity is, of course, easier. If procedures always follow a set order, or always follow after some particular member of an algorithm, these logical components of activity are stereotyped. If procedures and the transition from action to another have probabilistic features, the procedures are variable. It is hypothesized that the more time devoted to variable procedures, the more complex this process is. One can calculate measures of stereotyped and variable (changeable) components for executed activity and logical conditions.

We should note that the measure of changeable and stereotyped logical and executive components of activity are interconnected. The more logical conditions, the more they have outputs, and therefore the more changeable the executive components of activity are. However, in some cases, different logical conditions have an output that leads to the same operator of an algorithm. That is why we introduce different measures of "stereotypy" and "changeability" of logical and executive components of activity.

The time devoted to stereotyped and variable operators and logical conditions can be determined according to the following formulae:

$$t_{st} = \sum_{j=1}^{m'} P^o_{jst} \, t^o_{jst}; \quad t_{ch} = \sum_{j'=1}^{m''} P^o_{jch} \, t^o_{jch} \quad\quad (6.6)$$

$$l_{st} = \sum_{i=1}^{k'} P^l_{ist} \, t^l_{ist}; \quad l_{ch} = \sum_{i=1}^{k''} P^l_{ich} \, t^l_{ich} \quad\quad (6.7)$$

where P^o_{jst} and P^o_{jch} are the probability of the appearance of j-th stereotyped and variable operators; t^o_{jst} and t^o_{jch} are the performance time of j-th stereotyped and variable operators; P^l_{ist} and P^l_{ich} are the probability of the appearance of j-th stereotyped and variable logical conditions t^l_{ist}; and t^l_{ich} is the performance time of i-th stereotyped and variable logical conditions.

Accordingly, the measure of stereotyped and variable logical components of activity can be determined from the expressions:

$$L_{st} = l_{st}/L_g \quad\quad (6.8)$$

$$L_{ch} = l_{ch}/L_g \quad\quad (6.9)$$

where l_{st} and l_{ch} are the mathematical mean of performance time of stereotyped and variable procedures.

In the same way, we can determine the stereotyped and variable executive components of activity:

$$T_{st} = t_{st}/T_{ex} \qu\quad (6.10)$$

$$T_{ch} = t_{ch}/T_{ex} \quad\quad (6.11)$$

where t_{st} and t_{ch} are the mathematical mean of performance time of stereotyped and variable afferent and efferent operators.

There are two aspects impinging on the complexity of work activity. In the first, the sequence of actions or character of the decision-making process is predominantly determined by external stimuli or information provided external to the individual. In the second, the performance of actions and decision-making processes is determined largely by information retrieved from the long-term memory. The latter situation is more complex for the performer (Bedny, 1979; Konopkin, 1980). When one tries to evaluate the complexity of logical conditions one should distinguish those logical conditions performed based on memorized information from those based on external information. From this we introduce a measure of complexity for logical conditions that are performed based long-term memory:

$$L_{ltm} = l_{ltm}/L_g \qquad (6.12)$$

where L_{ltm} is the proportion of time for logical components of work activity depending mostly on memory and l_{ltm} is the mean performance time for logical conditions predominantly governed by memory.

Another important measure of the complexity of work activity is the measure of the workload imposed on working memory. This is the proportion of time for retaining current information in the working memory relative to the duration of task performance. The greater this proportion, the greater the workload imposed on working memory. During the evaluation of complexity, the time interval is connected with retaining information in working memory. Here we introduce two additional rules:

1. If, during the decision-making process the operator should perform sequentially more than three logical conditions, this time interval should be related to the fourth category of complexity. This rule is based on Zarakovsky (1966), who experimentally proved that the presence of three or more logical conditions in sequence sharply increases the mental workload of the operator and the probability of error.

2. If the operator should keep information in the memory that approaches the capacity of the working memory (four to five elements), this interval of time can also be related to the fourth category of complexity.

These rules permit us to not only determine the duration of the workload of the working memory, but also the level of complexity of time that is connected to the loading of the memory.

The measure that characterizes the workload of the working memory can be determined according to the following formula:

$$N_{wm} = t_{wm}/T_g \qquad (6.13)$$

where N_{wm} is the proportion of time in which current information is refined in the working memory and t_{wm} is the mean time for activity related to the storage in working memory of current information concerning task performance.

The complexity of afferent activity components can be evaluated depending on the detectability of perceived stimuli. When the operator receives information from stimuli with low detectabilty the task should be considered as more complex. That complexity is the proportion of time required for discrimination and recognition of stimuli with low-detectability features. The formula is:

$$Q = T'_\alpha/T_\alpha \qquad (6.14)$$

where T'_α is the time for discrimination and recognition of different features of the task approaching the threshold characteristics of sense receptors (low level of detectability).

Another factor affecting task complexity is the repetitive component of work activity. The more often the same elements of activity are performed during task execution, the easier the task is for the performer. For measuring these characteristics we introduce a formula (for logical conditions and afferent and efferent operators):

$$Z^l = t^l_{rep}/L_g; \ Z^\alpha = t^\alpha_{rep}/T_\alpha; \ Z^{ef} = t^{ex}_{rep}/T_{ex} \qquad (6.15)$$

where Z^l, Z^α, and Z^{ef} are the proportion of time for repetitive logical conditions and afferent and efferent components of work activity; and t^l_{rep}, t^α_{rep}, and t^{ex}_{rep} are the mean times needed to perform repetitive logical conditions and afferent and efferent components of activity.

The category of complexity of any member of the algorithm can be evaluated according to the 5-point scale that was described in section 6.5. If any one category of complexity predominates and exceeds 70% of that time for a particular task element, the general complexity of this component of activity belongs to this category of activity. In other cases, one averages the complexity value and places that value on the 5-point scale.

Let us consider how we can evaluate the complexity of time intervals connected with an active waiting period. These intervals of time can be encountered within or between tasks. For example, an operator performed one task and waited before starting to perform a second. In spite of the absence of external observed behavior, this time requires concentration of attention. The operator is ready to become immediately involved in the situation. This active waiting period can also be evaluated in complexity according to the level of concentration of attention during the waiting period. For quantitative evaluation of the complexity of this period of time we introduce the following rules:

1. If the waiting periods require a low, average, or high level of concentration of attention, they are described by the first, second, and third categories of complexity respectively.
2. If the waiting periods convey emotional stress (danger of trauma or accident), they are described by the fourth category of complexity.
3. If the waiting periods require keeping information continuously in the working memory, this requires increasing their complexity levels by one.

The existence of an active waiting period requires the introduction of additional measures of complexity calculated in the same manner as previously. One measure is called the "proportion of active waiting period." This measure can be calculated according to the following formula:

$$\Delta T_w = t_w/T \qquad (6.16)$$

where t_w is the mean time for an active waiting period and T is the total duration of task performance.

If an active waiting period continuously follows a particular element of the activity or task, it is a repetitive active period, measured by the formula:

$$W^{st} = t_{wst}/t_w \qquad (6.17)$$

where t_{wst} is the time for repetitive waiting components and t_w is the total duration of the waiting period.

If the waiting periods of time are identical according to their internal psychological content, they are repetitive. The proportion of repetitive active waiting periods can be calculated the same way.

Table 6.5 summarizes the various complexity measures.

6.7 APPLICATION OF SYSTEMIC-STRUCTURAL ANALYSIS OF WORK ACTIVITY FOR NEW DESIGN

This section is an example of how the theoretical concept described in this chapter can be applied to the design of an underwater unmanned vehicle (UUV). Figure 6.14 is a drawing of an underwater bulldozer recommended by Lobanov (1983) for work at a maximum depth of 20 meters. A pontoon equipped with a power source is on the water surface and the operator's capsule is underwater. The capsule and pontoon are connected with a tubular column with a scaled opening on the ladder inside its upper end. The operator's capsule has built-in ventilating equipment and an emergency exit.

TABLE 6.5
Measures of the complexity of Task Performance

Name of Measure	Formula for Calculation	Variables	Psychological Meaning
Time for algorithm execution	$T = \sum\limits_{i=1}^{n} P_i t_i$	P_i -- occurrence probability, t_i -- occurrence time of i-th member of algorithm	Duration of task performance
Time for performance of logical conditions	$L_\ell = \sum\limits_{i'=1}^{kll} P_i t_i$	P_i -- occurrence probability, t_i -- occurrence time of i-th logical conditions	Duration of decision-making process
Time for performance of afferent operators	$T_\alpha = \sum\limits_{R=1}^{q} P^\alpha t^\alpha$	P^α -- occurrence probability, t^α -- occurrence time of r-th afferent operators	Duration of perceptual workload
Time for performance of efferent operators	$T_{ex} = \sum\limits_{j=1}^{m\,o\,o} P_i t_i$	P_i -- occurrence probability, t_i -- occurrence time of j-th efferent operators	Duration of executive components of activity
Time for discrimination and recognition of distinctive features of task approaching threshold characteristics of sense receptors	$T'_\alpha = \sum\limits_{R}^{\alpha'\,\alpha'\,\alpha'} P_{r'} t_{r'}$	$P_{r'}$ - occurrence probability, $t_{r'}$ -- occurrence time of r'-th afferent operators, characteristics of which approach threshold value (required additional EF)	Duration of perceptual process connected with weak stimuli (closely to threshold characteristics)
Proportion of time for logical conditions to time for executive activity	$N_i = L_\ell / T_{ex}$	L_ℓ -- time for performance of logical conditions, T_{ex} -- time for executive components of activity	Relationship between decision-making process and executive components of activity

(Continued)

278

TABLE 6.5
(Continued)

Name of Measure	Formula for Calculation	Variables	Psychological Meaning
Measure of stereotype of logical processing of information	$L_{st} = l_{st}/L_g$	l_{st} -- time for stereotype of logical processing of information, L_g -- time for performance of logical conditions	Characteristics of inflexibility or rigidity of decision-making process
Measure of changeability of logical processing of information	$L_{ch} = l_{ch}/L_g$	l_{ch} -- time for changeable logical processing of information	Characteristics of irregularity or flexibility of decision-making process
Measure of stereotype of executive components of activity	$N_{st} = t_{st}/T_{ex}$	t_{st} -- time for stereotype of executive components of activity, T_{ex} -- time for executive components of activity	Characteristics of inflexibility or rigidity of executive components of activity
Measure of changeability of executive components of work activity	$N_{ch} = t_{ch}/T_{ex}$	t_{ch} -- time for changeable executive components of activity	Characteristics of irregularity or flexibility of exec-utive components of activity
Scale of complexity (a) algorithm (b) member of algorithm	X_r - level of complexity $(1, 2, . . 5)$	Based on level of concentration of attention during task performance (1 - minimum concentration, 5 - maximum)	Level of mental effort during task performance and performance of different elements. Unevenness of mental effort and critical points of task performance.

(Continued)

279

TABLE 6.5
(Continued)

Name of Measure	Formula for Calculation	Variables	Psychological Meaning
Proportion of time for repetitive logical components of work activity	$Z^l = t^l_{rep}/L_g$	t_{rep} -- time for performance of identical logical conditions	Characteristics of habitualness of information processing
Proportion of time for repetitive afferent components of work activity	$Z^\alpha = t^\alpha_{rep}/T_\alpha$	t_{rep} -- time for performance of identical afferent components	Characteristics of habitualness of perceiving process
Proportion of time for repetitive efferent components of work activity	$Z^{ef} = t^{ef}_{rep}/T_{ex}$	t^{ef}_{rep} -- time for performance of identical efferent components	Characteristics of habitualness of executive components of activity
Proportion of time for logical components of work activity depending largely on information selected from long-term memory rather than extero-ceptive information	$L_{ltm} = l_{ltm}/L_g$	l_{ltm} -- time for logical components of activity whose operational nature is predominantly governed by information retrieved from the long-term memory	Level of memory workload and complexity of decision-making process
Proportion of time for retaining current information in working memory	$N_{wm} = t_{wm}/T$	t_{wm} -- time for activity related to storage in working memory of current information concerning task performance	Level of workload of working memory and complexity of decision-making process

(Continued)

280

TABLE 6.5
Measures of the complexity of Task Performance

Name of Measure	Formula for Calculation	Variables	Psychological Meaning
Proportion of time for discrimination and recognition of distinct features of task approaching threshold characteristics of sense receptors	$Q = T'_\alpha / T_\alpha$	T'_α -- time for discrimination and recognition of different features of task approaching threshold characteristics of sense receptors	Characteristics of complexity, sensory, and perceptual processes
Proportion of active waiting period in total work process	$\Delta T_w = t_w / T$	t_w -- time for active waiting period in work process	Relationship between active waiting period and performance
Category of complexity of active waiting periods	$X_w - 1 \ldots 4$	Concentration of attention during waiting period (1 - minimum, 4 - maximum)	Level of mental effort during active waiting period
Proportion of time for repetitive waiting periods or work activity	$Z^w = t_{wrep} / t_w$	t_{wrep} -- time for repetitive waiting periods	Characteristics of habitualness of waiting periods
Measure of changeability of waiting periods in work process	$W^{ch} = t_{wch} / t_w$	t_{wch} -- time for changeable waiting components in work process	characteristics of irregularity of waiting periods
Measure of stereotype of waiting periods in work process	$W^{st} = t_{wst} / t_w$	t_{wst} -- time for stereotypy wait component in work process	Irregularity of waiting periods

Note. From Bedny (1987).

281

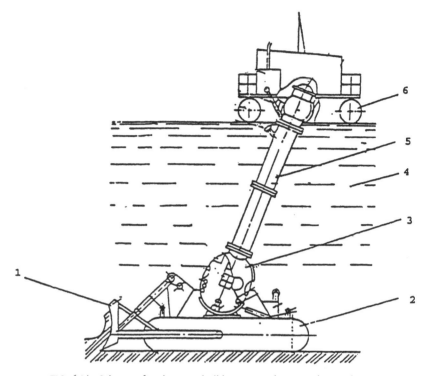

FIG. 6.14. Scheme of underwater bulldozer according to Lobanov (1983).

However, this does not guarantee the operator's safety. Despite the existence of flexible joints, any sudden disturbance or rough underwater surface can cause damage to the sealed opening in the column, and the emergency exit will become useless. Because of this it was suggested that a UUV be used. Three design alternatives for the UUV were considered. The first version was suggested by engineers, the second by American psychologists (according to Yastrebov, 1977; Yur'nev, 1975), and the third by Bedny (1987). Let us consider all of these versions.

6.7.1 Stage One: Qualitative Analysis

6.7.1.1 General Psychological Description. One way to improve the operator's safety is to place the control equipment on the pontoon or on shore and, where the vehicle must operate at a considerable depth, on a surface vessel stabilized in relation to underwater beacons. The idea is to replace the manned underwater vehicle with an unmanned, remotely controlled one (the UUV). This, however, complicates the control of the vehicle substantially by distorting the correlation between the axis of the operator's body and that of the vehicle. This distortion demands constant mental adjustments between these axes, which causes considerable mental fatigue. Therefore, to make the process of remote control easier, according to Yur'nev (1975) and Yastrebov (1977), consultants suggested using a rotating control panel located on a surface vessel or ashore to control the movements of underwater robots. Rotation of the control panel and its inclining and reclining were performed by the operation using feedback from the preceding motion.

However, operating such robots remains complicated. In particular, the UUV for construction purposes was not taken into consideration. Exploration of the continental shelf and the ocean depth have required the development of various underwater machines for laying deep-water pipelines and cables on the ocean bed. During this work, the movement of UUVs from one position to another is especially important. The problem is intensified by deteriorated visibility, caused by an agitated and murky mixture of water and sand particles, which complicates the process of control.

Because the operator handles the UUV by remote control using data from a display, the coordinate system of an operator is independent from that of the UUV. An operator may have problems selecting a right or left turn, or backward or forward movements. Because of this, the use of a display panel that changes its position concurrently with changes in the UUV's position was proposed. Using feedback, we reproduce only inclining and reclining movements of the UUV, just similar to the cases of underwater manipulators that are moving in the water.

The 360° rotation of the control panel should be determined not by feedback from the UUV but by active control actions of an operator who, simultaneously with movements of the UUV on the seabed, also rotates his or her display panel in the same direction and to the same degree. A rotating display was also proposed. The display screen has a scale mask and an icon (a marker) showing the current position of the UUV (see Fig. 6.15).

When the operator's panel rotates because the UUV turns, the display should rotate at the same angle but in an opposite direction. This allows one to associate the operator's coordinate system with the UUV's. The axes OX and OY will always coincide with the controlled panel axes. As an example, Fig. 6.16 shows three positions of the UUV and three positions as displayed relative to the four beacons on the seabed. This representation of information is very close to a real situation and does not need any recoding of information by the operator.

Using this approach and also data from an analysis of bulldozers' D455 (Komatsy Company, Japan) control panels, we offered a new version of control and display panels (see Fig. 6.17). A qualitative analysis of the UUV's control principles allows the ergonomist to consider three possible design alternatives: a stable panel and coordinate mask, a stable mask and rotating panel, and both panel and mask rotating. To select among the three possibilities it was felt necessary to evaluate the complexity of each alternative

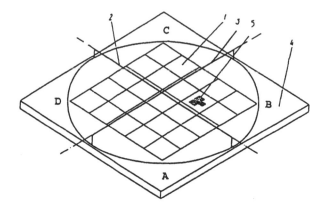

FIG. 6.15. Control board's indication. From Bedny (1987). 1—turning opposite direction display with coordinate lines; 2–3—index of coordinate axes of UUV; 4—control boards' horizontal panel; 5—coded image of UUV; A, B, C, D—points indicating positions of underwater electronic beacons. From Bedny (1987).

by using only analytical procedures. First of all let us discuss the initial conditions. The UUV is moving along a given trajectory (see Fig. 6.17) that includes all typical cases of its movements in given square. The speed is considered constant and equals 2.5 km/hr. The seabed beacons are located in points 1–4, bounding a square of 50 × 50 m² in size. Each quarter of a square has a number equal to the number of a seabed beacon.

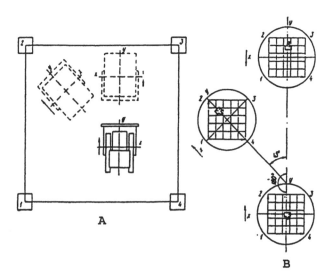

FIG. 6.16. Representation of the UUV on display screen. A—actual situation; B—image of the screen; 1–4 beacons.

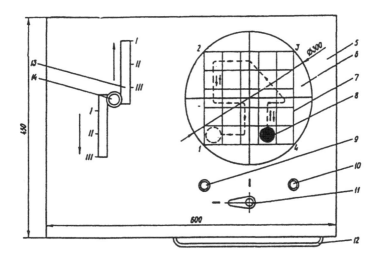

FIG. 6.17. Plan view of the control board. 1–4 + points indicating positions of underwater electronic beacons; 5—control board's horizontal panel; 6—control board's indicator; 7—coordinate lines; 8—coded image of UUV; 9—left turn switch; 10—right turn switch; 11—UUV's engine switch; 12—operator's hand rest; 13—sticks slot; 14—stick.

The UUV operator's task is described next, first using task elements, then in terms of activity elements (mental and motor actions and operations). According to the concept of design described in this chapter, the first stage requires verbal descriptions of the task with different levels of detail.

6.7.1.2 Broader Verbal Description of Task. The task consists of the following steps:

1. Move the UUV from its initial position in the fourth square to the turning point.
2. Stop the UUV and return it to the initial position by reverse gear.
3. Move the UUV to the first turn again and stop it.
4. Turn the UUV right at 90° and move to the second turning point.
5. Turn the UUV left at 135°.
6. Move the UUV along the given path to the third square.
7. Turn the UUV left at 45° and move along the given trajectory at the second quarter.
8. Turn the UUV at 90° so that its front side faces the operator.
9. Move the UUV to the next turning point.
10. Stop the UUV and switch into reverse gear.

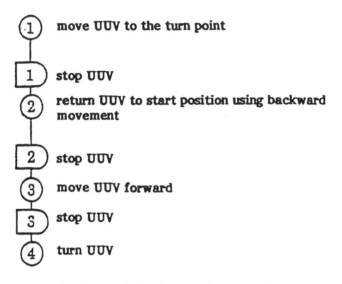

FIG. 6.18. Global symbolic model operative task.

11. Stop the UUV at the place of the previous turn.
12. Move forward again and stop the UUV at the point of the first turn.
13. Turn the UUV at 90° and move to the endpoint.
14. Stop the UUV and turn off the engine.

Based on the verbal description, a symbolic model was developed (see Fig. 6.18). (In Fig. 6.17 only the part of the task that ends by the first right turn at 90° is presented.) Based on the verbal and symbolic descriptions a more detailed verbal description of the task performance was developed.

6.7.1.3 Detailed Verbal Description of the Task (fragment).

1. Turn on the engine with the right hand, turn the shift stick to a moderate speed position;
2. Move hands back to initial position;
3. Wait until the UUV moves to the first turning point;
4. Reverse the shift stick to a moderate backward speed with the left hand;
5. Wait until the UUV gets to the turning point;
6. Turn the UUV around at 90° by pressing the right button;
7. Stop the UUV by pulling the transmission stick with the left hand.

The next step is to provide a detailed symbolic description of task performance.

6.7.1.3 Detailed Symbolic Model of Task. The model is given in Fig. 6.19 (that same part of the task). A comparison between Figs. 6.17 and 6.18 demonstrates that we can describe the expanded and more detailed components of the same task by using the same symbols.

6.7.2 Stage Two: Algorithmic Analysis

This stage involves the algorithmic descriptions of the task. From the algorithmic verbal description (which is not presented here) a logical expression of this task can be developed:

$$O^{\varepsilon}_1 \; O^{\alpha w}_2 \; O^{\varepsilon}_3 \; O^{\alpha w}_4 \; O^{\varepsilon}_5 \; O^{\alpha w}_6 \; O^{\varepsilon}_7 \; O^{\alpha w}_8 \; O^{\varepsilon}_9 \; O^{\varepsilon}_{10} \; O^{\alpha w}_{11} \; O^{\varepsilon}_{12}$$
$$O^{\alpha w}_{13} \; O^{\varepsilon}_{14} \; O^{\alpha w}_{15} \; O^{\varepsilon}_{16} \; O^{\alpha w}_{17} \; O^{\varepsilon}_{18} \; O^{\alpha w}_{19} \; O^{\varepsilon}_{20} \; O^{\alpha w}_{21} \; O^{\varepsilon}_{22} \; O^{\varepsilon}_{23}$$
$$O^{\alpha w}_{24} \; O^{\varepsilon}_{25} \; O^{\alpha w}_{26} \; O^{\varepsilon}_{27} \; O^{\alpha w}_{28} \; O^{\varepsilon}_{29} \; O^{\alpha w}_{30} \; O^{\varepsilon}_{31} \; O^{\alpha w}_{32} \; O^{\varepsilon}_{33}$$
$$O^{\alpha w}_{34} \; O^{\varepsilon}_{35} \; O^{\alpha w}_{36} \; O^{\varepsilon}_{37} \; O^{\alpha w}_{38} \; O^{\varepsilon}_{39} \; O^{\alpha w}_{40} \; O^{\varepsilon}_{41}$$

(6.18)

This scheme needs some explanation. It does not contain logical conditions because the sequence of its elements is determined unambiguously by the trajectory of the UUV. Afferent operators are excluded from the algorithm but an active waiting time was included. The visual afferential activity overlaps the periods of waiting state to a great extent, because they were designated as $O^{\alpha w}$. As the periods of waiting are classified only by the level of concentration of attention, all measures concerned with afferent operators have a value of 0. Hence, all measures of logical conditions also have estimated numerical values equaling 0. At the same time some quantitative measures concerned with estimates of active waiting time are also included.

6.7.3 Stage Three: Time Structure Analysis

Up to this point the ergonomics analysis used task elements as units of analysis. From then on only typical elements of activity were used as units of analysis. This means, for example, instead of using task element "move a stick" the element of activity "move-M" was used, or instead of task element "to take a meter reading," the typical elements of activity "reads," "eye focus-ET," and so forth, were used (the same task elements can be performed by different elements of activity).

Time activity structure can be displayed in the form of a table or a graph. As an example we present a fragment of a table in Table 6.6 (16 members of the algorithm out of a total of 41), which describes the temporal structure of an activity when controlling a UUV by remote control (third version).

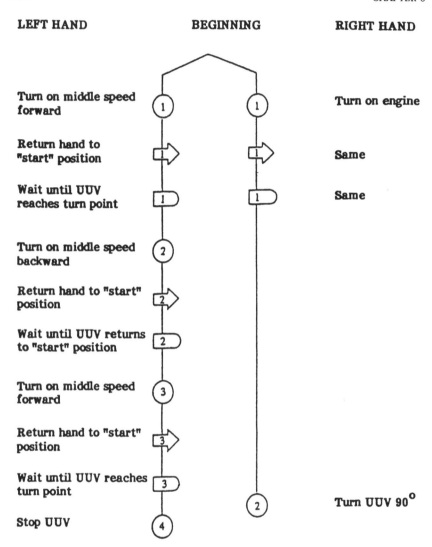

FIG. 6.19. Part of detailed symbolic model of the task.

6.7.4 Stage Four: Quantitative Analysis
of UUV Task Complexity

At this point, all qualitative and temporal data are transformed into quantitative measures that describe the complexity of the UUV's task performance for all three design alternatives (see Table 6.7). Not all measures of complexity showed differences between the three configurations of the UUV. Areas 1–3 (see Fig. 6.17) received contradicting information coming from an indicator because the relation between the operator's body coordinate

TABLE 6.6
Fragment of Time Structure of Task (Third Version)

Algorithm Members	Motor Activity				Mental Activity	
	right hand, Right leg	Time (sec)	Left Hand, Left Leg	Time (sec)	Description	Time (sec)
O^ε_1	R13A + G1A + M2.5A + RL1 + R13E + G5	0.7	R32A + G1A + M2A2.5 + M10B2.5 + RL1 + R43E + G5	1.65	-	-
$O^{\varepsilon w}_2$	-	-	-	-	Active waiting	21.4
O^ε_3	-	-	R43A + G1A + M10A2.5 + M4A2.5 + M10B2.5 + RL1 + R24E + G5	2.0	-	-
$O^{\alpha w}_4$	-	-	-	-	Active waiting	21.4
O^ε_5	-	-	R24A + G1A + M10A2.5 + M4A2.5 + RL1 + R43E + G5	1.86	-	-
$O^{\alpha w}_6$	-	-	-	-	Active waiting	21.4
O^ε_7	R13A + G5 + AP2	0.72	-	-	-	-
$O^{\alpha w}_8$	-	-	-	-	Active waiting	10.0

(Continued)

289

TABLE 6.6
(Continued)

Algorithm Members	Motor Activity				Mental Activity	
	right hand, Right leg	Time (sec)	Left Hand, Left Leg	Time (sec)	Description	Time (sec)
O^{ε}_9	RL2 + R13E + Gl	-	R43A + GlA + M10A2.5 + M2B2.5	1.24	-	-
O^{ε}_{10}	-	-	M2A2.5 + M10B2.5 + RL1 + R43E + G5	1.0	-	-
$O^{\alpha w}_{11}$	-	-	-	-	Active waiting	20.0
O^{ε}_{12}	-	-	R13B + G5 + AP2	0.6	-	-
$O^{\alpha w}_{13}$	-	-	-	-	Active waiting	15.0
O^{ε}_{14}	-	-	RL2 + R13E + G5	0.24	-	-
$O^{\alpha w}_{15}$	-	-	-	-	Active waiting	39.0
O^{ε}_{16}	-	-	R13B + G5 + AP2	0.6	-	-

axes and the UUV's did not match (for the first and second versions). This led to a more complicated decision when pushing the left (#9) or right (#10) buttons (see Fig. 6.17). Consequently, the displacement of a hand toward the left or right button, according to Rule 8 in section 6.3, must be classified as a microelement RB or even RC. In the third case, where contradiction between information coming from the indicator and operator's body coor-

TABLE 6.7
Quantitative Measures of Evaluating Task Complexity for Manipulation of UUV on Underwater Surface

Name of Measure	1st Version	2nd Version	3rd Version
Time for algorithm execution (seconds)	311.8	311.8	309.2
Time for performance of logical conditions	0	0	0
Time for performance of afferent operators	0	0	0
Time for performance of efferent operators	16.8	16.8	14.2
Time for discrimination and recognition of distinctive features of task at threshold level	0	0	0
Proportion of time for logical conditions to time for executing activity	0	0	0
Measure of stereotype of logical processing of information	0	0	0
Measure of changeability of logical processing of information	0	0	0
Measure of stereotype of operator processing of information	0	1	1
Measure of changeability of executive components of work activity	0	0	0
Scale of complexity (except active waiting period)			
(a) algorithm	2	2	1
(b) member of algorithm			
O^ε_9; O^ε_{10}; O^ε_{12}; O^ε_{14}; O^ε_{16}; O^ε_{18}	2; 2; 2; 1; 2; 1	2; 2; 2; 1; 2; 1	1; 1; 1; 1; 1; 1
O^ε_{20}; O^ε_{22}; O^ε_{23}; O^ε_{25}; O^ε_{27}; O^ε_{29}	2; 2; 2; 2; 2; 2	2; 2; 2; 2; 2; 2	1; 1; 1; 1; 1; 1
O^ε_{31}; O^ε_{33}; O^ε_{35}; O^ε_{37}; O^ε_{39}; O^ε_{41}	1; 2; 1; 2; 1; 1	1; 2; 1; 2; 1; 1	1; 1; 1; 1; 1; 1
Proportion of time for repetitive logical components of work activity	0	0	0
Proportion of time for repetitive afferent components of work activity	0	0	0
Proportion of time for repetitive efferent components of work activity	0.13	0.13	0.85

(Continued)

TABLE 6.7
(Continued)

Name of Measure	1st Version	2nd Version	3rd Version
Proportion of time for logical components of work activity depending of information selected from long-term memory rather than exteroceptive information	0	0	0
Proportion of time for retaining current information in working memory	0.83	0.83	0
Proportion of time for discrimination and recognition of distinct features of task approaching threshold characteristics of sense receptors	0	0	0
Proportion of active waiting periods in total work process	0.94	0.94	0.95
Scale of complexity of active waiting periods	2	2	1
Proportion of time for repetitive waiting periods of work activity	0	0.14	0.93
Measure of changeability of waiting periods in work process	0	0	0
Measure of stereotype of waiting periods in work process	1	1	1

Note. From Bedny (1987).

dinate axes is excluded, the element "move a hand toward a given button" is always RA (requires lower concentration of attention). As a result, moving the left or right arms to button 9 and 10 in the first and second versions corresponds to the second or third categories of complexity. However, the same motor actions in the third version always correspond to the first category of complexity. The same occurs when the operator manipulates stick #14 (see Fig. 6.17).

The existence of interfering information regarding the UUV's position meant that in the first and second versions the time to perform motor responses equaled 16.8 seconds as against 14.2 seconds in the third version (see Table 6.7). The complexity of control for the first and second versions equals two, and complexity for the third version equals one. Twelve members of the algorithm had a second category of complexity in the first two versions. In the third version all members of the algorithm had the first category of complexity (see Table 6.7).

Another measure is "proportion of time for repetitive response components of work activity" (Z^{ef}). In all three versions, the operator performed externally identical actions, but in the first two versions, the same actions required the operator to use totally different information. This means that cognitive regulation of the same motor actions was changed because the UUV's axes continuously changed their position in relation to the operator's own axis. Because of this, Z^{ef} equaled .13 for the first two versions. In the third version, because the operator's axis and that of the UUV coincided, the same motor actions were regulated by the same cognitive operations and the proportion of time for repetitive response components of work activity increased to .85. This significantly simplifies the control of the UUV (see Table 6.7).

In the first and second design versions, an operator must remember the position of the UUV's axis toward his or her own. This leads to a significant load on the working memory. An operator does not need this data except when performing O^{ε}_{14}, O^{ε}_{18}, O^{ε}_{31}, O^{ε}_{39}, and O^{ε}_{41} returning his or her hands to the initial position. The performance of O^{ε}_{41} does not demand memory when the left hand returns. This time equals RL + R32E + G5 plus the right hand's time (turn off the engine). These elements of activity take 2.91 seconds. From this, the proportion of time for retaining current information in the working memory for the first and second versions of the UUV, N_{wm}, equals .83. In the third design version, the operator does not need to keep the UUV's relative position in mind. Because of this, N_{wm} is zero. This is evidence of the sharply decreasing workload of the working memory for the last version. Similar calculations were made for other components of work activity (see Table 6.7). These demonstrated that the third design alternative was superior to the other two versions. Data that are present in this chapter demonstrate the systemic-structural analysis of work activity successfully solving complicated design problems by using only analytical procedures.

6.8 CONCLUDING COMMENTS

Chapters 1 through 3 described a complex theory of activity and a model of self-regulation. The present chapter describes the theoretical concept of analytical design that should be used to perform the behavioral design of a system. It is natural to ask what the connection is between the theory of activity and the suggested concept. The decomposition of holistic activity into separate hierarchically organized units during task performance, discovering their relationship, and discovering the goal, motives, and strategies of activity performance are all among the essential elements in activity theory.

In opposition, cognitive analysis used in Western engineering psychology suggests studying only psychic processes. This approach does not suggest dividing the task into separate units of analysis (e.g., see Wickens, 1992). Similarly, special units of analysis are not used in ergonomic study (see Kirwan & Ainsworth, 1992). As follows from chapter 4, the anthropocentric approach can be considered one of the more important approaches in the design of work activity. Chapter 6 has attempted to present this approach in a practical context. It makes use of various methods of cognitive psychology, particularly those that have branched off into task analysis. In addition to these methods, procedures derived from the theory of activity have also been used.

According to the activity approach, activity is not the sum of reactions to individual stimuli, but rather a system organized in a particular way and having a complicated structure. The structure of an activity has a logical and spatial-temporal organization of actions and operations that are performed by the operator. An analysis of this structure of an activity suggests various stages and levels of analysis on the basis of which an activity can be broken down into such standard units of analysis as actions and operations. The principle of systemic-structural approach and the requirements of decomposition of an activity into separate interconnected units are the conceptual foundations for the four analytical stages of design.

The systemic-structural organization of cognitive operations, their logical sequence, and mutual interaction depend on the goal of the activity to a great extent. The goal is the more important factor in determining how information is selected. Without understanding the goal, it is impossible to develop an algorithm of performance. Does this mean that specialists continually operate with such notions as goal, motive, and so on, during the design process? Obviously, it does not. For example, while developing the time structure of an activity, specialists pay attention first to mental and motor operations. Each operation has its own subgoal of which the person may be unaware. Instead, in order to analyze motives, specialists pay attention to the significance of actions and operations and their difficulty for the performer. From this, it does not follow that the goal and motive are ignored in the design process.

A flexible transition form one stage and level of analysis to another permits designers to continually reorient their attention from one component of the activity to another. At one stage of design specialists pay attention to the goal and motive, whereas at other stages they focus on the activity's operational structure. It is impossible to comprehend all components of the analyzed system at the same stage of design. Concentrating on the various aspects of an activity, which are important at a particular moment and are not connected with the goal and motive, does not deny their importance in the design process.

The complexity and hierarchical organization of an activity are also important factors that cause specialists to use the stages and levels of analysis in the design process. Another important aspect of design is to take the strategies of performance into consideration. The strategies of activity during task performance include the activity's structure that is developed within the framework of boundaries under external conditions and the dynamic internal criteria of evaluating task performance. Any changes in the strategy of performance immediately result in changes in the activity's structure. This means that the structure should always be described after taking the operator's strategies into consideration. Because of this, presenting the structure as a linear sequence of steps, which include perceiving information, interpreting information, forming hypotheses, making decisions, is too simplistic and sometimes incorrect. A person can change the sequence of these steps. For example, one can formulate a hypothesis and then formulate the goal of perceiving information, evaluate the possibility of achieving this goal, and gather information. As can be seen there is a complicated interrelationship between the goal, motive, and strategies.

The structure of an activity includes an algorithm of activity that presents a logical organization of actions and operations that lead to achieving the goal of the activity. An analysis of the structure of an activity suggests the following basic stages:

1. Determine the number of tasks that must be performed by the operator.

2. Analyze the task on a technological level when the typical elements of the task are used as units of analysis (decision-action diagram, timeline analysis, etc.).

3. Analyze the task on a psychological level when psychic function and the typical elements of the activity are used as units of analysis. This includes (a) cognitive analysis of a performed activity when human information processing focused upon decomposition of an activity into sensation, perception, memory, thinking, and so forth, and (b) activity approach when the goal and motive of an activity is analyzed followed by a study of the activity's structure when its typical elements are used as the units of analysis. The relationship between stages 2 and 3 is not rigidly sequential.

The central point in studying the structure of an activity is the decomposition of the activity into separate actions and operations, and during algorithmic analysis, dividing the task into operators and logical conditions. Dividing the complicated holistic system of activity into separate units should create the possibility of a reverse transition from the individual units to the holistic system (recombination). These are the basic requirements for the units of analysis used in the theory of activity. This fundamental principle was taken into consideration when developing the concepts addressed in this chapter. The basic units of analyst were task, action, operation, and member of algorithm. The activity was broken down into these units and then the reverse process occurred, which provides the integration of these units into the holistic time structure of the activity. This can be discovered in the description of the activity's time structure. The holistic time structure of work activity was presented as a logically organized system of operators and logical conditions that, in turn, consist of actions, and those of operations (motion or mental). This can be accomplished only after qualitative analysis, which includes a cognitive analysis of task performance (analysis of psychic processes, perceiving information, interpreting information, making decisions, etc.).

It should be noted that the idea of breaking an activity down into separate elements, which can be used as units of analysis, was first suggested by scientists at the beginning of the century who were working in a totally different field. This included Vygotsky (1956) in the former Soviet Union, and Gilbreth (1911) in the United States. In the first case, this idea was an important source of developing the theory of activity. In the second case, it was the basis of developing "time and motion study."

In order to extract separate actions, it is necessary to determine the subgoals of particular actions and the final goal of the task. The goal of the action integrates separate operations in a holistic subsystem, which is called "action," and the final goal of the task integrates separate actions into the holistic structure of an activity during task performance. The goal is the future desired result. However, the goal is only that result connected with the general motive of an activity. How can the motivational aspects be taken into consideration during the design process? The motivational aspects of work activity are considered at the first qualitative stage analysis. At the following stage, the gathered data are transformed into such indices as significance and difficulty of actions and task performance. It follows from the model of self-regulation of an activity and the model of attention, that these indices are intimately connected with motivation and level of concentration of attention on the performed activity and its elements. This permits one to conclude that a quantitative evaluation of task complexity takes not only the cognitive but also the motivational components of activity into consideration. A cognitive analysis of task performance is followed by extracting such units of analysis as actions and operations, which provide a description not only of psychic processes but also of the structure of activity, which can be changed depending on the physical configuration of the equipment and method of task performance. As can be seen, the Western task analysis techniques suggest decomposing behavior into stimulus and response categories and categories of human functions (e.g., perception, information interpretation, hypothesis development, decision making). In addition, the activity approach uses such units of analysis as actions, operations, and function blocks. This permits a more efficient use of the principles of cognitive psychology in the design process.

Another important component of the suggested design concept is evaluating the complexity of task performance. Because an activity is a process, motor and cognitive operations and their duration are used as units of measurement. Not only is the duration of different operations taken into consideration, but their qualitative characteristics, probability of appearing in the structure of the activity, ability to be performed sequentially and simultaneously, and logical sequence in the structure of activity are as well. Quantative evaluation of complexity utilizes a multiple system of measures that can be compared with each other.

Russian psychologists believe that the principles of cognitive psychology have not been properly applied to task analysis. Transforming the principles of cognitive psychology into task analysis is very difficult. To do this, a more precise language to describe the task in psychological terms is necessary. Using the customary psychological categories of perception, thinking, decision making, and so on, is insufficient. Russian psychologists believe that applying the units of analysis developed in the theory of activity permits specialists to transform cognitive psychology principles into task analysis more easily. They feel that the systemic-structural concept of design presented in chapter 6 is an integration of the Russian activity approach with that of Western scientists.

Learning and Training

7.1 OVERVIEW AND INTRODUCTION

The development of an efficient training process is required before determining training needs, and the most important component of determining training needs is task analysis (Lendy, 1989; Patrick, 1992), which was discussed in chapter 6. Based on this, we consider the problem that directly connects the theory of learning and the development of the training process.

The system of teaching in the former Soviet Union differed significantly from that in the United States and the West generally. Any learning concept, if it was to have government support, had to be practically applicable and could not contradict Marxist ideology. Russian theorists have criticized the Western approach to learning as being based on behavioristic principles derived from animal studies.

Vocational training for operators in the USSR had both theoretical and practical elements. For those who controlled complex automated systems, a different, problem-oriented system was used. In all cases training was intimately related to work activity theory and goal formation. Although a very primitive form of learning is based on association principles (sensory and motor learning), the goal orientation of work activity theory makes it impossible to reduce learning to mere association. Cognitive learning based on goal-directed actions represents a higher level; the more advanced form of cognitive learning (centered on abstract relationships) is termed *intellectual learning*. The lower form of sensory, motor, and sensorimotor learning can become cognitive when consciousness and goal are involved. Because of this, associative and cognitive learning are interconnected.

Learning can be both incidental (unpremeditated) and goal directed. In the latter, what one learns depends on the goal the learner has. External activity is not required for learning. There are two interconnected components of activity: *executive* (directed to transformation of the external environment) and *gnostic* or *explorative* (gathering information about that environment). Problem-solving ("discovery") learning can be contrasted with algorithmic learning (a general method of developing heuristics for particular classes of task). The first is part of introductory learning; the second requires task performance.

Different types of motivation (intrinsic and extrinsic) affect the learning process. The student performs various functions from relatively passive perception of information provided by the teacher to independent search for and use of information and task performance under the teacher's supervision.

From the Russian standpoint the major elements of teaching are learning tasks, actions, goals, motives, feedback, self-control, algorithm, and strategies. Concepts such as stimulus, response, reinforcement, and reward are seldom employed. A basic learning principle is the attempt to individualize (adjusting the speed and method of information presentation to the student's ability to acquire it based on his or her prior achievement and personal features). Individualization must reflect strategies of goal achievement for the individual.

There are various types of knowledge. Declarative knowledge is in the form of images, concepts, and propositions and may be subsumed under meta-knowledge. Procedural knowledge transforms information in long-term memory. Practice transforms knowledge into skill. Declarative knowledge is very often at a higher level of generalization than procedural knowledge. There are four stages of knowledge acquisition: familiarity (recognition), copy (reproduction), productive (knowledge use), and transformation (solution of new problems). Russians also presume two levels of skill. The possession of a high level of skills is the capability to use the first level of skills and knowledge in a practical situation.

Practice produces changes in activity structure, related to integration and control of the action. This is also associated with the individual's selection of performance strategies. The learning curves for individual components of the skills should be combined with learning curves for holistic skills being learned. Russian theorists utilize a concept termed *deautomatization* in contrast to the automatization that occurs when learning a skill; the former is the transfer of action control from unconscious physiological mechanisms to central conscious ones, the reverse of what occurs in skill acquisition. Skill acquisition is reflected in the stabilization (reduced variability) of performance. This is measured by a coefficient of dispersion.

Gal'perin's (1959, 1966) theory of learning based on activity principles is described in this chapter. It emphasizes the unity of inner psychic states and behavior, and the internalization of material actions. Any action includes orienting, executive, and control elements. The theory assumes that action has certain basic dimensions: shape (nature), generalization, completeness, and degree of mastery. The latter two relate to automatization in skill acquisition.

7.2 THE ACTIVITY VIEWS OF LEARNING

The Russian theory of learning, which is the basis of its teaching methods, differs in some respects from that of the West. Some difficulties arise because of differences in terminology. Moreover, there is a greater interaction between the theory and practice. Any concept of learning, if it is to have the government's support, must appear to be practically applicable. In the past the concept could not contradict Marxist ideology because the theory of learning was considered a practical application of psychology not only for teaching, but also for ideological conditioning. What the situation is at the present time is difficult to say.

The combination of theory and ideology is of course foreign to the West. Moreover, some Western psychologists (e.g., Goldstein, 1974) point out that there are wide differences between learning theory and practical teaching. A major difference in orientation arises also because Russian psychologists believe that associative learning theory makes use of behavioral concepts based on the same mechanisms that drive animals. Among these critics are Leont'ev (1977), Rubinshtein (1959), Talizina (1975), and Yaroshevsky (1985). Their criticisms of behaviorism are founded on the presumed failure of Western theory to deal with fundamentals, such as motive, goal, language, and conscience, that are essentials in the Russian concept of activity.

One can evaluate the behaviorist paradigm against practical situations. Ormrod (1990) provided the following example of the behavioristic approach: "Many newer cars sound a loud buzzer if the keys are still in the ignition, when the driver's door is opened; removal of the keys from the ignition is negatively reinforced because the buzzer stops" (p. 53). What would happen if, instead of a buzzer, one used beautiful music? Maybe in this case the driver would close the door and "key-leaving" behavior would be maintained by the positive reinforcement of the beautiful music. Taking the key would, from the behaviorist approach, entail self-punishment.

Whether one uses the buzzer or music, the issue is having the driver take the key. What Russian theorists believe is that behaviorism fails to understand that the essential element in this concrete action is the goal of removing the key from the ignition, regardless of the means of doing so. The buzzer only recalls what the driver has to do. But in behavioristic learning theory conscious goals do not exist.

Another example is found in Wade and Tavris (1990) in the following question and answers:

A prisoner is released from jail early because of good behavior. This is an
example of:
a. Positive Reinforcement
b. Classical Conditioning
c. *Negative Reinforcement*
d. Punishment (p. 185)

For these authors the correct answer is negative reinforcement of improper
behavior. But prisoners who are being released not only remove the negative
stimuli of prison, but at the same time have an opportunity to reenter society,
all of which are very powerful positive incentives. In such cases behaviorist
theories fail to capture the causal dynamics of the practical situation.

Any theoretical concept of learning, in order to demonstrate validity,
should demonstrate its ability to be used for teaching and instruction. Under
the term *teaching*, we understand that the activity of the teacher and the
student is integrated in a holistic system and provides the formation of a
new experience in the student.

With regard to vocational teaching (training of operators and blue-collar
workers), this process can be divided into theoretical (lecture) and practical
(task performance) components. The progression is from the general to the
specific, from the student's relatively passive reception of information to his
or her more active participation in the specifics of task performance. This
is a general pattern found in both the East and West.

Russian scientists highlight two notions: the training system and method
of training. The training system is understood as the ·method of preparing
the content of the training process. This includes organizing the material
into separate units, as well as grouping and creating a sequence through
which the student acquires it. The systems approach is very important in
developing the training system, emphasizing the significance of analyzing
the interrelationships among knowledge and skills and the method of organizing them. There are two major vocational training systems in Russia.
One is the "operational-complex" system of vocational training. The other
is the "problem-analytical" system (Batishev, 1977). The first system was
developed at the beginning of the century and was widely used throughout
the world. This system was well known as the Russian system of vocational
training. The major notions of this systems are operations and complex. At
the first stage of training, the students acquire separate operations that are
associated with skills that are involved in the performance of separate elements of production operations. For example, the student learns how to use
a particular tool. At the final stage of training, the student performs a holistic
production operation. She or he also studies how to perform different operations in changeable conditions.

The problem-analytical system (Batishev, 1977) is used in the education
of operators who regulate process control facilities (e.g., chemical plants)

and other automated manufacturing plants. In this system the major training elements are not production operations but rather task problems. Because the trainee's function is not to control operations (the computer does this), but to monitor the system's adherence to parametric limits or to select alternative modes of operation, and to diagnose what has gone wrong when it inevitably does, the analysis of the problem inherent in the system malfunction is what the student is required to learn. The trainee studies this problem first theoretically and then in actual practice, usually in a simulator. Each problem is divided into smaller units called "situations" or subproblems that, when combined, form the total problem. The trainee studies first the individual situation and then the problem as a whole.

Russian psychologists emphasize that learning must include goal formation, consciousness, and speech. They believe that Western behavioristic learning theory is oriented largely around the formation of associations between stimuli and responses; and they believe associative learning (characteristic, they say, of animal learning) can produce only reactive behavior. They believe that such reactive behavior prevents the individual from developing the flexibility needed for goal-directed activity in response to differing situations. The human does make use of associative learning at a primitive level (involving trial and error), primarily for sensory learning (the ability to differentiate stimuli and acquire perceptual skills), motor learning, and sensorimotor learning.

According to Russian scientists (Asratyan, 1966) and some Americans (e.g., E. J. Green, 1962), operant conditioning utilizes the same mechanisms discovered by Pavlov. Classical and operant conditioning are both associationist. Russian scientists feel that cognitive (goal-directed, conscious) learning plays the major role in human learning and that associative learning mechanisms have only an auxiliary character.

Speech is, of course, involved in cognitive learning. Words become controlled and regulate learning in the same way as the external world controls and regulates performance.

All learning is multilevel. Learning at lower levels (using the mechanisms of classical and operant conditioning) can be transformed into higher level (cognitive) learning because of the involvement of conscious activity. Awareness that a goal has been or can be attained can attach itself to reinforcement of sensory, motor, and sensorimotor levels of learning. Russian theorists believe that there is a conscious component in all learning; and to the extent that this consciousness can be reinforced, the lower level of learning assumes some of the attributes of the higher. There are at least two kinds of cognitive levels of learning. If the initial kind of cognitive learning involves primarily consciousness of goal attainment (or failure), a more complex type (called intellectual; Itelson, 1972) involves logical relationships (particular-general, concrete-abstract, class-subclass, and functional relationships (goal-tools, cause-consequence, quality-quantity, and action-result). In this case we are not talking about sensing the relationship between particular objects and the actions one takes in relationship to these, but about an understanding of relationships among objects, events, and phenomena in general.

As was shown by Vygotsky (1962), scientific concepts reflect the essential relationship and structure of different things and phenomena that often cannot be directly comprehended. A special system of symbols for designating these things and phenomena must be employed. This is language and syntax—the method of connecting words into a meaningful sentence. According to Vygotsky, words and language enable one to substitute symbols for direct sensory experience. The sociohistorical theory of mental development (Vygotsky, 1960) suggests that individuals learn through interaction with other members of society. The language with which the person operates reflects the experience and knowledge that was developed by society over a long period of time. Acquiring the concepts designated by the various words reflects the knowledge accumulated by humankind over its history.

By operating on the environment and objects in the environment, a person discovers essential but hidden associations and relationships. These operations can be performed not only externally, but also internally—on images and concepts. This may explain in part the development of superordinate concepts within the individual. Mental activity is like task performance in the external environment and may, over time, substitute for the latter. This concept is important for Russian theories of learning.

The distinction of the associative and cognitive levels of learning is somewhat artificial, because in reality, these two levels of learning are closely associated. In early childhood associative learning undoubtedly plays a more important role. But with adulthood the reverse occurs.

Russian theorists believe that the cognitive level of learning as conceptualized in Russian psychology differs from that of the West because they believe this level of learning is based on activity principles involving not only cognitive but also behavioral (physical) and motivational aspects. Moreover, they do not consider associative and cognitive learning as distinct entities but as interconnected levels within the activity. Learning performed on the activity level (e.g., involving consciousness and goal formation and attainment) includes also the associative level of learning. This can be seen as a hierarchically organized process that involves what has been termed *incidental learning*. For example, a person performing a task accidentally learns how to perform a different set of actions. In this case, the person acquires experience without deliberate learning efforts. Although this unpremeditated learning may, in some cases, be efficient, in most cases, premeditated (goal-directed) learning is more efficient.

For example, P. I. Zinchenko (1961) studied the efficiency of memorization under premeditated and unpremeditated conditions. Subjects had to classify different cards (varying pictures and numbers). The subjects were instructed to organize the cards either by pictures or by numbers. Those who were instructed to organize by pictures memorized the numbered cards very poorly, although their performance with the picture cards was much better. The reverse was true of those who organized by number. Zinchenko concluded that a person can easily memorize those things that are included in his or her goal. This experiment also demonstrated that memorization was dependent not only on the individual features of the stimulus, but on the way the material was used. Memorization is influenced by motives, goals, and the method of performing the activity.

Earlier, two components of activity were postulated: executive components directed to the transformation of the external environment, and gnostic or explorative components of activity, the goal of which is to gather or transform and comprehend information about the external world.

Gnostic or explorative components of activity are especially important in learning. It can be divided into external, practical explorative activity, which is interconnected with internal, mental activity, for example, assembling and disassembling of different objects, changing their position and relationship, and studying their features. This kind of activity can also be internal or mental where a person operates with words, statements, symbols, and images. In order to make it easier to perform this activity, a person can use external symbols or tools such as instructions, graphics, or equations.

Learning in Russia is conceptualized as having a specific structure: instructor and student goals; a training task; certain required student actions, goals, motives, feedback, and body of information to be imparted; and instructor control and evaluation (Davidov, 1972). Notions such as stimulus, response, reinforcement, and reward are seldom employed. The goal of the training situation is for the student to develop a general method of solving a class of problems or performing individual tasks. The instructor is supposed to provide conditions that will permit learning. The trainee should possess a precisely formulated goal of activity and be motivated for its achievement.

Particular attention is paid to developing the ability of the student to independently formulate and solve different problems. This is why multiple-choice questions almost are not used in the Russian school system. Some information need only be recognized or reproduced by the students, whereas other information should be actively used by the student in practice. Special attention is paid to mental development, which is considered not as simply absorbing knowledge, but as the ability to acquire knowledge independently, the ability to apply and transfer it to a different situation, and generate new knowledge based on existing knowledge. This is why much effort was devoted to developing "discovery" learning in the education process, and particularly in vocational training (see, e.g., Kudryavsev, 1975; Matyushkin, 1972).

Discovery learning is that in which the student is confronted with problems that require a search for unknown laws, methods of actions, and applicable rules. In contrast to the usual problems solved by students, in which all the information needed to solve the problem is provided, in discovery learning, the problem task involves a discrepancy between task conditions, the student's knowledge, and problem requirements. The student is required to resolve the discrepancy by discovering new information and applying it. The approach was strongly influenced by Bruner (1960).

In contrast to discovery learning, Landa (1974, 1976) developed a theory of learning that utilizes an algorithmic method. He created a general method of developing algorithms and heuristics for solving a particular class of tasks. Learning is based on applying general algorithms and heuristics directed to shaping general strategies of the thinking process. However, the student must also acquire specific skills to perform routine tasks. Sometimes there is a contradiction between the goal of developing a general method of thinking (which requires one set of teaching methods) and the goal of developing more specialized professional skills (which requires a different set of methods). In this case, a combination of teaching methods is necessary. This combination is called "developing" teaching (Kalmikova, 1979).

Whatever training method is used, there are at least two stages: introductory (theoretical) followed by task performance. The discovery method is used in the introductory stage where the student, under supervision, discovers different methods of task performance, develops these, and selects the preferred one. After these methods are learned, the student must follow them strictly (Bedny, 1981). Obviously, feedback is essential to the learning process. The final stage of learning involves evaluation of the outcome of task performance, either by the instructor or by some machine measurement.

The student has both intrinsic and extrinsic motivation. Interest in the training process, curiosity, and the desire to acquire a special skill are examples of intrinsic motivation. Reward, punishment, and competition are examples of extrinsic motivation.

Students perform various functions in the learning process (Itelson, 1972): relatively passive perception of information, active independent search for and use of information, and task performance. In the first case, the student performs actions based on imitation, reproduction, practice, rules, and standards. The teacher uses lectures, presentations of prepared information, explanations, and demonstrations. The second function involves self-instruction. In this situation, the goals and interests promoted by the students are very important. In the third case, the students are considered to be the object of pedagogical guidance. The instructor organizes the situation such that it shapes the required motivation of the students. The instructor formulates different problems and tasks with the student, discusses how they can be performed, develops a plan of action, and emphasizes the significance of information.

Any kind of teaching suggests active involvement on the part of the student. But this need not be overt as when mathematical problems are solved. External behavior is not always involved in the learning process. Leont'ev (1977), Gal'perin (1955), and Piaget (1952) demonstrated that thinking involves various mental actions: analysis, synthesis, discrimination, abstraction, generalization, orientation, selection, classification, coding, and recoding.

An analysis of the preceding material demonstrates that the major notions that are used in the study of learning are learning task, actions, goals, motives, feedback, self-control, algorithm, heuristic, strategies, and the like. Such notions as stimulus, response, reinforcement, and reward are seldom used. The speed at which a skill is acquired changes over time, tending to reach an asymptote.

Kojin, Tsukerman, and Pakhomov (1961) studied how operators acquired the skills of reading different instruments. After the first session the trainees had improved their performance only slightly. However, at the start of the second session they immediately began to perform better than at the conclusion of the previous session. During the following lessons, the same thing happened. Performance significantly improved from one lesson to another rather than within the lesson.

Kojin et al. (1961) introduced the notions of accumulation and realization processes. In accumulation, performance cannot be significantly improved. The accumulation processes should be transferred into realization processes (into performance). These data corresponded to American cognitive psychology. According to Lindsay and Norman (1992) and Norman (1976), the structure of the long-term memory has a dynamic and integrative nature. The system continually modifies itself through active interaction with the environment. Our understanding of new data continues to be expanded even though the data themselves will never be directly encountered again.

Our knowledge of the same data continually changes. Memory schemes can be developed and tested not only consciously, but also unconsciously. This continued evaluation of the stored knowledge in the memory influences the process of acquiring new information and skills. This is well known in gymnastics during the acquisition of motor skills. Sometimes a gymnast cannot perform a particular action for a long time. However, after a period of not attempting to perform it, the gymnast begins to perform this action almost immediately upon his or her return to it. This accumulative stage can be compared with what we know in cognitive psychology as the process of elaboration which, after a period of time, can influence performance.

One of the basic principles of teaching is the principle of individualization, which is also studied by American scientists. For example, the speed of presenting information to the student can be changed depending on the individual's achievement. In the former Soviet Union, a tremendous effort was devoted to classifying people according to their individual features. This study was begun by Pavlov (1927) and followed by Teplov (1961) and Nebylitsin (1965). Merlin (1973) and Klimov (1969) introduced the notion of individual style of activity, which is a specific strategy of activity determined by the nature of the person. The individual style can be conscious or unconscious. People may use the same strategies determined by different individual features of personality (Bedny, 1976; Bedny & Voskoboynikov, 1975). Practically speaking, however, it is difficult to discover these individual features.

7.3 KNOWLEDGE AND SKILLS

During teaching, the student acquires knowledge about phenomena in his or her field of specialization, and develops skills to perform various tasks. Knowledge and skills interact with each other (Platonov, 1970; Rasmussen, 1986). Knowledge is information encoded in the long-term memory in various ways, for example, verbally, as images, propositions, and productions (Klatsky, 1975; Velichkovsky, 1982). Verbal information is defined as a form of knowledge that represents an object as a set of its characteristic features (Landa, 1984).

People memorize the meaning of what they have perceived. According to J. R. Anderson (1985), Gagné (1985), and others, meanings are stored as propositions. Propositions reflect the relationship between different objects and are units of knowledge considered as a separate statement or assertion; they can be judged either true or false. Psychologists call these declarative knowledge, examples of which are definitions, theorems, laws, and rules. Concepts and propositions are interrelated but not the same. A student can acquire the correct concept of an object; for example, can list its characteristic features, but may not be able to give a correct definition of it. Landa (1984) categorized declarative knowledge as in Fig 7.1.

Knowledge may also reflect or subsume other knowledge and thus we have meta-knowledge. Individual items of declarative knowledge, without being organized hierarchically as a set of meta-knowledge, are insufficient and cannot usually be employed in practice.

Procedural knowledge is also extremely important, specifically in vocational training. This type of knowledge is associated with mental operations that transform information (images, concepts, and propositions) in long-term

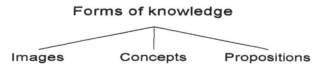

Forms of knowledge

Images   **Concepts**   **Propositions**

FIG. 7.1. Declarative knowledge according to Landa.

memory. Procedural knowledge is also needed in performing a physical action or operation.

Information in memory about how to act in various situations is known as production (J. R. Anderson, 1985). Activity theory and American cognitive psychology categorize knowledge as in Fig. 7.2 (Landa, 1984).

As one can see from Fig. 7.2, declarative and procedural knowledge are independent; if one has declarative knowledge, it does not mean that one has knowledge about procedures to be performed in a task to achieve a goal. The relationship between the two types of knowledge must be learned.

Procedural knowledge, which is required to achieve a goal, is not sufficient for a trainee to perform this procedure. One can see this in gymnastics, where the gymnast knows the technique, but cannot perform it without extensive practice. Practice transforms a knowledge into a skill.

Because information in memory is structured, the efficient organization of learning material by the instructor and the student becomes extremely important. Certain organizations of training material are more or less efficient than others. The organizing principle must be logical and reflect a causal-consequence relationship. The more organized the material, as in a lecture, the more the student will acquire from the lecture. Isolated items of information are more likely to be ignored by the student or later forgotten.

A. M. Novikov (1986) introduced the following classification of declarative and procedural knowledge: (a) knowledge about the object of an action (declarative knowledge): directed by sensed knowledge about objects (based on perceiving and imagining of objects); phenomenological (descriptive) knowledge about equipment, facilities, and operations; knowledge about the general principles of arrangement and operations of individual objects; knowledge about the natural-scientific basis and principles of organization and operation of technical objects; and (b) knowledge about actions (procedural knowledge): directly sensed knowledge about the performance of an action (based on perceiving and imagining of objects); phenomenological (descriptive) knowledge about actions with objects; knowledge about general rules (algorithms) of action with specific objects under different conditions; knowledge about the scientific basis and general principles of performing technological processes and method of organizing labor.

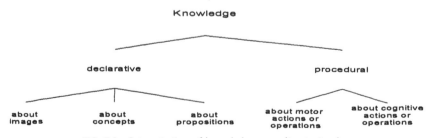

FIG. 7.2. Categorization of knowledge according to Landa.

Depending on the purpose of the teaching process, the level of generalization in the student's knowledge can be different. The higher the level of generalization, the easier it is to transfer knowledge and use it under different conditions. The level of generalization of declarative and procedural knowledge can be different, as declarative knowledge very often is at a higher level of generalization than procedural knowledge. For example, in vocational schools, theoretical teaching is at a much higher level than practical training.

Generalization and transfer require the extraction of critical, salient features of an object or situation from those that are nonessential, followed by determination of how these essential and nonessential features vary from one situation to another.

Bespal'ko (1977) classified the levels of knowledge acquisition as follows:

1. The first level is knowledge familiarity and is associated with the recognition of various objects, phenomena, processes, and features, for example, general knowledge about the organization of a company.
2. The second level is knowledge copy; the student can exactly reproduce tasks she or he performed before and use information in the same way she or he did before.
3. The third level is productive knowledge; the student can use his or her fund of information in different situations. For example, the operator can perform a diagnostic task or repair equipment.
4. The fourth level is knowledge transformation, associated with the creative process. The student can use information to solve new problems by developing new goals and methods of work.

It is assumed that the instructor is aware of the individual level to which his or her teaching is addressed.

One of the main purposes of teaching is to develop the student's capability to think. Thinking is not knowledge, but what one does with that knowledge. Landa (1982) defined thinking as a structure organized as a set of mental operations carried out in order to solve problems and make decisions. The principles for organizing these mental operations may vary, which influences the character of the thought process. One can have a large repertoire of images, concepts, and propositions in one's mind, but only a small repertoire of mental actions, and cannot therefore know which mental actions to apply to which knowledge.

Expert performers are much more aware than novices of the knowledge they use in solving problems and the physical actions they execute. This is why instructors who teach novices teach mostly knowledge and physical actions but not mental operations. Teaching a student to think in order to solve problems and make decisions requires a special method of teaching, which we discuss later. The degree of generality of the thinking process depends on how wide the class of problems is to which method can be applied. Depending on whether the method enables one to solve an individual specific problem or a class of problems, the method of thinking will be more or less specific or general. The notion of a dichotomous general versus specific method of thinking is relatives rather than absolutes. For example, a person may successfully solve a wide range of mathematical problems. In this case, we can tell that this person has a general method of thinking in mathematics but does not possess it with regard to nonmathematical problems.

Other important units of acquired activity are skills. In the Russian language, two words exist: *naviky* and *umeniya*. In English, they both translate to mean "skills" even though in Russian they indicate a first and second level of skills. The first level is actions (motor, sensory, and mental) developed for a particular level of automatization. They can be performed more easily, quickly, and economically with a high level of efficiency. One of the important characteristics of these skills is the development of the motor or mental structure that integrates individual motions and operations or separate actions into a holistic system or pattern. These skills can be efficiently used in a particular situation. For example, many actions connected with driving a car by an experienced driver are performed quickly and efficiently with a low level of awareness and concentration of attention. However, if something happens, the awareness of these actions as well as the level of attention can be sharply increased.

The second level of skills has much in common with the first. These kinds of skills consist of an individual's ability to organize knowledge and the first level of skills into a system, and efficiently use it for performing a particular class of tasks or solving a particular class of problems.

This is also related to the novice–expert continuum. The novice performs individual skills that are not very well integrated. The expert integrates the individual skills into a holistic activity that automatizes and submerges the individual skills and can efficiently use them in different situations.

From the standpoint of cognitive psychology, there are two ways to process information, conscious control and automatic. Conscious control processing requires individual attention, conscious thought, and mental effort. Automatic processing occurs with little effort and without conscious attention. The control process becomes automatic through repetition and practice. During the acquisition of skills, the student pays less attention to how to control his or her actions and pays greater attention to the goal and result of his or her actions. However, the student can transfer his or her attention from the result or product to his or her own actions if some deviation from the required standard performance occurs.

During this process, the student performs a meaningful and perceptive evaluation of his or her actions. A meaningful evaluation of his or her own performance is associated with conscious analysis and evaluation of actions and their result. The individual's perceptual evaluation of his or her achieved skill level depends in part on perception of such physiological phenomena as muscle tension and the subjective feeling of mental effort and emotional state. Perceptual control of motor skills can be divided into external or exteroreceptive and internal proprioreceptive and vestibular. Perceptual control passes from external to internal motor mechanisms. Petrovsky (1986) and Platonov (1970) suggested skill acquisition phases as in Table 7.1. From the psychological point of view, skills can be divided into three basic groups: sensory (visual, auditory, tactile, etc.), mental (calculation, reading, drawing, solving production problems, etc.), and motor (skills connected when performing motor actions). This classification has conventional features because

TABLE 7.1
The Phase of Developing Skills

Stage of Developing Skills	Character of Skills	Purpose of Skills	Specificity of Performed Actions
Familiarization	Grasping and conceptualizing actions	Comprehending method of performance	Distinct understanding of goal but vague understanding of method of performance; rough errors during performance
Preparation (analytical)	Conscious unskilled performance	Acquiring separate elements of actions; analyzing methods of performing them	Distinct understanding of performing action; imprecise and unstable performance; many unnecessary motions or mental operations; high level of concentration; narrow field of concentration; poor control
Standardization (synthetical)	Automatic performance of actions	Combination and integration of different components of action	Increasing quality of performing different motions and mental operations; their integration; elimination of unnecessary components of action; transfer of attention from action to result; improvement in control of performance; transfer to internal mental or kinesthetical control
Variation (situative)	Flexible adaptation to situation	Acquiring conscious self-regulation of action	Flexible, expedient performance of action; control based on special sensory synthesis; intellectual synthesis (intuition)

Note. From Platonov (1970).

all skills involve cognitive, sensory, and motor processes. When one tries to relate skills to a particular group, one pays attention to what component has a major importance.

Gil'buhk (1979) suggested a more detailed system of classification. Depending on the psychological content of actions involved in the corresponding skill, he classified skills as sensory, perceptual, mnemonic, thinking, imaginative, and motor.

In an earlier chapter, it was pointed out that cues can be derived from displays or from sensory impressions. It is possible, therefore, to categorize skills in responding to these cues as instrumental or noninstrumental. Any production operation, task, or problem can be performed in a different way, which suggests involving and using different knowledge and skills. In some cases, the officially required and actually used skills can be different. In other cases, the method actually used by a worker can involve more efficient skills than the officially required ones, which is why, after special study and analysis of the actually performed method, it can be recommended as the officially required procedure. For example, Galaktionov (1978) and Manwaring and Wood (1985) discovered that operators, while performing their jobs, acquired new ways of performing tasks.

Klimov (1969) discovered that weavers used an individual style of activity or different strategies of task performance. These strategies very often depended on the individual features of the weaver. For example, the weavers who possessed low mobility of the neural system (mobility of neural processes measured by the speed of alternation of conditioned reflex) used different preliminary adaptive actions to compensate for their actions. On the other hand, the quicker weavers did not use these actions and performed according to their ability to react quickly. Both groups worked efficiently and were unaware of their different strategies. Some strategies were applicable to a number of workers whereas others were more individualized.

In 1950, in the former Soviet Union, Kovalev's (1950) method was widely used. He studied how workers performed production operations, after which he attempted to combine different methods into one. However, in some cases, it was discovered that it was impossible to create a new method based on these principles because different elements interacted and interfered with each other.

7.4 BASIC LAWS OF SKILL ACQUISITION

Currently, there is, of course, no single universally accepted method of teaching. The various types of learning, such as rote memorization, concept learning, learning how to solve problems, and skill acquisition, are employed as the situation dictates. We consider in this and following sections only those types of learning and aspects of skill acquisition that are important in the professional field.

Obviously, skills are not acquired without practice. Repetitive performance of the same activity results in changes in its structure. These changes can be described as the following:

1. *Changes in the method of performing actions:* The individual operations are integrated. Unnecessary action components (unnecessary because increasing skill renders them unnecessary) are progressively eliminated. Performance speed is increased and activities formerly performed sequentially can now be performed simultaneously.

2. *Changes in method of sensory control of actions:* Visual control of motor actions is replaced by kinesthetic control. The student can now evaluate the relationship between individual parameters of the action. Size and distance are more adequately estimated.

3. *Integration of components of activity:* The individual's attention is transferred from the perception of his or her own actions to the output of the action. Actions are evaluated not only consciously but also intuitively.

4. *Integration of components of activity:* The individual prepares for (anticipates) subsequent actions, while performing preliminary ones. Transfer form one action to another is performed automatically without having to make the transition consciously.

Changes in the structure of a skill during practice can be explained by the individual's selection of strategies and method of regulation; less efficient strategies are eliminated. Strategy selection may be conscious or unconscious. A change in strategy can be discovered by examining performance indices such as error and time. If the performance indices are sufficiently sensitive, one will be able to see minute changes in the measures used; these reflect strategy changes. Because strategies are peculiar to the individual, one individual's learning curve will be slightly different from another's. However, in combining individual learning curves one can discern general types of curves. In one, the curve is negatively accelerated when a skill is mastered quickly during initial learning stages, after which the process slows and approaches asymptote. In another, the curve is positively accelerated when the skill is mastered slowly but picks up speed with time.

The first type of curve is one in which rote memorization and trial and error dominate, such as with sensorimotor skills. The second type is associated with skills requiring understanding, in which improvement proceeds slowly, but, after a trainee comprehends the method of performance, performance sharply improves. Sometimes a learning curve can be described by a mathematical formula that permits one to predict the process of skill acquisition and define the time required to achieve a stable level of performance.

Errors during skill acquisition emerge randomly. This is why the probability of their appearance can be calculated according to the following formula:

$$P = m/n \qquad (7.1)$$

where m is the number of correct performance trials, and n is the number of trials.

Figure 7.3 presents a learning curve that demonstrates the probability of trials without error. The formation of a skill does not begin at the zero point because the trainee has some preliminary knowledge and skills that permit him to perform some trials without error. P indicates that the probability of trials without error has increased and approached one. However, this level will never be achieved because accidental factors can influence performance.

The shape of the learning curve is affected by the nature of the task, the idiosyncratic features of the trainee, and the method of training. Averaging curves permits one to discover the general features of the skill acquisition process. At the same time, this can result in a loss of some important information about the particulars of skill acquisition (Venda & Ribalchenko, 1983). This is why, during the training process, the instructor will use both individual and average learning curves.

The learning curve has irregular features (peaks and troughs, plateaus). These fluctuations can be explained by not only accidental factors, but also changes of performance strategies. The plateaus demonstrate that using the strategies is no longer effective in improving performance. The troughs in-

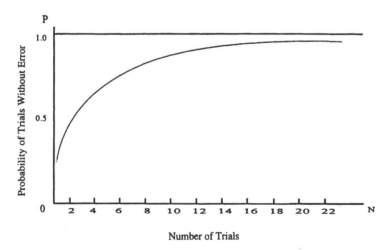

FIG. 7.3. The change improbability of trials without errors as function of number of trials. From Lomov et al. (1970).

dicate a worsening of performance as the trainee attempts to transfer to a new strategy. The new strategy is tested by the trainee and if evaluated to be more efficient it is used; inefficient ones are discarded. Using a new strategy very often requires more time initially and produces more errors. Any strategy has a series of stages: formulation of the strategy; its implementation, evaluation, correction, and perfection. In this way the acquisition of a skill is a self-regulation process. Where the strategy selection process is conscious, the student develops a set of possible strategies and deliberately selects one as best, based on conscious or unconscious criteria.

Another method of analyzing strategies at different stages of learning is in terms of the time structure of the learning performance (Bedny, 1987). For example, Fig. 7.4 shows the four stages of performance that are required to install pins into the holes of a pinboard. The pin's position in the hole depends on the following: if the pin is fluted and whether it is grasped with the right or left hand. Any line segments designate the duration of the motor or mental components of the activity. On Fig. 7.4, 1' and 2' designate the movement of pins by method M22C (at a distance of 22cm, with maximum concentration of attention); 3' and 5' designate identification of the decision-making process when skill is not sufficiently acquired; 1 and 2 designate the movement of pins by method M22B (at a distance of 22cm with average concentration); 3 and 5 designate the decision-making process when the skill has been acquired (in this situation, microelement EF is required); 4 and 6 designate turning pins at a 90° angle (microelement T90S); 7 and 10 designate moving pins by method M10C (at a distance of 10cm with maxi-

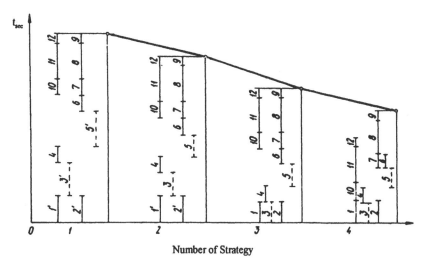

FIG. 7.4. The changing of time structure of activity during skills acquisition process. From Bedny (1987).

mum concentration); 8 and 11 designate putting pins in a hole (microelement P2SE); 9 and 12 designate release (microelement RL1).

The first strategy is characterized by the existence of motor pauses, the sequential recognition of signals and decision-making processes, the longer duration of the cognitive components of activity, a sequential turning of pins, and a high level of visual control of any movements. The second strategy has a similar allocation of different elements in time. The distinguishing factor is a reduction in the duration of cognitive components (duration of elements 3 and 5 is less than that of 3' and 5'). As a result, the duration of motor pauses is also reduced. The third strategy is characterized by a reduction in visual control when moving pins with both hand from a container to the holes. As a result, elements 1 and 2 require less concentration during their performance and become shorter in duration (elements of activity 1' and 2' are performed by method MC substituted by elements 1 and 2, performed by method MB). The cognitive components of the activity (element 3) are combined with motor elements 1 and 2, and element 4 (turning one pin) is executed immediately after elements 1 and 2. As a result, the general duration of a task is more reduced. The last strategy involves a further reduction of motor pauses because the elements of activities 10, 11, and 12 are performed immediately after completing element 1. Element 6 (turn pin) is combined with element 7 (moving pins to the hole). From Fig. 7.4, the students transfer to the last strategy not directly but through intermediate strategies.

It was discovered that the more simple the task was (e.g., when no pin was fluted), the number of intermediate strategies was reduced. This means that the more complicated the situation, the more intermediate strategies should be used. An analysis of Fig. 7.4 shows that the time structure demonstrates how the different components of an activity are allocated in time and how they are changed during skill acquisition. The transformation from one strategy to another is based on the student's past experiences and an evaluation of the intermediate results either consciously or unconsciously. This means that the skill acquisition process has a self-regulative nature.

At the beginning of practice, motor and cognitive skills are regulated consciously and require a considerable effort of attention. After extensive practice, skills can be performed faster, not under conscious control, and with less effort. This process is known as automatization. However, in continuous practice, a distortion of skills can sometimes be observed. This process is known as de-automatization. For example, a gymnast learns a complicated skill that later can be lost. Losing a skill can have either a stable or temporal character or in this instance, a stable or temporal de-automatization. In the first case, the skill is lost forever, whereas in the second case, the skill can be recovered either spontaneously or through special exercises. Bernshtein (1966) and Il'in (1986) hypothesized a "meaningful" and "kinetical" structure of motor skills. The meaningful structure of skill includes the image of situation, the goal of the action, and the program of performance (the first two obviously stem from self-regulation theory). The kinetical structure is characterized by spatial, temporal, and force parameters. The meaningful structure belongs to the psychological aspects of skill acquisition whereas kinetical is related to its biomechanical aspects. These are relatively independent but, at the same time, they can influence each other. For example, a trainee may correctly imitate an action, but when she or he attempts to perform that action with maximum effort and speed, the meaningful structure can be destroyed. If a person does not perform an activity for a long period of time, the skill temporarily disappears (extinction). Pavlov (1927) showed that extinction is not disappearance. Under certain circumstances, a response that disappears can return (spontaneous recovery). Sometimes, under emergency conditions, an undesired response that had been unlearned will reappear involuntarily. Extinction can be considered as one type of de-automatization of skill.

After a skill has been automatized, any attempt to regain conscious control of the skill for whatever reason may inhibit the skill—an example of de-automatization. However, the loss of a skill happens more often when the skill is first acquired by rote repetition. If, from the beginning, the trainee tries to consciously acquire different components of the skill after automatization, attempting conscious control of the skill does not result in de-automatization.

Transfer of and interference between skills can also be explained on the basis of self-regulation, depending on similarities and differences in the content of components of different function blocks. For example, task conditions can be different whereas task goals are similar. In this situation, achieving the same goal requires different program performance. For example, braking a car sharply on a dry road results in immediately stopping the car. When braking sharply on a slippery surface, the car can spin out of control. Because of this, a driver speeding toward an intersection where the light changes unexpectedly to red should react in a different way depending on the road conditions. In an emergency situation, the interference of skills can emerge. For example, instead of braking smoothly, the driver can brake sharply.

The speed of skill acquisition also can vary depending on the criteria employed. Thus, if the criterion is a specified speed of task performance, this can be achieved more rapidly than if the criterion is performance without error (Lomov, 1984).

Bedny and Zelenin (1988) suggested using multiple learning curves, one that describes acquisition of task performance as a whole, and others describing the individual elements of task performance. For example, Fig. 7.5 demonstrates two curves, curve **a** reflecting the acquisition of a production operation as a whole, curve **b** reflecting the acquisition of one of the components of the production operation. Multiple learning curves permit one

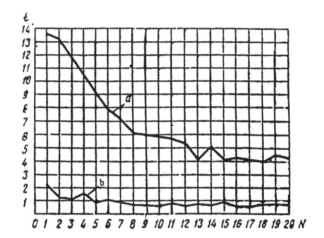

FIG. 7.5. Learning curves during turning of part on lathe. From Bedny and Zelenin (1986). A—curve of acquiring of task performance; B—curve of acquiring of element of task "locating of part into three jaw check."

to discover the speed of acquisition of individual subskills and how this is related to overall performance. This may be important in understanding the structure of the task.

Another method that permits one to analyze the structure of skills is based on a comparison of the performance of experienced persons with that of inexperienced trainees. This method includes an analysis of the changes in the process of skill acquisition when individual skill components are rearranged. For example, when a trainee acquires the skill of turning parts on a lathe (Bedny, 1981), it was discovered that the same component of the skill will be performed with a different speed, depending on the position of this component in the structure of the task. The less experience a worker has, the more the rearrangement of the skill components influenced his or her speed of performance. The expert in the same situation did not display the same speed variability. This can be explained by the fact that an experienced worker acquires multiple strategies that enable him or her to shift when task structure changes.

To study motor and perceptual-motor skill acquisition it is possible to make use of instrumentation such as video cameras and eye-movement recording devices. For mental skills, instrumentation is not useful because the skill acquisition process is covert; but procedures such as verbal protocol analysis, the problem behavior graph (Newell & Simon, 1972), interviews, and error analysis will be helpful.

An experimental method of studying the acquisition of cognitive skills is to vary the structure of the problems by introducing obstacles to the trainee such as limited information, introducing additional feedback, or eliminating feedback at different stages of skill acquisition. These variables can be introduced as part of a comparison of expert–novice performance.

An important criterion to determine when a skill is mastered is the stabilization of its structure (reduction of variability). When a motor skill is mastered, the trajectory, speed, acceleration, and time intervals between separate phases of the motion are similar when the action is repeated. This kind of skill is usual in experienced workers. For a trainee, however, the opposite is true. One can use the coefficient of dispersion of major parameters of actions during repetitive performance for the indices of stability of skills (A. M. Novikov, 1986). This coefficient can be calculated based on the analysis of time in which the action is performed. Another approach is based on the calculation of temporal parameters of different components of an action (A. M. Novikov, 1986). This can be calculated in the following way:

$$V_t = \left[\sum_{i=1}^{n} (T_i - \overline{T}) / N\overline{T} \right] \times 100\% \qquad (7.2)$$

where T_i is the time performance of an individual action or its particular part, $\overline{T}$ is the average time performance of action or its part, and N is number of cycles.

In the same way, the coefficient of dispersion can be determined according to force criteria.

$$V_f = \left[\sum_{i=1}^{n} (f_i - \overline{f}) / N\overline{f} \right] \times 100\% \qquad (7.3)$$

where $\overline{f}$ is the average value of effort, and f_i is the effort during the performance of a particular action or its part.

In the same way, we can determine the coefficient of dispersion of amplitude of an action (spatial parameters).

$$V_a = \left[\sum_{i=1}^{n} (A_i - \overline{A}) / N\overline{A} \right] \times 100\% \qquad (7.4)$$

where A_i is the individual value of amplitude, and $\overline{A}$ is the average value of amplitude.

A comparison of these coefficients permits one to analyze the stabilization of skills according to these dimensions. Stabilization based on the dimensions is an irregular process. Stabilization can first be observed according to the spatial structure of skills (V_a), then temporally (V_t), and then according to effort (V_f).

In an experiment performed by A. M. Novikov (1986), four groups of students in a vocational school assembled electronic devices. Each performed this task for a different length of time. The first group worked for 3 days, the second for 6 days, the third for 9 days, and the fourth for 12. During the experimental study, the following parameters were measured: (a) the performance of time operations, using a stop watch, and (b) the spatial parameters of the trainees' arms. These were measured at the work bench, the bench surface being marked with squares. Arm position was recorded on film and measured by the squares covered by the arm.

The stability of actions was analyzed according to V_t and V_a. Figure 7.6 describes time to complete the assembly. The curve in Fig. 7.6, part A is smooth, which makes it difficult to determine the exact time at which a skill was stabilized. Using the preceding formulae, spatial parameters began to be stabilized after the second day (Fig. 7.6, part C), temporal parameters after the fourth day (Fig. 7.6, part B). Stabilization of the skill was achieved in 5 days.

After a month and a half of not performing this task, a second experiment was conducted. The trainees again performed the task, with the results shown in Fig. 7.6, part D. In Group 1, the skill was not preserved—the trainees performed as if they had not practiced. In Group 2, when the stereotypy level was high enough, the trainees demonstrated a high level of retention, as did Groups 3 and 4. From this, one can conclude that the criteria for completion of the motor training process can be the stabilization of actions and that the typical learning curve is not as informative as data based on stabilization criteria.

Russian theorists attempt to explain the variability of skill acquisition in a different way. For example, such specialist in the physiology of work as Vinogradov (1966) explain the variability of skills as the result of the irradiation of neural processes and uncoordinated separate actions. However, this fact can also be explained by the self-regulative process, which suggests the existence of explorative activity based on which a student selects strategies of performance and corrects them based on feedback influences (Bedny, 1981; Bedny & Zelenin, 1989; A. M. Novikov, 1986). Both concepts are correct. The first concept reflects the fact connected with the absence of corresponding neural and psychological mechanisms of monitoring actions. The second is based on the necessity of forming these mechanisms according to the self-regulation process.

The learning curves for individual trainees have great diversity at the beginning of training but then begin to converge. The greater the diversity of the trainees, the more divergent the individual learning curves.

If the training method is not efficient, the variability among students is increased, the speed of skill acquisition is decreased, and the stabilization of skills is achieved at a lower performance level. Much depends on the

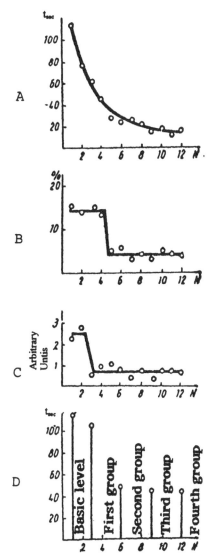

FIG. 7.6. Skill retention as dependent of duration of practice. From Novikov (1986). A—time spent for task; B—stability of time structure of task performance; C—stability of spatial structure of task performance; D—time spent for task after break in performance.

complexity of the task. For example, Fig. 7.7 displays different curves for varying task complexities.

One methodological problem that instructors encounter in training is defining the lengths of chains or operations that trainees can efficiently learn in one sitting. The lengths of the chains determine the size of the steps. In simple situations, integral methods of instructions can be used because students acquire all their elements in a single instance. Other methods involve a step-by-step method of acquiring multioperation procedures. In this situ-

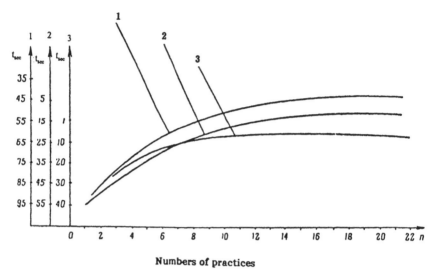

FIG. 7.7. Learning curve for tasks with different complexity. From Lomov et al. (1970). 1—pilots' actions restarting engine after accidental shut-off during flight (more complex task); 2—pilots' actions adjusting engine after RMM dysfunction (average complexity of task); 3—pilots' actions following drop in oil pressure (less complex task).

ation the trainee practices the first operation, until mastery, after which she or he practices the second operation until mastery. Then she or he will perform both operations together until mastery and so on. This paradigm is well known as distributed practice and whole–part learning (McCormick & Ilgen, 1985; Shaporinsky, 1981).

Currently, there are five combinations of part and whole in learning: simple-sequence, progressive-sequence, uncompleted-sequence, holistic method, and some combination of the preceding. Simple-sequence requires the learner to perform separate operations, and after acquiring them, the learner performs the whole task. The progressive-sequence requires the student to preform the first operation, and after mastering it, he or she performs the first and second operations together, then first, second, and third, and so forth. The method of the uncompleted sequence requires that the instructor extract the more complex operations of a task, which are then learned separately: Following this the problem is presented and solved as a whole. The holistic method suggests that the student learn the whole task or solve the problem in one step.

All of the preceding methods have some weaknesses and benefits. For example, the method of simple-sequence permits the student to acquire the individual components of the task more efficiently. However, they may have difficulty in integrating these individually learned operations. The method of progressive-sequence also has its weaknesses in that different operations performed by students may not be executed the same number of times. The amount of repetition and time spent practicing separate operations depends on the position of the operation in the sequence. The holistic method complicates task performance for the inexperienced student and does not permit him or her to concentrate on individual operations. The selection of a method of teaching depends on the nature of the task, the extent to which is proceduralized, and the goal of the teaching process. For example, if task operations are closely interconnected, then the holistic method becomes more desirable. However, if the holistic method is too complicated for the student, it is better to use the simple-sequence method.

7.5 THEORY OF STAGE-BY-STAGE FORMATION OF MENTAL ACTIONS AND SKILLS

For a long time, an associationist theory of learning based on Pavlov's conditioned response concepts dominated the former Soviet Union because the government favored Pavlov. After Stalin's death in the mid-1960s, new theories of learning based on activity theory and stemming from Vygotsky's work were developed. One is called the theory of stage-by-stage formation of mental actions and skills and was developed by Gal'perin (1959, 1966). This theory, which was successfully used for teaching elementary- and high-school children, was also used in adult training.

An important statement of Vygotsky's (1960) is the principle of the unity of psychic processes and behavior. He emphasized the importance of speech as a mediating mechanism. In the same way, tools mediate between the worker and the performance of his or her task. Sight of the tool becomes a symbol that mediates between the worker and the performance. According to Vygotsky, sign systems regulate human behavior; that is, the worker learns and remembers that she or he must do so and so with that tool. Speech and tools are thus both mediating mechanisms. The manner in which they mediate is through the process of internalization, which is essentially the transformation of external activity with a physical object into a mental operations with symbols. Internalization as a general process was also explored by Piaget (1952), Leont'ev (1957), and Gal'perin (1959). For Vygotsky, learning is produced by the internalization of activity during social interaction. Leon'tev and Gal'perin felt that internalization is primarily associated with work activity. For them speech is used as a mediating device in human activity. Consequently, work activity and its verbal aspects are the more important conditions for internalization. Contrary to Piaget, the Russian scientists emphasized that thought cannot be reduced simply to logic, although the latter are important components of thinking. Thinking also includes formation of goals, assumptions, the development of plans, and the extraction of significant task conditions, all of which are linked with emotion and motivation. According to Vigotsky and his followers, the most important way of analyzing mental processes is to examine their origins, including ontogenesis and phylogenesis. For example, they conducted research that was aimed at how culture influences mental processes. Russian psychologists are also interested in examining what they call "microgenesis" or how a psychological process develops in individuals in a relatively short period of time, for example, during an experimental session. Presumably all of this is directly associated with skill acquisition.

The major notion is that internal activity depends on external activity. The psychic process is not only a system of images and knowledges but also a system of actions and operations integrated by motives and goals. Because psychic activity has many similarities to external activity, a major unit of analysis becomes not "stimulus reaction" but action. From this follows an important principle according to which knowledge can be acquired only through the system of a student's actions (motor and mental).

The more important informational components of an activity, which are required for the correct regulation of actions and to perform without errors, are called reference points by Gal'perin (1969). An adequate mental presentation by the student of the required reference points in the new material will permit correct execution of the entire task by going from one point to another. The notion of reference points is close to that which Landa (1974) called "indicative features of the algorithm." In a broader sense, any information, properties, qualities, and relations between objects that permit a correct analysis of the task and its objective conditions can be considered reference points. When the student becomes capable of discovering the required reference points in the new material or task, it permits him or her to perform a new task. The system of reference points, based on which the student can develop a mental plan of performance, is the orienting basis of an action. The orienting basis of an action can be considered information that is used by the student to regulate his or her own actions. Because of this, an analysis of information presented to the student and that which, in reality, he or she used, is an important component of Gal'perin's theory.

According to Gal'parin (1969), any action (mental or motor) includes in itself orienting, executive, and control elements. Action is always regulated by an image of the environment and an image of the action and is directed to achieve a conscious goal. The internal plan and image of the action that is used by humans to perform the action or the "orienting basis of activity or action" can be considered the more important regulative mechanism of activity. These psychic components should be distinguished from objectively presented systems of reference points that will permit correct execution of new actions.

The executive components of action permit the transformation of objects or information in the environment. Executive components can be material (with real objects) or ideal, mental.

The control aspect of the action tracks the performance of the action and compares output with the standard of success (i.e., feedback). Hence the action can in part be conceptualized as a guidance system.

Talizina (1975) stated that in order to provide efficient guidance in the learning process, it is necessary to point out the goal of guidance (what must be achieved during the learning process), discover the initial state of the guidance process (what initial knowledge and skills the student possesses), determine the program of control actions that permit a transfer from one level of knowledge to another (the teacher's control of the student's learning activity), supply feedback (provide the student with information about the achieved result), and supply corrective mechanisms (provide corrective influences from the teacher). These concepts are based on cybernetic foundations that are very important to Russian scientists, who believe these concepts are very useful in training.

An action has basic characteristics, such as the level at which it is fulfilled (shape), the amount of generalization, the completeness of the operation accomplished, and the degree of mastery.

The level of an act can be either material or its material representation or sign (the materialized form of action). The material level involves the manipulation of external objects. When we mention material representation, the student performs actions with a scheme, drawing, pictures, models, and so on. The mental level of an act involves internal mental actions or operations. Mental actions, like material actions, have their own orienting basis of action, control, and executive components. Perceptual actions also exist. According to Gal'perin's (1969) theory they are more complex than material. This kind of action emerges as a result of the transformation of material actions, and involves, for example, such processes as endeavoring to perform calculations about objects by visualizing them in the head.

Mental actions are performed internally with concepts, images, and propositions. It is assumed that mental actions were first performed with physical objects and then internalized as described previously. Either a part of the action or the whole action can be related to describe the aforementioned form of actions. For example, the adding of numbers in the mind is the purely mental form of actions. However, if the person performs the production operation by using different tools, then the object of action is material. Executive parts of these actions are performed in external or material form. Orienting the basis of action and control components of action can be performed in perceptual form based on audible speech or mental form. Mental actions also have their orienting basis of action, control, and executive parts. Orienting and control components of action can exist in perceptual, external verbal, and mental form.

Generalization of the action is simply what Western psychologists know as the process of generalization and transfer of knowledge and skill. However, in Gal'perin's (1969) theory generalization is considered as a special feature of the mental action itself. If the action can be correctly used in a number of similar situations, or in a new situation, it means that this action possesses the dimension generalization.

Another important feature of the action is its completeness. Galperin (1969) extracted the process of abbreviation of mental acts as a distinct action dimension. Thus, initially actions depend on mental support operations that, as the skill is mastered, are progressively eliminated; this is spoken of as "abbreviating" the action. Talizana (1975) suggested that the orientation basis of action is abbreviated first and becomes unconscious before other action elements are eliminated.

The next action dimension is the degree of mastery. This feature, well known in psychology, is the process of automatization. Mastering and abbreviation of action will be manifested as increasing speed of performance, reduced error, and reduced variation in performance (stability). According to Gal'perin (1969), unconscious automatization is undesirable because it reduces understanding of the underlying principles of actions and inhibits generalization. To master an act means not simply to memorize action. It means the ability of the students to correctly use this action in different situations.

The degree of generalization, the degree of completeness, and the method of the presentation to the student affect the orienting basis of actions. These criteria permit extractions of three basic types of orienting basis of actions.

The first type is characterized by incomplete content of orienting basis of action. The system of reference points extracted by the student independently and the orienting basis is not complete. This tends to produce learning by trial and error.

The second type is characterized by the existence of all required reference points for correct performance of action. However, the orienting basis of actions is presented to students in readymade form. Moreover, the orienting basis of action is presented in a specific form that is suitable only in particular cases. Transfer and generalization become difficult. In any new cases the student should study again.

The third type of orienting basis also is presented in general form. However, these orienting points can be useful for all problems of a particular class. The orienting basis is developed by students independently and according to a particular plan under supervision of the teacher. In this case (according to Gal'perin, 1969), trials and errors as the method of acquiring knowledge and skills are eliminated.

An example of presenting the orienting basis of action in general form is a situation when a trainee acquires general algorithms for developing technological processes. If the student knows these algorithms, he or she can independently develop the technological process for any particular part. Extraction of the orienting basis of action should be performed based on the principle of typicalization of the technological process and the correct selection task or production operations that must be performed by the trainee (Bedny, Zgidelev, & Figanov, 1972).

The student's activity occurs in a sequence of stages. These stages represent the level at which a mental act is fulfilled.

The first stage of knowledge and skill acquisition is called the formation of an orienting base for a task and actions. At this stage, the student receives information about the goal of a task or problem or the goal of an action. The student also receives all information about reference points in the new material. As indicated previously, the system of reference points is considered to be the orienting basis of actions. This information is used by a student to regulate his or her actions. Usually, this information is presented to students through written or oral instructions. The system of orientation in a task or action must be presented to the student in such a way that would permit the student to efficiently perform not only a particular action but any other action that belongs to the class of which the particular action is an instance. In other words, the orienting basis of an action should present generalized data and principles. Because of this, a student studies how to correctly extract the general characteristics of a phenomenon.

At the next stage, the student should acquire the skills of how to efficiently use the reference points in different situations. The student learns how to use the orienting system by doing. At this stage, the trainee performs particular actions and regulates them according to the information involved in the orienting basis. Actions are performed accompanied by audible speech. Physical actions are combined with verbal actions. For example, at preschool age, a student learns arithmetic by moving objects and simultaneously counting aloud.

At the third stage, actions are performed based on only verbalization without direct support from objects. The student essentially talks to him or herself. Verbal actions include all operations involved in the performance of particular actions. Using material actions and combining them with audible speech permits the teacher to carefully control what the student performs. All actions performed by the student are under his or her conscious control.

At the fourth stage, the trainee performs actions with external speech but in a whisper. From the beginning, this speech has a complete character. This means that the student should precisely and completely describe all operations involved in performing mental actions. The efficiency of performance is evaluated only according to the achieved result.

At the fifth stage, the student performs verbal actions based on internal speech. Actions are abbreviated, become unconscious to the student, and seem to appear simultaneously. The student is oriented only to the ultimate goal. At this stage, the transformation of external audible actions to an internal plane occurs. This is not simply "speech minus sound," but a significant reconstruction of speech. The audible form of speech becomes the audible image of a word. This image retains some of the variable articulation and kinesthetic patterns on which it is based. However, the structure of articulation in internal speech is significantly changed and cannot be considered audible speech. Hence, in the learning process, external speech to oneself is gradually mastered and spontaneously transferred into internal speech to oneself. Here, the "verbal formula" of an action appears in an abbreviated form.

At the last stage, the teacher controls the final action output. The major principle of Gal'perin's (1969) idea is that the acquisition of mental skills is a process of internalization when an initially material form of actions is later performed as a mental operation. During this process, the external control of actions transforms into internal control and is considered to be an act of attention (Gal'perin, 1966). During all of these stages, actions are reduced, abbreviated, and performed automatically.

During the training process, actions and activity as a whole change their characteristics. If we talk about motor actions, only particular parts of these actions can be internalized. The orienting basis of an action and the control part of an action belong to them. The executive component of an action (motor component) can change only its own structure but still performs externally.

One of the major principles of Galperin's (1969) current theory is that of guiding a student's activity. Gal'perin first introduced the notions of orientational, executive, and control parts of actions in the theory of activity. From this it follows that an action can be considered the guiding subsystem of an activity. The student controls and regulates his or her own actions under a teacher's supervision. Gal'perin emphasized the importance of consciously mastering different skills. This theory has been applied successfully in teaching school students, elementary-school children in particular, but evidence exists that this theory can be successfully applied to teaching adults, in particular the human-machine system (Kondrat'eva & Titova, 1965; Zarakovsky et al., 1974).

7.6 CONCLUDING COMMENTS

For many years, associationist concepts of learning, based on Pavlov's conditioned reflex theory, dominated the former Soviet Union. Association is a form of conditioned reflex. However, in the middle of the 1960s a concept of learning emerged based on activity theory. This concept, suggested by Gal'perin (1959), was called the stage-by-stage theory of developing mental acts. If, in the associative learning concept, the major units of analysis are associations, the major units in Gal'perin's theory were actions (which could be either external-physical or internal-mental). This theory does not deny the existence of associations; however, the occurrence of an association is considered the result of an organized system of actions performed by the student.

Russian scientists reject association when the process of its formation is simplified to the schema, stimulus-response-reinforcement. They emphasize the importance of goal-directed activity in the formation of external actions and internal mental operations. As has been discussed in this chapter, this process transitions the student from external operations (by manipulating physical objects) to an internal mental form. Gal'perin (1959) considered this process of transforming an external physical activity into an internal psychic one to be a process of internalization.

The Russian system of teaching is based on the principle of the instructor precisely guiding and controlling a student's actions when the latter perform tasks and solves problems. Actions are considered to be a guiding system that contains orienting, executive, and control aspects.

From the activity point of view, learning is not considered only to be either a function of memory or a means of strengthening associations by rote repetition of reactions. Learning is, however, a process of internalization or a reorganization of the internal dynamic structure of an activity.

Other Theories of Learning
From the Activity Viewpoint

8.1 OVERVIEW AND INTRODUCTION

This chapter describes learning theories based on activity principles and develops a self-regulation theory of learning. Examples of research that illustrate certain aspects of the self-regulation learning theory are presented.

Landa's (1974) algo-heuristic theory, which is a predecessor of the self-regulation, sees learning as problem centered. He distinguished between two classes of problems, one that invariably can be solved by algorithms, the other that can be solved (but not invariably) by heuristics (i.e., methods that are discovered in the course of problem solution). In the algo-heuristic theory the student searches a "field" of conceptual categories and discovers the correct response by examining and discarding incorrect categories, eventually arriving at a solution by the process of elimination. This process involves consideration of hypotheses and selection of one for testing by comparison with an already available concept classification system. If the selected hypothesis tests satisfactorily, it is accepted; if not, the student reverts to a previous stage of problem solving, develops and selects another hypothesis, and tests it, continuing until a procedure leads to a correct solution. The algo-heuristic methodology is reductionistic (to elementary operations) and holistic (because it is problem centered, the student cannot solve the problem in a series of steps but must work the problem through to completion). Because of the preceding characteristics, the algo-heuristic theory is particularly appropriate for complex cognitive learning. It has many characteristics in common with the self-regulation theory of learning, which is discussed next.

The self-regulation theory of learning asserts that self-regulation is the basis for all learning, including stimulus-response association and conscious and unconscious learning. The relationship between conscious and unconscious components of learning can be understood only by analysis of the relationship between the goal and the performance outcome. Learning in the self-regulation framework is conceptualized as the development not of stimulus-response connections but of performance strategies, a process that has much in common with the problem solving of the algo-heuristic theory. Exploration (gathering of data from the environment and analysis of the problem situation) is a critical feature of self-regulation learning because it is necessary if an appropriate strategy is to be developed. Learning is the transition from one performance strategy to another. Stages of learning are conceptualized as processes of working through preliminary (incorrect) performance strategies to more developed ones. According to self-regulation theory of learning the specialist should think not in terms of types of learning, but of levels of learning. Learning based on associationist processes is a type of learning, because it does not consider a conscious goal. Motivational and social factors are extremely important in self-regulation learning.

A series of examples is provided to illustrate certain aspects of the self-regulation theory. Special attention is paid to something called the "orienting basis of action," which is the extraction of significant aspects of the problem environment to which the student should respond. Changes in performing during learning are viewed as reconstruction of a performance strategies. Because all learning is essentially problem solving, the more complex the learning task (the more difficult the problem), the more preliminary strategies must be developed and tested. An important aspect of skill acquisition is the transformation of unconscious aspects of self-regulation into conscious ones.

8.2 THE ALGO-HEURISTIC THEORY OF LEARNING

The algo-heuristic theory of learning and instruction (AHT) was developed by Landa in the former Soviet Union in the middle of the 1950s where it soon created a strong following. After Landa's articles and books had been translated into a number of European and Asian languages, the AHT received a wide recognition also in many other countries. Landa emigrated to the United States. In 1974 and 1976, respectively, his books, *Algorithimization in Learning and Instruction* and *Instructional Regulation and Control: Cybernetics, Algorithmization, and Heuristics in Instructions* were translated into English. In the United States, his theory and methodology were labeled by his American peers "Landamatics" (see, e.g., *Encyclopedia of Psychology*, 1984; Reigeluth, 1983). His basic idea is briefly considered in this section.

AHT describes a system of operations or actions that turn knowledge into skills. Landa (1974) distinguished between two classes of problems. The first, algorithmic problems, are those that can be solved by using an algorithm, which means that by following a specfic prescription (a set of instructions), one can solve any problem that belongs to a particular class. The second, non-algorithmic problems (semialgorithmic, semiheuristic and heuristic, which were described in chap. 1), require special instructions (heuristics) for their solution. These can increase the efficiency of solving problems. But they do not guarantee the solutions. By using special procedures and rules one can decrease the degree of uncertainty of instructions for solving nonalgorithmic tasks.

Algorithmic and nonalgorithmic tasks and processes often interact. For example, algorithmic methods can be used as subprocesses in solving nonalgorithmic problems. Moreover, problems may originally be considered nonalgorithmic. However, if one finds a method of solving these problems the problems will become algorithmic.

Landa (1974) made a distinction between two kinds of algorithms in the learning process: algorithms of transformation and algorithms of identification. The first specifies the processes geared to the transformation of objects or information; the second specifies processes of relating an object or concept to a class of objects or concepts. Algorithms of identification provide an affective tool for describing actions involved in the process of concept formation. The study of concept formation has a long history in Russia, beginning with Vygtosky (1956). Like Piaget, he began with preschool children and discovered that they used trial and error in acquiring concepts. This was because they were unable to differentiate essential from nonessential features of the stimuli presented to them. Identification algorithms enable one to make concept acquisition and formation a well-structured and effective process. Transformation algorithms were described in more detail in chapter 6. This chapter focuses on identification algorithms.

Until the mid-1950s, the notion of an algorithm was considered purely mathematical. Analyzing how people solve nonmathematical problems, Landa (1974) discovered that people often used nonmathematical algorithm processes similar to mathematical processes. Because each algorithm is applicable to all problems belonging to a specific class rather than to any single problem, it represents a general method for solution. Landa introduced the notion of a human algorithm viewing mathematical algorithms as a particular subclass of the human algorithm. Later, with the advance of computers, the notion of algorithm began to be used in computer science. Algorithms for programming computers were called computer algorithms. Although computer and human algorithms both guide the process of solving problems, the former do not necessarily model the ways in which humans solve problems. Elementary operations in computers are static; mental operations evolve and are dynamic. The same algorithm can be used by different computers, but this is not always true of humans. The fact that humans and mathematical algorithms have something in common has led Landa and other Russian theorists to an intense interest in the algorithmic approach to thinking, learning, and instructions.

An efficient identification algorithm requires one to develop an effective concept definition of objects belonging to a certain class. This in turn requires discovery of such objects' features (attributes) that will permit a rapid, precise identification of the objects. The relationships between attributes, on the one hand, and the class to which an object belongs, on the other hand, can be expressed in the "if-then" logical form. Landa (1974) showed that attributes of objects belonging to certain class can be connected by different logical connectives and thus have different logical structures. This can be expressed, for example, in proposition of the type: if some object x has attributes a and b or d then it belongs to class m; if it has attributes a and b and not d then it belongs to class p; if it has attributes a but not b and c, then it belongs to class s, and so on. These relationships can be described by notations used in symbolic logic:

$$a(x) \text{ and } b(x) \rightleftarrows x\varepsilon M$$
$$a(x) \text{ and } \overline{b}(x) \rightleftarrows x\varepsilon P \tag{8.1}$$
$$\overline{a}(x) \rightleftarrows x\varepsilon S$$

where $a(x)$ means that object x has an attribute a, $\overline{b}(x)$ means that object x does not have an attribute b, $\rightleftarrows$ above means "if, and only if, -then," and sign ε means "belong to a class."

Algorithms are often viewed as rules. Landa (1974) maintained that this is to a certain degree true, but that it is very important to understand the specificity of algorithms as rules. According to Landa, algorithms are rules of how to apply rules rather than the basal rules themselves. In other words, algorithms are rules of higher order. Landa illustrated this in the following way: Consider a simple rule, "If there are conditions a and b, then, and only then, do M." Many students know rules related to certain disciplines or professions (e.g., grammar rules, physics rules) but make errors in their applications. According to Landa, this is because they do not know *what to do* with a rule, and with the object of its application, in order to apply it. Here are some of the actions to be performed in order to apply the rule properly:

1. Isolate the first condition indicated in the rule (in our example, condition a).
2. Check to see if this condition is presented. If no, do not perform the action M. End. If yes, go to item 3.
3. Check to see if the second condition is present (in our example, condition b). If condition b is presented, do M. If condition b is not present, do not do M.

In other words, it is an algorithm of applying the rule that describes the operations that everyone has to perform on the rule and on the conditions indicated in it in order to apply the rule correctly. Of course, this algorithm can also be called a rule, but it is clear that this rule differs from the basal rule in that it prescribes what to do with it in order to apply it. One should know an algorithm of applying the rules in order to be able to correctly use them.

Landa (1982) indicated that very often algorithms are confused with flow charts and that the algorithmic approach to learning, instruction, and performance is wrongfully viewed as flow charting. He pointed out that each algorithm can be presented as a flow chart (albeit not only as a flow chart), but not every flow chart is an algorithm. Educational training and other materials abound with flow charts that look like algorithms. However, in most cases they are not because they lack some, or all, of the characteristic features of algorithms (e.g., the precision and unambiguity of instructions, and the instructions' ability to completely specify and determine the actions to be performed). It is possible to flow chart all kinds of processes, procedures, and instructions, including those that are imprecise, ambiguous, vague, and hard-to-understand-and-follow.

From Landa's (1974) approach follows a natural sequence of steps for developing algorithms to be used in instructions. One must first define the task and subtask (essentially perform a task analysis). Then it is necessary to uncover algorithmic processes (largely unconscious) used by expert performers in performing the task and its subtasks. It is especially important when there is a significant involvement of mental operations. The next step is to explicitly describe unconscious algorithmic processes as algorithms that would serve as prescriptions for what novices should do in their minds in order to perform as experts. It is especially important to break experts' processes into relatively elementary mental operations.

In mathematics, a very limited number of the most simple actions (operation) that all normal humans can perform (such as recognition and differentiation of simple symbols, substitution of one symbol by another one, etc.) were selected as elementary. These operations are called formal because they deal only with very simple objects' symbols that clearly differ by their form (shape). In mathematics, only those instructions are considered algorithmic that address these simple formal actions. With nonmathematical algorithms, Landa (1974) pointed out the situation is much more complex. Nonmathematical algorithms have to address operations that are geared not only to symbolical objects but to all kinds of objects, including mental (images, concepts, propositions). What kind of operations and instructions should be considered elementary, or sufficiently elementary, in activities that are not limited to dealing with mathematical symbolic objects?

Landa (1974) said that the criterion of elementary of such operations can be only psychological. He offered a simple definition: An operation can be viewed as elementary (or sufficiently elementary) if a person is able to perform it without asking others or himself how to do it. Correspondingly, an instruction is elementary (and therefore algorithmic) if it is addressed to such an operation. For example, if a person knows how to give an object (say a pen) to another person, then this operation is elementary for a person and the instruction "give a pen" is an elementary (algorithmic) instruction. On the other hand, if, when asked to find an adverb in a sentence a person were not be able to do it or were to ask "how," then this operation and its corresponding instruction would not be elementary (algorithmic) for that person. The advantage of this criterion is that it is an experimental criterion and finding out whether an operation (and instruction) is elementary or not can be accomplished through a relatively simple test.

Because humans who perform operations, and to whom instructions are addressed, are learning and developing creatures, it is obvious that what is elementary for a person at one stage of his or her development may be not elementary at another, earlier stage. On the other hand, an operation (and instruction) that may be elementary for one person may be not elementary for another person, and vice versa. From this follows one of the fundamental theses in Landa's (1994) theory, that the notion of elementary of operations and their corresponding instructions is relative. If the notion of operations and their corresponding instructions is relative, then the notion of nonmathematical algorithms is relative as well. A system of instructions that may be an algorithm for a person at one stage of his or her development may not be an algorithm at another stage. And, one and the same system of instructions may be an algorithm for one person and not an algorithm for another one.

Not all problems are algorithmic in nature. For many problems it is impossible to devise algorithms; general solution procedures do not exist. For these, Landa used heuristics, which are defined in a general sense as methods "encouraging the student to discover for himself" (Stein, 1980). A series of nonelementary operations, or elementary operations that cannot be performed in a uniform way under the same conditions, is defined by Landa (1976, 1984) as a heuristic process. A prescription describing these operations is a heuristic prescription. Presumably the instructor can develop these prescriptions on his or her own, or can formalize heuristic solutions discovered by students; and these can be transmitted to later student generations.

The essential difference that distinguishes heuristic from algorithmic instructions is that the former cannot guarantee a problem solution. The heuristic may work, but not always. One of the major functions of heuristics is that they decrease the degree of uncertainty in searching for a solution to a problem. Depending on the degree of uncertainty contained in the heuristic instructions, Landa (1976, 1978) divided all nonalgorithmic processes into semialgorithmic, semiheuristic, and heuristic.

One of the major objectives of AHT is to develop general cognitive skills that will enable the student to find solutions to a great variety of problems.

According to Landa (1976), intelligence is nothing other than a set (system) of general methods of thinking. From this it follows that intelligence can be taught by teaching systems of general operations, which make up general methods of thinking.

The following is an example of the general algorithmic method of thinking. Objects' attributes can be connected via a logical conjunction *and*, or a conjunction *or*, or their combination. Figure 8.1 is an unpublished example from Landa of a general method of determining whether an object belongs to a certain task when the objects' attributes are connected via the conjunction *and*. Figure 8.2, on the other hand, is an example of a general method of identifying the belonging of an object to a certain class when the object's attributes are connected via the logical disjunction *or*. If one contrasts the concept of "angle" and "adjective," it will become apparent that their attri-

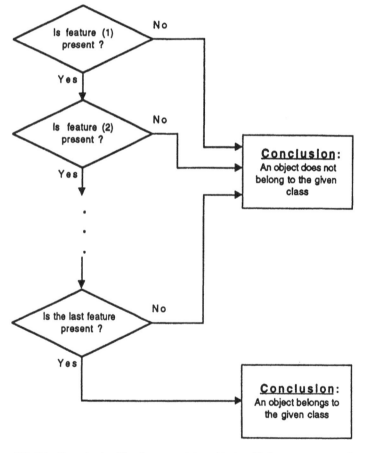

FIG. 8.1. Generic algorithm for recognizing objects with features connected by AND (conjunctively) according to Landa.

butes are connected via the logical conjunction *and*, which determines the same (identical) procedures of their identification. The process of identification will differ only in its content but not in its structure.

One of the basic procedures of AHT is the "breaking down of a complex, unobservable cognitive processes into more elementary, but also unobservable cognitive operations that could be unambiguously executed by learners" (Landa, 1983, p. 173). There are no easy methods for achieving this. Landa described some methods for what he termed "penetrating experts' mind" to discover the mental operations of which mental processes consist. Several of these are widely used in research involving thought processes:

1. Perform concurrent or retrospective verbal protocol analysis (Ericson & Simon, 1984), during or after task performance by anyone (expert or novice) performing the task.

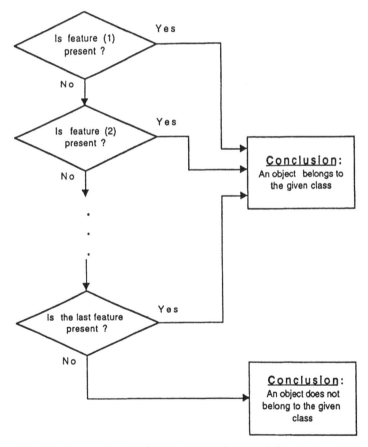

FIG. 8.2. Generic algorithm for recognizing objects with features connected by OR (disjunctively) according to Landa.

2. Cross-examine an expert as to how he or she typically solves the problem; cross-examine a novice about task performance difficulties.

3. Ask an expert to train a novice in the performance of a task, while observing the training process. A variation of this is to ask the expert to develop a set of instructions for training the novice.

4. Compare novice–expert differences in task performance.

5. Analyze errors made by subjects during task performance.

6. Infer the mental operations involved in task performance.

7. Change the conditions of task performance from those under which the task was previously learned and measure task performance to see how it has changed; observe subject performance, and/or questioning the subject about the difficulties he or she encountered under the changed conditions. Learning the algorithm or heuristic does not mean that a person can immediately perform the task requiring it. Practice in actual execution of operations prescribed in algorithms or heuristic is required.

In the framework of the AHT, of special interest is the process of actuation of knowledge units and mental operations in the course of thinking. As a rule, people have in their minds thousands of knowledge units and mental operations. However, when they solve problems and make decisions they use (apply) only some of them. To be able to apply them, they must first be actuated in the mind. Failure to actuate the needed knowledge units(s) and/or operation(s) makes it impossible to use (apply) them in the course of solving a problem, and thus leads to failure to find a solution.

Knowledge units and mental operations can be actuated in three basic ways: (a) externally by outward stimuli (tangible objects and/or instructions), (b) internally by other knowledge units and or mental operations, and (c) internally by verbal self-instructions as to what to do in order to perform some task or achieve some goal.

Actuation in cases (a) and (b) is based on associations formed in previous learning and experience. Actuation in these cases occurs by associations, involuntarily, that is, automatically. Processes actuated by association are passive processes because they occur without any special effort, that is, by themselves.

Actuation is case (c) is different. A situation or a problem may automatically actuate certain knowledge units and mental operations that may be inadequate for finding a solution. However, if a person knows what to do in order to find out a solution and gives her or himself a corresponding self-instruction, this self-instruction may be able to actuate such physical and/or mental actions that were not actuated automatically by association.

For example, consider a student who is trying to solve a geometric problem. Perception of a diagram of an isosceles triangle automatically—by association—actuates in the student's mind the knowledge of three attributes of isosceles triangles (out of five existing and in principle known to him or her). These attributes, however, do not lead to finding a solution. Then the student asks her or himself a question (self-question): "What are the other attributes of the isosceles triangles?" (This self-question performs the function of a self-instruction: "Find other attributes"). The self-question (self-instruction) actuates mental actions of voluntary search in memory for an isosceles triangle's attributes that did not come to mind by association involuntarily, by themselves.

Thus, self-instructions actuate certain physical or mental actions that in turn actuate certain knowledge (images, concepts, and/or propositions) that were not actuated by association but were necessary to use (apply) in order to solve the problem. According to Landa (1984), self-instruction and its corresponding mental or physical actions are thus able to overcome associations. Overcoming associations is a prerequisite for solving any kind of nonstandard, creative problem, because associations actuated by the problem conditions do not lead to the solution.

Processes involving actuation of knowledge and actions by self-instruction that lead to overcoming knowledge and actions actuated by associations are active processes.

The ability of self-instruction to actuate knowledge units and actions, not actuation by association, lies at the foundation of self-guidance, self-regulation, and self-control. Landa (1978) stated that one of the major objectives of instruction is to form processes of thinking (i.e., systems of mental operations). That is, instruction does not involve knowledge of how to think, but rather knowledge of algorithms and heuristic. What then is the function of algorithms and their knowledge?

The major function of algorithms (and heuristics) is that they explicitly describe all the necessary operations and their systems and thus serve as models and blueprints of what is to be formed in the minds of students and trainees. Having a precise blueprint of the processes to be formed, the instructor can develop all the required operations and all their systems in the minds of learners, in a purposeful, systematic and reliable way.

Algorithms can serve as models and blueprints of processes to be learned not only for instructors but also for students. Knowing the algorithms, students in many cases can follow them without any instructor and thus develop the required mental processes on their own.

Another important function of algorithms is that they make learners aware of the processes of thinking in general and of their own mental operations and thought processes in particular. But once a learner is conscious of what he or she should mentally do and how to do it in order to solve defined classes of problems and perform, he or she can voluntarily regulate and direct his or her own processes of thinking.

Landa (1976) indicated that knowing the algorithm—that is, knowing what to do—does not make the ability of doing. To develop this ability, it is necessary to convert the knowledge of operations to be performed into actual operations, that is, an algorithmic process. For algorithmic processes to be efficient, they have to have certain characteristics, the most important of which are the smoothness and speed of their flow.

The algorithmic process cannot be considered well formed if the performer must think, at each step, what to do next, "stumbles" in carrying out operations, or performs them slowly. According to Landa (1978, 1983), there are two conditions for achieving smoothness and speed performing an algorithmic process: The instructions of the algorithm must be internalized and the operations automatized.

The internalization of instructions consists of transforming the external instructions of an algorithm into internal instructions, that is, self-instruction. The automatization of instructions consists of changing the psychophysiological mechanisms of their actuation and regulation, specifically, changing the source of their actuation.

In algorithm-based learning when students are exposed to explicit algorithms, the source of actuation of operations changes in the following way:

- Phase 1: Operations are actuated by the external instructions of the algorithm. Because of this, the learner is aware of the operations she or he performs and they are performed consciously. During this phase instructions are involuntarily memorized (internalized) and turn into self-instructions.

- Phase 2: Operations are actuated by self-instructions. Because operations are still actuated by instructions, the learner is aware of the operations he or she executes and they are still performed consciously. However, during this phase two kinds of direct connections are formed: (a) connections between the action component of operation (e.g., isolate) and the objects at which operations are directed (e.g., attributes of a triangle), on the one hand, and the operations and the requirements of problems and tasks (e.g., proof that a triangle that has such and such attributes is similar to a triangle that has such and such attributes) on the other hand, and (b) connections between operations themselves.

  Once these connections have been formed, the operations can now be directly actuated by problem and task conditions. Self-instructions are no longer needed and disappear.

- Phase 3: Operations are actuated by problems and task conditions. Because operations are no longer actuated by any kind of instructions, awareness of the operations starts to perform, and may begin gradually to wane. With the passage of time, the algorithm instructions get gradually forgotten (because they are to not utilization) and the operations cease to be conscious. Only when Phase 3 is reached is the desired effortless flow of operations achieved. At this phase the process of actuation and carrying out of operations becomes automatized.

Automatization of operations and processes is nothing other than the formation of direct connections between operations and their objects, and between operations themselves, that create the basis for the transition from one source of their actuation (instructions and self-instructions) to another source (objects and other operations).

In the course of going through the phases, the operations change from conscious to unconscious. Landa (1976) called their operations postconscious to distinguish then from unconscious operations that never reach the level of consciousness. Such operations are preconscious. In conventional instruction where explicit algorithms are not taught, most unconscious operations are preconscious, whereas unconscious operations formed on the basis of algorithm and then automatized are postconscious. The quality of unconscious postconscious operations and processes is much higher than the quality of unconscious preconscious ones.

One of the drawbacks of conventional instruction and training is the fact that students are taught to solve specific problems and perform specific task but typically are not taught general methods. This manifests itself in well-known situations in which students can replicate the instructor's reasoning and solution, but are nonplussed when they are confronted with new problems (often even of the same type).

Material presented in this chapter demonstrates that the formation of knowledge and mental skills cannot be reduced to the formation of different associations. The basic notions of the theory are mental actions (operations). Mental operations are viewed as real actions that are performed on mental objects (knowledge) and their connections, rather than actions performed on tangible, material objects. Moreover, in many cases mental operations enable one to overcome the associations that were formed in the previous learning and experience, which makes it possible to solve problems that were not encountered in the past experience and do not outwardly resemble them.

8.3 SELF-REGULATION THEORY OF LEARNING

Another theory of learning based on the concepts of activity was developed by Bedny. The self-regulation process discussed in chapter 2 has several stages. In the first, task performance is quite stable; that is, skills are well established and there are no new conditions to which they must be generalized.

A second stage involves transition. A strategy learned in a particular situation must now be performed in another situation with which the performer is also quite familiar. There are few problems here. In another transition, however, the strategy must be transferred from the accustomed situation to one with which the performer is not familiar. As one can see from the position of self-regulation, learning is a transition from one well-known strategy to a new strategy or adaptation of a well-known strategy for a new situation.

The notion of strategy is fundamental to the self-regulation theory of learning. In this section strategies are treated as plans for goal achievement that are responsive to external contingencies, as well as to the internal state of the system. The notion of strategy is intimately associated with the notions of algorithmic and heuristic methods of performance. The concept of strategy is more comprehensive than that of either algorithm or heuristic. Strategies are typically a complex integration for deploying different algorithmic and heuristic methods of performance. Strategies have a dynamic and adaptive character enabling changes in the approach to goal attainment as a function of external and internal conditions of the self-regulation system.

The processes described previously can be considered from the standpoint of self-regulation, a concept that is not completely novel to Western psychologists. For example, Bandura's (1977) theory of observational learning has been referred to in previous chapters. Bandura saw self-regulation as a linear sequence of stages—observing oneself, judging oneself, rewarding oneself, and regulating oneself. However, self-regulation cannot be presented as linear sequence of steps. Russian theorists see self-regulation as having a recursive, loop structure with function blocks having forward and backward connections. Moreover, the self-regulation process, as described by Anokhin (1962), Bernshtein (1966), and G. A. Miller et al. (1960), involves a hierarchical organization. For example, during the analysis of motor acts, Bernshtein (1947) described five levels of self-regulation in which a higher level of self-regulation can manage a lower level. Each level is connected with a particular part of the neural system. Any lower level of self-regulation involves a lower level of physiological control.

The highest level of regulation is connected with the activity of the cerebrum and with verbal-logical thought. This level of regulation only exists in humankind. It is responsible for symbolical coordination, which is connected with thinking and consciousness. The second level of regulation guides more molecular subtasks and is particularly responsible for developing and guiding sequences of motion actions. A third level regulates the movement of bodies in space. A fourth level of regulation is responsible for the contraction of individual and groups of muscles. This level also evaluates the reactive forces of different motions. At the fifth level, muscle tone is regulated. Only the first two levels are associated with consciousness, but awareness is not always necessary, because the self-regulative process can function without it. If, however, behavior is directed at a deliberate goal, consciousness is required.

The Russian theorists Anokhin (1955), Bernshtein (1947), Bedny (1981), and Bedny and Zelenin, (1989) asserted that the basis for any kind of learning, even that of stimulus-response association, is the self-regulation process. This is because to them the basis of all behavior, even that which is physiological, such as the salivary conditioned response of Pavlov's dogs, is self-regulation. They denied a simple, straightforward association between stimulus and response, even in the most apparently simple learned behaviors. Thus, even when they accept that humans are capable and make use of associative learning, for them self-regulative processes are the foundation of such learning. For example, conditioned response can be explained according to the model of self-regulation developed by Anokhin (1955; see Fig. 1.1). It is possible to derive complex behaviors from very simple associative building blocks because the latter are progressively assembled into larger action units. The brain is assumed to develop a model of a required objective future state, compare stimulus situations and action outputs, evaluate actions, and develop new programs of performance.

This requires one also to consider the relationship between conscious and unconscious learning. Both can be involved in the same learning performance. Learning derived from association without conscious knowledge of logical interrelationships between phenomena is based on reinforcement that can have informational and motivational functions, although the informational function of reinforcement is limited because there is no understanding of the stimulus-response relationship.

Relationships between conscious and unconscious components of learning can be understood only through the analysis of the interrelationship between a goal and a result. Tikhomirov (1984) outlined three kinds of outcomes:

1. A conscious goal is achieved.
2. During goal-directed activity persons achieve an inadvertent accessory outcome that is originally unconscious but can become conscious.
3. An outcome is achieved based on involuntary behavior that is involved in the goal-directed activity; the accessory output always remains unconscious.

Accessory outcomes that were not associated with the goal and cannot be expressed verbally are always unconscious to the individual. However, this result still influences the process of regulation of activity. Very often it is possible to provide transformation of the outcome into one that directly associates with a goal. In this case the outcome becomes conscious (Ponomarev, 1976). The attainment of a conscious outcome is considered a conscious level of self-regulation. When one achieves an accessory outcome, its results can become conscious by reorienting one's attention. In this case, the incidental learning is subsumed in the content of the goal-directed activity.

This is transferred from an unconscious level of self-regulation into a conscious level. In the third case, outcomes and processes of self-regulation are always unconscious. In the first case, persons always make inferences about logical and functional interrelationships between functions and goals. In the second case, it becomes possible only after reorienting one's attention to these accessory results.

One can talk about a cognitive level of learning (as it is understood in Russian psychology) only in those cases when the learning process is directed to achieve a conscious goal. From this it follows that learning can be voluntary and involuntary, and conscious and unconscious. For example, a child can play with toys and acquire different skills and knowledge unconsciously without any voluntary efforts. However, major types of learning by humans are considered to be conscious. This type of learning suggests existing consciously established goals. It implies a conscious level of self-regulation that has complex interrelationships with an unconscious level of self-regulation. This last level of self-regulation can be considered as an associationist level of learning. As an example of a transition from an unconscious to a conscious level of learning, there is a transfer from unconscious explorative behavior (associative level of learning) into conscious explorative activity (cognitive level of learning). This activity is connected with one's conscious goal via the comprehension of a particular situation that emerges in explorative activity.

Association can also appear as a result of a consciously developed action during its automatization. In this case, we are talking about the cognitive level of learning. From the self-regulation point of view, association is one of the results of the learning process. In order to understand why one or another association was formed during the learning process, it is necessary to study the self-regulation process that creates those associations.

The relationship between conscious and unconscious aspects of learning was demonstrated in an experiment by Gordeeva and Zinchenko (1982). They studied micromotions of the right hand when a person tried to move a "light spot" on a TV screen from one target to another, by means of a lever. These micromotions were registered by special devices. Although the micromotions were unconsciously performed and evaluated by the subject, they still influenced the formation of the subjects' internal images of their motor acts. The existence of these micromotions is evidence of an unconscious associationist level of self-regulation that cannot be transformed into conscious learning. At the same time subjects were aware of what they were trying to accomplish and tried to correct their performance, obviously conscious self-regulation. Changes in instructions and modification of task performance resulted not only in changes in conscious strategies of goal attainment but also in micromotion changes. Thus, during the learning process instructors can influence the unconscious level of self-regulation.

This can occur in a cognitive learning task just as it does in motor learning. This question was studied by Bedny (1981). He paid attention to the fact that a teacher guides a student's activity. However, a teacher's guide should be developed by taking into consideration that the thinking process of a student has self-regulation features. One method that permits consciousness of the different components of a problem in different ways is known as "reorientation." Introducing different instructions, prompts, reformulation of the goal, and so on, results in the student's becoming conscious of different components of the problem in different ways and developing different strategies of solving this problem. All of these data were the subject of discussion in chapter 3. In Landa's (1976) theory, this can be achieved by breaking down mental processes into relatively elementary mental units, which a person can perform consciously. From the self-regulation point of view, this type of mental operations (actions) can be regulated consciously by students (they can evaluate their method of performance, correct it, etc.). In general, we can say that by changing the system of teaching, we can change the relationship between the conscious and unconscious components of learning a wide enough range. Of course, these changes have their limits. However, even in those cases when the unconscious components of the learning are impossible to transfer to the conscious level, there are enough possibilities to influence the unconscious components of activity through the conscious level of self-regulation.

In one study Gal'perin and Sachko (1968) analyzed a student's skills dealing with the use of a vice for fixing wood pots. If the student secured the vice too tightly to the pot, the pot was damaged. If the student secured the vice too loosely, the pot was unstable. The instructor used a verbal expression such as "Fasten the pot more tightly or loosen the grip on the pot, by turning the lever on the vice." These instructions were ambiguous. The student could not transfer verbal instructions to muscular efforts and the instructor could not find the appropriate words to explain what the student should do. In such cases, the student acquired the correct action only after a long trial-and-error period. The use of a special measuring device with a pointer eliminated the difficulty after a few trials. The researchers interpreted these results as the process of internalization but it was actually self-regulation. In this case, the student developed a subjective criterion of success based on the objective criterion presented by the measuring device. When conscious strategies are inefficient and one does not know how to change them, one starts to use unconscious strategies of self-regulation. The student without the special measuring device unconsciously either clutched the pot too tightly or too loosely; he or she unconsciously corrected his or her actions until the required muscle effort was attained to secure the pot appropriately. This required a long period of training. When the instructor used the measuring device with the pointer, unconscious motor processes were transferred to a conscious level.

There are other examples from gymnastics. The skills involved include many cognitive components that may be conscious or unconscious. As in the pot example, the trainer and the student have difficulty translating verbal descriptions into muscle coordination and required space orientation. The difficulty is exacerbated because of the holistic nature of the gymnastic act. Because components are closely interconnected, it is difficult to deconstruct the act to enable the gymnast to practice an individual component.

When the trainer uses instructions such as "Straighten arm upward" or "Turn head left," she or he influences the gymnast's behavior on a conscious level of self-regulation (the gymnast consciously performs the action) while at the same time exerting an influence on an unconscious level. For example, when the gymnast turns her or his head, this immediately influences muscle tension which by trial and error is corrected unconsciously.

The development of complex motor actions requires the self-regulation process, and the connection between stimulus and reaction is not straightforward. For example, often a gymnast acquires an incorrect technique that becomes automatized. Because of this, the orientational and the evaluative components of the self-regulative process are weakened. The gymnast cannot change the technique of a particular element in spite of the trainer's instructions because programmed components of self-regulation become dominant, which enables the gymnast to consciously control and regulate movements. One procedure to solve the problem is to "weaken" the automatization. The trainer starts to employ varying recommendations that may even be incorrect, but that induce the gymnast to perform the particular element in a different way, which destroys the embedding. This is an example of influencing the unconscious level of self-regulation by means of conscious mechanisms.

The self-regulation theory refers not to stimulus-response connections but to performance strategies and their regulation. Automatization, which consists of eliminating conscious regulation of movement, can prevent learning of more complex tasks or even performing more simple tasks. For example, if a gymnast over a long period of time practices salto layouts—a particular gymnastic movement—with a 360° vertical turn (see Gaverdovsky, 1987, Fig. 8.3), it is very difficult for that gymnast to learn salto layouts with a 720° vertical turn. The opposite can occur if, over a long period of time, a gymnast practices a salto layout with a 720° vertical turn. Later he or she cannot perform a salto with a 360° turn, which is much easier and was performed well before.

Another example is also interesting. A tumbler is to perform a salto layout with a 360° turn during a competition. Because he had performed the salto with only a 720° turn over a long period of time, he performed it that way during a competition instead of with the 360° one. He did this twice and was unable to accept the fact that he had. Indeed, he was not even conscious of \

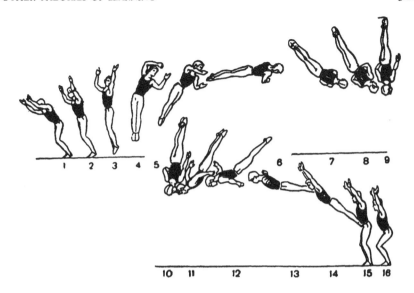

FIG. 8.3.   Salto layout with 360° vertical turn.

the additional turn. The stimulus from the environment automatically excites the executive (implementing) components of self-regulation, inhibiting the performance of less well learned actions. One solution to this problem is to vary strategies of performance during training (including analysis of the situation, goal formation, processing and evaluating results), and thus increase the likelihood that the gymnast can voluntary regulate his or her own actions.

Certain recommendations can be made. In attempting to learn more complex actions, the trainees should attempt to perform simpler actions under more complex conditions. For example, while the gymnast is attempting to learn a double salto, she or he should periodically perform a single salto but land on a higher surface than the original starting point. Or, an operator can learn to perform a task at a more rapid pace than is actually required; this guarantees that he or she can perform the task at the lower required pace.

To the self-regulation theory, associationist and cognitive levels of learning have complex interrelationships. Piaget (1952), in analyzing the question of the function of trial and error, discovered that animal actions do not have totally accidental characteristics and depend on the predevelopment of a cognitive scheme. Both correct and incorrect actions give animals information required for changes in behavior. He observed the behavior of his son, Laurent, at the age of 10 months; the child intentionally varied his actions in order to observe the similarities and differences in the resulting consequences. Analysis of the varying consequences of correct and incorrect actions enables one to separate action causes from the coincidental (correlations) (Lindsay & Norman, 1992). The more complex an unpredictable situation is, the more important exploration becomes. In the absence of any organized information, the individual will rely on contingencies of reinforcement or punishment when trying to reach desired goals. Under these circumstances, humans possess no advantage over animals in reliance on information derived by trial and error. It is possible that any self-regulative system in a state of uncertainty initiates random, unsystematic behaviors.

Linhart (1970) studied explorative reactions in the child to discover how operant reactions develop. For this purpose he used an aquarium with artificial fish and a rubber pump, which the child squeezed to force air into the aquarium. As a result, the fish and the water began to move. However, the air could be forced into the aquarium only when a red light was turned on. The child was not given this information. If the child squeezed the pump when the red light was off, no movement of fish and water occurred. The movement of the fish and water served as a reinforcement for the red light. Linhart measured the frequency, strength of squeezes, and latency (between the bulb "on" and squeeze) and discovered that playing gradually transferred into learning. The child tried to discover why in some cases after squeezing, the fish did not move. Gradually, the child developed positive conditioned reactions to squeeze the pump only when the light was on and the negative conditioned reactions not squeeze the pump when the light was off. This learning process proceeded through three stages, the first involving increased frequency and strength of squeezes independent of whether the light was on or off. Linhart called this stage the explorative phase of learning whose purpose was to examine the situation and consequences of one's own actions. In the second stage, although the frequency and the strength of squeezes when the light was on continued to occur at a high level, the frequency and strength of squeezing when the light was off decreased. The child came to understand when its actions were successful. However, these reactions were unstable. The number of errors gradually decreased over time, but did not entirely disappear. In the third stage, precise and stable reactions emerged. By questioning the experimenter and analyzing his or her answers, the child tried consciously to explore and manipulate its environment. The child tried to find the causes of its erroneous actions. The important factor was not the illumination of the light but the movement in the tank. These results could be interpreted to mean that the child's own actions regulated his or her behavior in accordance with the desired goal. From this study, it follows that the basis for conditioned reactions is the self-regulation process, which provides adequate strategies for task performance. Learning can be considered a transformation of strategies of performance that correspond to a particular phase of learning. The more complex a task, the longer it takes the individual to find a truly effective strategy, which means that there may be a series of attempts to develop one, only to revise or discard it. These initial attempts are called intermediate strategies.

Venda and Ribal'chenko (1983) also conceptualized learning as a process of transition from one strategy to another. This suggests that learning should be organized in accordance with phases of skill acquisition, each of which in turn represents the development of preliminary strategies and the transition to more refined ones.

The learner is particularly sensitive to the influence of feedback in the beginning phases of learning. However, in the last stages of learning, the role of feedback sharply decreases and the learner begins to work in accordance to the program of activity she or he has developed. To introduce conscious feedback into automatized skills can weaken them. Although feedback is important, one must also consider the importance of the method of orientation to the task. Orientation and evaluation processes in self-regulation are interconnected, but they perform different functions. Their structure is produced by different function blocks. Because of that, orientational and evaluative components of actions cannot be considered as a holistic unit as was suggested by Gal'perin (1966).

From the specificity of the learning process, it is very important to separate such relatively stable orientational components as "goal" and "formation of task" from the dynamic components of orientation, such as "subjectively relevant task conditions." This last function block demonstrates that the system of information presented to the students should be changed depending on the stage of the learning process. This means that the orienting basis of an action is continually changed during learning.

The model of activity presented to the student as a standard must be changed depending on the stage of the learning process. During learning, the student uses different temporal components of an activity that cannot be considered erroneous. These components have an intermittent character and are not included in the final stage of performance. All of this is evidence that the system of instructions should be changed depending on the stage of the learning process. Based on this, Bedny (1981) formulated the principle of dynamic orientation at different stages of skill acquisition. Dynamic orientation includes changes in the method of presentation of indicative features or reference points in the task, and new material, change in feedback, and conscious control of performance. This requires a dynamic system of instruction to be applied to the learning process.

A conscious level of self-regulation does not mean that all components of the activity become conscious. The relationship between conscious and unconscious levels of self-regulation may change depending on different stages of learning. The relationship between conscious and unconscious levels of self-regulation may change depending on different stages of learning. Orientation and evaluation processes in self-regulation are interconnected, but they perform different functions. Their structure is provided by different function blocks (see Fig. 2.6). Because of that, orientational and evaluative components of actions can not be considered as a holistic unit as was suggested by Gal'perin (1966).

When the learning task is not clearly conceptualized (in instructions or in the student's mind), the student explores the problem by attempting to formulate the goal and extract what appear to be the significant aspects of the problem. Following this, the student evaluates the difficulty and significance of the problem; this influences the student's motivation to solve the problem. Later the student formulates a hypothesis about the nature of the problem solution as the basis for a program that is tested by implementing the hypothesis.

The test of the hypothesis is completed when the student examines the consequences of his or her actions; this is the evaluative component of self-regulation, which includes feedback. If the result is negative for the hypothesis, the student regroups and develops another action program, unless she or he perseverates or can think of no other solution. The theory of self-regulated learning suggests complex relationships among explorative, motivational, executive, and evaluative components of activity.

The processes just described are compatible with Lindsay and Norman's (1992) three phases of learning: accretion, restructuring, and tuning. In the first phase, new information is gradually accumulated and stored in memory schemata. In the second phase, the memory structure must be modified, because in most cases the newly organized schema is not completely appropriate for the new material. Accretion, restructuring, and tuning can occur over and over again. From the point of view of theory presented in this section, Lindsay and Norman's stages are essentially self-regulation processes that occur at a conscious or unconscious level.

In summary, existing conscious and unconscious levels of self-regulation involve both associationist and cognitive learning processes. Associationist learning is essentially unconscious and is regulated by task outputs. Cognitive learning is controlled by a conscious goal and the relationship between that goal and the output. Because of this, outputs not related to the goal will lack meaning.

From the self-regulation point of view, theorists do not think in terms of types of learning but levels of learning: the associationist level and the cognitive level. Laws governing the lower associationist level cannot be totally fitted into the higher cognitive level. Association learning is a lower, more primitive type or level of learning than that of cognition. Associationist learning is connected with unconscious, nongoal processes, cognitive with conscious, goal-related processes. Nevertheless, both levels of learning depend on self-regulative processes and interconnected.

The more difficult the process of attaining a given goal, the more necessary intermediate strategies become, and therefore, the number of trials to criterion and errors are increased. In other words, the strategies or rather their instability cause trials and errors to increase. The subject learns by changing his or her strategies of activity. The strategies are essentially hypotheses that must be tested, and the errors made during the learning process may also represent quasi-experiments that generate information about the adequacy of the strategy hypotheses. This is a very novel way of conceptualizing the problem of error and deserves to be examined in greater detail by self-regulation theorists.

The theory of stage-by-stage formation of mental activity asserts that it is possible to develop mistake-free learning (Gal'perin, 1966; Talizina, 1975). However, any precise instructions, procedures, and input influences cannot predict what can happen at the output of a self-regulation system, which is the student, because it is impossible to completely eliminate errors (Bedny, 1981; Landa, 1974).

Errors in the learning process have informational features. They can be restricted, but to totally reduce them is impossible. For the purpose of knowing how to perform actions correctly, students should know what is incorrect. A. M. Novikov (1986) discovered that, in some cases, presenting a complete orienting basis of actions to the student gives a negative result. He demonstrated the result of an experiment where a trainee learned how to apply exact effort to particular controls when using additional visual orienting points on an oscilloscope. The trainee very quickly performed the required motor actions without errors. However, when the information from the oscilloscope was removed, the trainee could not perform the required actions correctly. When the instructor began to use discreet feedback instead of the oscilloscope, which included turning on a bulb only during a period when the trainee performed errors, it resulted in an increase in the efficiency of the training process.

In the second case, trainees were given incomplete information about how to perform actions (incomplete orienting basis of actions). This increased the efficiency of learning. In another example, A. M. Novikov (1986) demonstrated the efficiency of using special simulators to develop the skill of filing. This simulator can present information about the required trajectory of the file on the oscilloscope's screen. It was discovered that the more efficient method of training was one where a student worked without a simulator previously and performed incorrect actions. Later, when the required information was presented on a simulator, they could compare erroneous and correct actions. This increased the efficiency of the training process.

All of these data are evidence that trainees should independently evaluate correct and incorrect actions and compare them. This helps them to correct their actions and develop appropriate strategies of task performance. The pace of task performance during the training process increased not only because of an increase in speed of performance, but mainly because of a change in the strategy of performance (Bedny, 1981). Testing different strategies, their corrections and transformation provided by the mechanisms of self-regulation, and errors all play an important role in this process. Errors also play an important role in cognitive skills. An analysis of how a person solves different problems also gives evidence that a person learns by transforming strategies and analyzing errors. Learning how to solve different problems can be presented as a process of promoting different mental hypotheses, their examination, correction, and performance.

In the stage-by-stage theory of learning, the process of mastering skills is presented as the transformation from conscious to unconscious because of automatization. According to the self-regulation concept of learning, this is a two-sided process. The trainee ceases to be aware of some components of activity while becoming aware of others. This usually happens when the trainee alters his or her attention to different components of the task at the various stages of learning. The trainee actively attempts different strategies of performance, transforms them, and, in the process, transforms his or her own understanding of what the required standard of performance is.

In human learning, conscious control and evaluation of the outcome of an activity is important. This conscious control and evaluation of the process of performance, when the student can independently examine and regulate his or her own activity based on feedback influences, is called self-control. From the pedagogical point of view, self-control is considered to be a system of skills (sensory, motor, and thinking) that are required for the planning, regulation, and performance of a task by the student. These skills can be automatized, which means that the student is not always conscious of them. Self-control skills are a particular system of actions that can be shaped during the learning process.

The theory of internalization plays an important part in the conceptualization of learning (Gal'perin, 1959, 1966; Piaget, 1952; Talizina, 1975). This theory is widely recognized in the former Soviet Union, although not all Russian scientists accept it. A detailed consideration of this fundamental problem is beyond the scope of this book. Here, we only want to mention that some scientists have written on the idea of the independence of internal mental activity from external (e.g., Lindsay & Norman, 1992; Linhart, 1970; Rubinshtein, 1958). Independence does not mean that these two kinds of activity do not influence each other.

Scientists have also written on the idea of the independence of internal mental activity from external stimuli (e.g., Lindsay & Norman, 1992; Linhart, 1970; Rubinshtein, 1958). According to Rubinshtein, external motor actions are only tools for the cognition of the relationship between objects. He wrote that internalization cannot be considered a transition from external to internal mental activity, but is a process of transition from one type of existence to another. Independence does not mean that these two kinds of activity do not influence each other.

According to the self-regulation concept of learning (Bedny, 1981; Bedny & Zelenin, 1989), external practical and internal cognitive actions are interconnected. Because of the existence of feed-forward and feedback interconnections between these action types, the possibility of their comparison and evaluation enables active formation of both external and internal actions. Linhart (1970) wrote that it is not a process of transferring from external behavior to an internal plan of performance, but rather the formation of internal operations that have their own central features. External motor actions can be considered a support for the performance of internal mental actions. During the gradual acquisition of mental actions, external support is less necessary and, finally, they can only be performed on a mental plane. During the simultaneous performance of external and internal actions, they actively shape and correct each other, but are not transferred from an external to an internal plane. This causes changes in the strategies of mental operations. The character of these changes is, of course, influenced by external activity. The existing cycles of feedback influences regarding external motor activity influence the structure of mental operations, which, in turn, influence external motor activity. Because of this, we can use the term *comparison* of mental and motor actions during the learning process.

This is one of the important self-regulative mechanisms of transforming the strategies of activity during learning. The transformation of strategies of activity during learning is not an internalization, but rather a process of reconstruction.

From this it follows that learning is considered not as an internalization as thought by Gal'perin (1959), Piaget (1952), and others, but rather a process of actively forming separate actions and strategies of performance. In this dynamic process, the relationship between the conscious and unconscious components of learning, both internal and external activity, is continually changed.

Tikhomirov (1984) criticized Gal'perin's (1966) theory because he did not discover the specificity of the goal in the structure of the orientational basis of activity. This is why when we analyze the orientational components of learning we use the model of self-regulation presented in chapter 2.

From the theory of self-regulation, the orientational components of an activity involve the following function blocks: "goal," "formation of task," "subjectively relevant task conditions," "assessing the task's difficulty," and "assessment of sense of task." There is a complicated relationship among them. This means that the orientational components of learning have a complicated structure.

In Gal'perin's (1966) theory, the process of verbalization plays an important role. However, not all components of the learning process can be verbalized. In the self-regulation concept of learning, not only the verbalized but also the nonverbalized aspects of learning are important and place an emphasis on the dynamic of the teaching process. The learning process can be described as the individual stages through which a learner must pass. In Galperin's theory, the dynamic components of learning are totally ignored.

One of the general laws of learning is the law of effect: An action that leads to desirable outcomes is likely to be repeated in similar circumstances. According to the self-regulation theory of learning, actions that lead to accepted goals have a greater probability of being repeated in similar circumstances. One cannot overemphasize the profound implications of the difference between "goal" and "outcome" in this context. An outcome is merely the result that occurs; a goal is an image or conceptual representation of a future desired outcome.

As has been pointed out, without a relationship to a goal, the performance output has a merely incidental, transitory effect and cannot be interpreted meaningfully by the individual. In other words, the outcome has no significance until it is associated in the student's mind with a goal of some sort; hence the importance of the stage in learning and task performance of goal formulation.

In the activity concept of learning goal formation, motivational and social factors are extremely important. The method of assigning goals can affect the process of training blue-collar workers. This was shown by Bedny's (1981) study, whose purpose was to discover how introducing different goals to the student influences the training process.

The system of training students in the extensive system of vocational schools in the former Soviet Union is highly centralized and standardized. In consequence, there are some constraints in that training. For example, the instructor can use a time standard for performance only when the student can perform production operations at the final stage of practice and presumably possesses the required skills and knowledge. This is based on the assumption that introducing a time standard early in the training can decrease the quality of work because the student does not yet have the required knowledge and skills. However, the vocational school encounters problems when the student completes the program and goes to work in a plant where he or she is unable to perform to quantity requirements in a specified time period. This reduces productivity, lowers the salary of young workers, and causes dissatisfaction and increased turnover rate. The problem of how to increase the productivity of young workers was the motivation for Bedny's (1981) study conducted with blue-collar workers in various industries. Preliminary observations of students in an apprenticeship discovered that, in most cases, the students did not know how much time they should spend on a particular job. The instructor paid attention only to the quality of the output. A stop watch study discovered that students spent a great deal of time unproductively and did not organize their work in time. Students had poor motivation and the instructors spent a great deal of effort disciplining them. After being trained without time limits, the students could not increase the pace of task performance in production conditions.

The transition to a faster pace of performance required reconstructing the structure of skills (Bedny, 1975). The skill of performing the same task at a different pace requires different skills. It was hypothesized that introducing a goal to the student, which included a time standard, could increase performance quantity without decreasing work quality. The results of the study with students who performed bench and assembly work reveal that introducing a time standard increased not only quantity but also the quality of work (Bedny, 1981; Bedny & Zelenin, 1986). In addition, introducing a time standard reduced unproductive time by increased work motivation, improved discipline, and reduced fatigue.

A similar result was achieved in a study of first-year students milling special bolts. This task was performed individually for a period of 2 days. Because the work was monotonous, productivity was usually reduced on the second day. It was suggested that presenting a precise quantity goal to the students could increase the work productivity and reduce monotony.

The study was conducted for approximately 3 weeks. On the first day, the students worked as if under regular conditions, without the requirement to produce a particular number of pieces (the time standard was absent). On the second day, the students performed the same task, but had to produce 30 pieces per day. The students did not know they were taking part in an experiment and were being observed (single blind study), which eliminates the "Hawthorne effect."

The results demonstrated that the introduction of a standard increased productivity sharply. The differences in productivity are statistically significant. Figure 8.4 compares the number of units produced on the first day without a time standard (curve A) with that produced with a time standard (curve B). Not only did productivity increase, but also the variation in productivity changed. When the time standard was introduced, productivity quickly rose, maintained a high level for a longer period, and later decreased.

Although the introduction of a time standard had a significant impact on performance, this occurred only when it was introduced to first-year students. If the student had worked without a time standard for a long time and had almost finished the training course, the time standard was ineffective. This can be explained by the fact that students developed their own performance standard as a function of peer comparison. Because the objective time standard did not exist for these students, they did not evaluate themselves by such a standard, but by a standard developed informally within the group

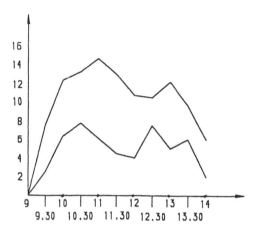

FIG. 8.4. Dynamics of productivity during the shift. From Bedny (1981). A—with a time standard; B—without a time standard.

(see Bandura's formulation). Introducing the time standard at the beginning of the training influenced the informal social standard developed by student.

From the standpoint of self-regulation, a goal that contains objective data about standard requirements performs important cognitive and motivational functions. A comparison of the time standard with actual output increases the efficiency of such activity function blocks as "subjective parameters of a successful result," "subjective parameters of admissible deviations," and "information about interim and final results." From a motivational standpoint, the goal, which now includes a standard of achievement, has a positive influence on the motivational aspect of self-regulation. According to the activity model, the individual can either adjust to that goal or change it, based on feedback derived from the function block "making a decision about a correction."

This process permits an objective goal to be converted into a personal goal of self-assessment. Individuals continually transform the solution of practical problems into self-assessment evaluations. This enables people to create self-beliefs about their efficiency. In the case of students, this is important because sensitivity to one's own achievement plays a more critical role for students than for experts.

8.4 SOME EXAMPLES

This section describes some studies that attempt to show that learning theories based on activity principles can explain certain behaviors. These examples cannot, however, totally validate these theories, if only because each theory contains assumptions that are not susceptible to measurement, and implications that are measurable but immensely difficult to do so.

The range of situations described in these examples is broad: from the training of an operator to reorganize different stimuli to pilot performance in aircraft to tractor engines to production processes. The emphasis on the latter is because Russia (under whatever name it was called), for cultural, historical, and industrial reasons, always emphasized production processes in its research.

One must begin with the theory stage-by-stage formation of mental actions and skills. We consider a situation where training must be given with different signs and symbols. This method was developed by Goryainov (according to Zarakovsky et al., 1974). Training was performed according to particular steps and levels suggested by Gal'perin (1966).

The first step (preliminary) of the training process was considered the formation of the orienting basis of the action. For this purpose, the teacher developed special cards. On each card, a particular sign or symbol was depicted. Special written instructions were developed for each card. The instruction described the orienting basis of each action, the executive and control components. This required careful task analysis procedures. Each card was compared by the student, with the picture of a real object. The trainer named the category or type of object that was coded by the symbol. Based on these data, the student developed a general image of the symbols as a holistic sign system.

The second step involves the student performing action at different levels. The first level is the materialized form of actions. The cards with symbols were placed before the trainee, who had to independently group the symbols according to category, type, subtype, and functional purpose, and select from the whole alphabet of symbols particular kind of symbols with general features or particular symbols. The trainee drew pictures of different symbols in a specified order, first depicting the major component and then the additional component(s). Before the trainee performed these tasks, he or she explained aloud what major and additional parts of each symbol is, their relationship and how they influence the meaning of the symbol, and so on.

The second level involves verbalization without cards. Here, the operator performs the following tasks: (a) Talk about system classification of symbols in given alphabet, (b) talk about additional parts (features) of symbols and how they influence classification of these symbols to a particular class, and (c) describe aloud the differences between different symbols that are verbally presented by the instructor. Name, using memory, all the symbols that belong to the alphabet.

The third level involves the mental (internal) performance of actions. At this stage the operator performs tasks without verbalization and tools, performing various calculations with symbols, grouping them according to different criteria. He or she can also make drawings of the symbols according to verbal description by others. The final level is connected with using simulators when different lighted symbols are displayed on the simulators.

The second example has to do with the training of pilots. This example demonstrates the importance of the orientational components of activity. One of the important components of orientation for a pilot is his or her image of flight. The pilot, of course, can fly based on displayed (instrument) information without having a sufficiently developed image of flight, but under these circumstances, flight reliability is reduced. The flight image is particularly important when the pilot acquires flight skills largely in a ground simulator. Although he or she has a simulated visual display (external view of the environment) and certain physical stimuli, such as engine noise, the pilot in the simulator learns how to orient him or herself primarily by instrument information. Russian aviation psychologists feel that the increase in learning in ground simulators requires a method of training that produces a more valid image of fight and "a sense of feeling" about the aircraft (Vorona, Aleshin, & Safronov, 1984).

Formation of a flight image depends on particular system of actions as well as a system of indicative features that engender the image. At the first stage of learning it is very important that trainees learn how to select indicative features relevant to a particular situation. From the perspective of self-regulation, the more important function blocks providing the image of flight include "goal," "formation of task," and "subjectively relevant task conditions." The specific relationship among these function blocks determines the relationship between the relatively stable components and dynamical components of trainees' orientation during the learning process. The relatively stable components are "goal" and "formation of task." The dynamical components are "subjectively relevant task conditions."

Krukov and Kremen (1983a, 1983b) developed a method of training that illustrates this point called "method of critical points." Critical points of a flight may be treated as a certain spot of the flight profile that is associated with the transition from one stage of flight pattern to another (see Fig. 8.5). Each critical point has its own logical structure of indicative features. The noninstrumental information, such as vibrations, acceleration, and so on, also plays an important role. Depending on the task performed by the pilot, the logical

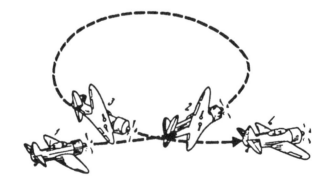

FIG. 8.5. "Critical points" in a loop. From Kremen and Krukov (1983).

structure of indicative features in the same stage may be modified. Trainees study how to extract appropriately indicative features (relevant instrumental as well as noninstrumental information) and constitute a coherent structure. Essentially, this process is a diagnostic one. At the first stage, training is performed on ground conditions. For this purpose the instructor can use slides, movies, actual signals, vibration, noise, and the like. The trainee should interpret the situation and describe the image of flight verbally, while manipulating the simulation materials. At the second stage the trainee studies how to perform similar tasks in real flight together with instructors. The strategy of the pilot's behavior is described as a logically organized system of perceptual, mental, and behavioral actions in which he or she is involved. Based on the evaluation of correct and incorrect actions and their organization, pilots acquire strategies for performing this diagnostic task efficiently.

An experimental study by Zavalova et al. (1986) compared the performance of pilots trained using the traditional method of gathering information from displays with performance in flight of those pilots who used the new training organized around critical points. The latter increased their ability to spend more time gathering information outside the cabin (Fig. 8.6). Time for gathering displayed information from the control panel was reduced 14% to 21% without reducing efficiency of task performance.

Gil'bukh (1979) studied the use of an oscilloscope in training to indicate when an amount of pressure required of a student for pressing a lever was achieved. With this device, excellent training results were achieved; but after transition to the real equipment, which lacked an oscilloscope, performance deteriorated. When an indicator was installed in the learning situation that illuminated only when the required pressure was attained, performance in the real situation improved.

Although one might think on the basis of Gal'perin's (1966) theory that the visual information provided by oscilloscope in training should improve

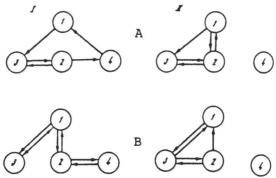

FIG. 8.6. Eye-movement strategies of a pilot. While in a loop (A) and third circle (B). From Kremen and Krukov (1983). I—experimental group; II—control group; 1—out of cabin space; 2—center of dashboard; 3—left part of dashboard; 4—right part of dashboard. Arrows show the eye movement.

performance, what was overlooked was that in the operational situation kinesthesis was the foundation of the orienting basis of the required action. The oscilloscope provide only complete visual information and ignored kinesthetic regulation of actions. When the indicator was used, the orienting basis of action was incomplete. However, efficiency of training increased because the students regulated their actions based not only on visual information, but also on muscle efforts. Because the orientating basis of actions was incomplete, students performed not only correct but also incorrect actions. Based on comparison of visual information from the indicator and muscle effort, students can compare correct and incorrect actions and determine the required muscle efforts. The point of this example is that the information presented to students should activate the required mechanisms of self-regulation.

Landa developed a training program for the Du Pont Company. This company faces problems with training and performance—both within the company and at the customer's plants. Preliminary studies show that the training of new technical services representatives is too long (up to 1 year), unreliable, and expensive. The existing manual does not provide detailed guidance to their customers' operators (employees of the manufacturer's customers). It contains a lot of theoretical and practical information, but is in no way considered to be a performance guide that tells an operator in a step-by-step, algorithmic manner how to perform each of the tasks and solve each of the problems. The manual does not serve as a performance guide or training manual because it is written in the logic of the subject matter and not in the logic of learning the subject matter. Another problem is that training is carried out on a one-to-one basis by another technical representative who shares with the trainee his or her knowledge and experience. This process is too long and unreliable. Due to inadequate training, customer operators perform at less than expert level, which results in big economic losses. The main objectives of the study were:

1. Identify a troublesome diagnostic and treatment problem that creates difficulties in learning and performance and leads to producing defective cables.
2. Uncover algorithmic processes underlying the ability of an expert.
3. Devise an algorithm that would describe the process of an expert and serve as a guide for trainees, novices, and nonexperts. This system of algorithms is called a *master algorithm.*
4. Test this system of algorithms with Du Pont's operators.
5. Experimentally identify how they perform in comparison with a master algorithm (how they are more or less effective).
6. Develop new versions of algorithms that combine features of a master algorithm with those that are attained in experimental studies. This

algorithm is superior to all individual algorithms used by different operators (because of this, that algorithm is called a *super expert algorithm*).

7. Present the algorithm in the form of an algorithm performance guide,
   and make a cost efficiency analysis.

On the basis of the operators' better solutions, the original algorithms of
experts were converted into super expert algorithms. For example, Fig. 8.7
presents one kind of super algorithm. When developing a super algorithm,
one must ensure that the description of any function block of the algorithm
can be comprehended in the same manner by different operators. Here, we
present an example of an algorithmic description of task performance for
blue-collar workers, however, the theory is more often applied for training
white-collar workers.

In the self-regulative theory of learning, the process of developing skills
is represented as a series of stages of searching for and developing suitable
strategies of activity. These strategies require the organization of individual
activity components into multidimensional activity structures that may not
all be acquired simultaneously. Sometimes the process of acquiring a skill
must be divided into individual components and must be mastered separately. However, it is not always possible to divide a complicated skill into
separate components. In this case, the instructor should use holistic method
of training. Contradictions emerge between the necessity of using the holistic
method of training and the inability of the student to comprehend all com-

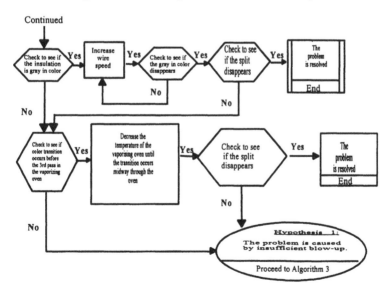

FIG. 8.7. Algorithm for diagnosis and treatment of temperature profiling
according to Landa.

ponents of skill mastery. Dynamical orientation is also important in this case. The instructor should emphasize student attention on those components of skill that are more important in a particular phase of training. In this case the holistic method of training is combined with instruction that emphasizes student attention on those components of skills that are more important in a particular phase of training (Bedny, 1981; Bedny & Zelenin, 1989). From the self-regulation point of view, this is associated with the process of extracting the "relevant task condition."

A. M. Novikov (1986) used this principle during training of blue-collar workers in a vocational school. He studied the hand chipping of metal. The parameters involved in this skill included the precision and force of hammer use on a flat chisel, the method of grasping the chisel, the pressure exerted, duration of work without pauses, muscular bioelectricity, and flexibility of the elbow. He discovered that all individual skill parameters could not be mastered simultaneously and that the interrelationships of those parameters to each other sometimes worked against each other. Thus, mastering one skill parameter sometimes resulted in weakening another parameter. In spite of the same instructions given to all students, they actively searched for more personally suitable methods of task performances.

Changes in the performance of skills are explained as resulting from reconstruction of activity strategies. One recommendation from A. M. Novikov's (1986) study was to use dynamic instructions that change depending on the stage of skill acquisition: The students' attention can be reoriented and recommendations about self-control can be introduced. For example, in the beginning the student is afraid that he or she might hit the left arm with which the flat chisel is held. Because of that the recommendation was made to install a rod in the vice at the same angle as the flat chisel. This device freed up the left hand from holding the chisel. The student can regulate hammering by the use of visual and kinesthetic information. In the second stage, a firing pin with a special surface is used. This results in sharper deviation of the hammer after incorrect hitting. This information is more helpful in directing the students to evaluating hammering more kinesthetically. In the third stage, the student uses a protective device that protects the left hand when holding the rod with the regular firing pin, which enables the student to hammer the rod with the same force and transfer his or her attention to the point of the rod fixed by the vice (expert workers should look at the object fixed by the vice, not by the firing pin). Other stages of the training process were developed in the same way. Applying a dynamic method of orientation to the students with a holistic method of training sharply increases the efficiency of the training process.

In another experiment, Khodikina and Portnoy (1985) taught students how to use a keyboard. They used different methods: The students altered between typing separate letters and their combinations, and were required to transfer attention from precision to speed and the reverse, and from free pace to a strictly required pace. Because of these variations, the speed of mastering skills increased more than two times over than traditional method.

The point of the study as it relates to the self-regulative theory of learning is that it shows the importance of a dynamic orientating basis of student performance. The activity goal may contain requirements that cannot be taken into account simultaneously by students. In these cases, instructions should be changed at different stages of learning.

Chebisheva (1969) and Bedny (1981) demonstrated that variation of pace performance of a task provides formation of an awareness of time and an opportunity to correctly regulate task performance in time. As pointed out previously, exercises performed at a faster pace then really required have a positive influence on performance.

For successful training, the objectively administered instructions and subjectively accepted instruction must match each other. It is, therefore, important not only to provide a dynamic orientation to the student, but to enable transformation of objective instructions into corresponding self-instructions. As has been emphasized previously, activity theory draws a distinction between objective stimuli and the perception, understanding, and acceptance of those stimuli.

The orienting basis of motor actions includes certain motor images, including kinesthetic, vestibular, acoustic, tactile, and visual, and awareness of time. Three kinds of information are used in the creation of motor images. One kind of information is based on the instructions provided by the instructor, another is based on the observation of others, the third is information that can be gathered during the student's task performance from his or her internal sources. It is important that these information sources agree on the meaning of the information they provide.

Venda and Ribal'chenko (1983) studied the training of electric power station operators, using various methods of presenting information in a mimic or representation of the station components. There were three methods of presentation, one in which all information was always presented to the operator, another in which the components in the mimic describing individual station operations were illuminated when these operations were performed, a third that reflected an algorithm of task performance by using a command information graphic display. The psychologist unexpectedly turned off vital equipment such as air and dust fans, feed water pumps, fuel lines, and the like. The operators had to recognize these faults, diagnose them, and perform required responses. Time performance, errors, motions, verbal responses, eye movements, and so forth, were registered.

The subjects were divided into three groups. The first group used the first method of presenting information, the second groups used the second, and the third used the third method of presentation. The first method requires a complete analysis of the emergency situation. The second method helps the operator to see and follow the links between the displays and controls. The third method frees the subject from analysis and decision making. In this situation, the final solution can be directly implemented.

The most efficient method was that of the algorithm. The more effective the method of presentation, the fewer intermediate strategies had to be used by the student. The conclusion reached was that the more complex the task, the more intermediate strategies are required, the strategy being a method of attempting to solve the problem. Figure 8.8 presents operator performance as a function of the method of information presentation. The third curve is smoothest because fewer intermediate strategies were developed by operators (only one strategy C_3 was used). It also reflects much less time for task performance. Curve 2 reflects the employment of two strategies (B_2 and C_2). And curve 1 demonstrates the operator used three strategies (A_1, B_1, and C_1).

The horizontal line in Fig. 8.8 demonstrates that the time performance of a task can be similar (points a, b, c, d). However the strategies of performance at all of these points are different. From this it follows that the performance time of the task is insufficient to draw a conclusion about the operator's level of experience. For example, at points a and d, the operator performed the task at the same time, but at point d, the operator has more experience than at point a.

Activity theorists make a special point of attempting to demonstrate that learning can be viewed as an active regulative process, because this principle is at the heart of the activity theory of learning. They emphasize that an important aspect of skill acquisition is the transformation of unconscious components of self-regulation into conscious ones. This is particularly important for the development of sensory-perceptual skills. The training process

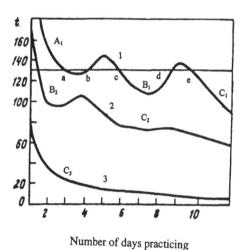

Number of days practicing

FIG. 8.8. Performance of control room operators as a function of method of information presentation. From Venda and Ribal'chenko (1983). A, B, C—method of information presentation.

includes description of indicative features of task and developing actions leading to identification of elements required for task performance. Often the process of formation of these kinds of actions takes a long time because they are formed through extensive, unconscious practice. Thus the relationship between conscious and unconscious processes of self-regulation becomes important in this case. Transformation of an unconscious level of self-regulation into a conscious one may be achieved through a method called "comparison of different actions" and especially through the process of verbalization.

Kalikinsky (1968) illustrated this point. His study deals with training students to diagnose malfunctions in tractor engines by listening to the noise the engines produce—for example, grinding, clicking, determining loudness, localization, and the like. In order to develop this diagnostic skill, different sounds were recorded onto a tape. In addition these sounds were reproduced on an oscilloscope. Acoustical stimuli varied and student could compare them by using acoustical and visual information. The student compared the acoustical sounds with the visual patterns on the oscilloscope and was instructed how to correlate the sounds to the oscilloscope patterns. Following that, the students attempted to recognize sound without visual information. If sounds were not correctly recognized then the visual information would be repeated. Verbal descriptions for the sound stimulus were introduced where possible. Then trainees were expected to recognize and differentiate the sounds, as well as describe verbally where possible. If these procedures were performed incorrectly then the preliminary procedures were repeated. This enabled the students to regulate consciously the perceptual and mental actions of the diagnostic tasks.

The training process may be represented as several stages. The first entails learning to discriminate the various sounds in conjunction with the associated visual information. The second entails assigning a diagnostic meaning to the sound. The third is the performance of a diagnostic task under operational conditions. The relationship between these stages is not strictly sequential and depends on the result achieved by the student.

Similar methods were used for developing sensory and perceptual skills by Kharkin and Bedny (unpublished). Analysis of this study showed that associations of visual stimuli with acoustical stimuli and than with verbal responses are not simply a result of their pairing as it is described in associated theories of learning. The experiment demonstrated that the trainee actively selects identificational features of stimuli, compares them, and organizes them in a system based on perceptual, thinking, verbal, and motor actions. Activity is analyzed and corrected according to the goal of performance, feedback, subjective criterion of success, and so on. This demonstrates that associative learning has a complicated recursive structure.

The assignment of a verbalized meaning to nonverbal stimuli, such as sounds, is of course very common. The American coauthor of this book had personal experience of this when, in his first professional job with the U.S. in 1951, he was required to learn sonarman skills. These involved not only the need to discriminate underwater sounds produced by enemy vessels (e.g., motor torpedo boat, destroyer), but also to interpret the visual display of an amorphous, highly unstructured sonar pip as a vessel or a natural ocean object. Western psychologists might, of course, interpret the assignment of meaning to nonverbal sounds and lights as simple association of stimuli. However, as indicated previously, learning by simple association is viewed by activity theorists as a primitive form of self-regulation.

One component of training is learning by observation, either as part of a more holistic training process or as a relatively independent one, in which case it has its own goals, motives, and actions. However, observation as an independent process is possible only in simple situations. Learning by observation is significant for the social learning theory of Bandura (1977). For example, a new employee learns how to perform a specific task by observing an experienced employee. The behavior to be learned must be overt, of course, which makes observation inappropriate for cognitive learning tasks.

Some activity theorists view learning by observation as based on self-regulation mechanisms. They see the process of observing activity as one of interpretation based on the observer's developing interpretative strategies of performance.

In the study done by Bedny (1979), subjects performed tasks that involved inserting pins into holes. Actions and their sequence as performed by subjects may be precisely observed. Thirty subjects were divided into two groups: One used individual training by direct experience, and the other used training through observation. Two components of skill acquisition were discovered. One group of components was connected with a certain sequence of actions. The other group was associated with a method of performing particular actions (strategy of attention, method of grasping pin, sequence of the decision-making process, etc.). The first group of components may be learned by observation. However, the second group of components cannot be learned through observation because they are cognitive in nature, or are otherwise inaccessible to observation. Among the latter there are those that are potentially subject to direct observation only through performance or through direct supervision. Learning by direct experience was, as one might expect, significantly superior to that of learning by observation. This suggests that observation can enhance skill acquisition when it is part of a larger training program, but cannot be substituted for direct experience. Learning by observation play a more important role in social behavior. However, leaning by observation cannot be reduced to memorization and subsequent reproduction. Observed events are actively interpreted by the subject and later the subject may demonstrate totally different behaviors than those initially observed.

THE MAJOR THEORETICAL CONCEPTS

The following are the major theoretical concepts of learning that are based on the activity principles:

1. Learning is a process of sequential changes of cognitive executive and motivational aspects of activity that emerges from external conditions and the student's own actions. The student as a result of training acquires a knowledge of reality that differs from his or her pretraining knowledge of reality. She or he also acquires general principles of thought required to solve problems, and a system of actions required to perform different tasks.

2. In contrast to cognitive psychology, which breaks down complex mental processes into separate stages for the purpose of explaining and describing the structure of mental processes, the theory of learning based on activity principles divides mental processes into mental actions or operations that can be used by learner in training process. The complex system of knowledge and mental processes dissects into their components knowledge units and mental operations in such a way that they become elementary for the leaner and can be used for acquiring new knowledge and skills. The notion of elementary knowledge units and mental operations is relative and can be changed as a result of learning and cognitive development. Dividing mental processes into their components depends on instructional objectives and the stage of development of learners' mental processes.

3. Reflexes, associations, and mental processes or their particular stages are not used as units of analysis in the learning process. Rather actions and mental operations and their systemic organization are required to solve problems or particular types of tasks. Activity is considered a self-regulative system with a hierarchical structure. A student evaluates the consequences of his or her action and changes the method of performing an activity based on external and internal criteria, past experience, and the significance of the task.

4. Learning is not only a process of strengthening associations based on the memorization and selection of required information, but also is a complex process of functional and cyclic changes in the structure of mental operations and activity.

5. As a result of self-regulation, the strategies of activities performed by students change during the learning process. The more complex the acquired activity, the more intermediate strategies (in the sense of preliminary strategies leading to the most efficient one) will be utilized by the students. Learning as a transition through stages and strategies requires the student's dynamic orientation. Learning acquires a sequential phase character. As learning proceeds, the number of intermediate strategies is reduced. Only

in a simple situation, when the student has adequate past experience, can she or he acquire knowledge and skills by bypassing intermediate strategies.

6. Human learning has a multilevel organization. One level is involved with the unconscious aspects of self-regulation and is related to associative learning, which in turn is based on blind trial and error, reinforcement, and reward, and regulated by behavior consequences. The associative level includes unconscious sensory, motor, and sensorimotor learning.

The second level of learning involves a conscious level of self-regulation controlled by a conscious goal and is characterized by conscious analysis, selection of data, recognition of problems, development, testing, and selection of hypotheses, the evaluation of performance, and correction of activity in accordance with the established goal. Both levels are interconnected and influence each other. By changing instructions one can change the conscious strategies of student's activity, and through this, indirectly influence the unconscious level of self-regulation. By reorienting the student's attention, it is possible to change the relationship between two levels of self-regulation and transfer some components of unconscious associative learning to the cognitive level.

7. In the initial stage of learning, internal and external explorative activity is especially important, because, based on it, a student analyzes the situation and reorganizes and corrects the structure of his or her activity. The comparison between actions and their consequences corrects behavior. As skills are acquired, external explorative activity is reduced and activity is performed in an internal plan.

If a student has enough training and/or experience, explorative activity is immediately internal. In external exploration the student learns by doing, whereas in internal exploration she or he learns through observation, comprehension, interpretation, and evaluation of the mental result. As skill is acquired the role of explorative activity and feedback is reduced. Program components of self-regulation play a great role.

8. During learning the emotional-motivational aspects of self-regulation change. At the first stage of learning, frequent transition from evaluating the result of an activity to evaluating a student's own ability to perform the task is observed. This process is conversion of the goal to obtain some task outcome into a self-evaluative or egocentric goal. This aspect of self-regulation is an important source of the internal motivation to learn. During the gradual acquisition of knowledge and skills, the frequency of these transformational processes is reduced, as well as the student's emotional sensitivity to their own results.

9. In learning, feedback is important, but self-regulation cannot be reduced to the effects of feedback. In self-regulation orienting, programming, executive, and evaluative components are as important. Based on feedback,

the consequence of an activity is compared with the goal, using success criteria. This enables the reformulation of strategies that are changed into more efficient ones. The role of feedback depends on the stages of learning; it changes in intensity and content. Extrinsic feedback influences the process of formation of intrinsic feedback. In the training process, extrinsic feedback should activate the process of forming correct intrinsic feedback, but not be substituted for it. In solving this problem, it is important discover the relationship between the conscious and unconscious level of self-regulation.

10. The process of acquiring knowledge and skills is considered a process of either internalization, when the actions are performed according to the particular levels and provide a transformation of material or materialized actions to a mental plan, or as a process of active formation of actions based on the mechanisms of self-regulation, which provides a complicated reconstruction and developing of external actions and internal mental operations.

11. One important principle of developing cognitive skills is the step-by-step methodology of forming complex mental activity from its component elements (actions or operations). These principles provide direct connections between systems of mental operations and verbal instructions, empowering in many cases students' conscious awareness of their mental operations and their system's nature. Systems of instruction presented to the students depend on the type of cognitive activity. Four types of processes have been referred to under cognitive activity: algorithmic, semialgorithmic, semiheuristic, and heuristic. These made it possible to build a unified system of instructions that may be used for development of the whole spectrum of mental activity.

8.5 CONCLUDING COMMENTS

In the West it is common to separate theories of learning from theories of instruction. Theories of learning emphasize the description and explanation of qualitative changes of behavior as a function of different kinds of practice. They frequently neglect the issue of training design. In contrast, Russian psychologists treat instruction as technology grounded in a theory of learning. Technology of instruction in some instances may be similar. However, their theoretical basis and method of explanation may differ. The successful technology of instruction often is based on intuitive decisions of teachers. Theories of learning in such instances enable us to explain these successes and facilitate their dissemination. According to Russian scientists a theory of learning that does not lead to a design of instruction is inadequate.

Analysis of the material presented in chapters 7 and 8 exhibits the fact that there are various theories of learning in the former Soviet Union. Until the 1950s only the associationist concepts of learning existed. All of them were derived from Pavlov's theory of conditioned reflex, which had official government sanction. Following the death of Stalin all theories of learning then stemmed from the principles of activity. Related to these theories are the theory of stage-by-stage formation of mental actions and skill, the algo-heuristic theory of performance, learning, and instruction (AHT), and the self-regulation theory of learning. Fundamental to these theories are such notions as knowledge, mental and physical actions, goal, motive, strategies, and the like. The activity perspective distinguishes tangible material objects and mental objects, that is, knowledge (images, concept, and propositions and the connections). Material objects may be transformed by physical actions. Mental objects are transformed by mental actions or operations.

The major task of teaching is the formation of goal-directed actions and their system (strategies), not merely associations. Associations that are formed in the learning process are not sufficient for acquiring information of goal-directed activity. Actions and strategies of activity empower the learner to actively change the situation, as well as his or her own behavior to attain the goal, thereby transcending the limits of merely adapting to a changing environment. In this regard the motivational processes are especially important for establishing goals, the ability to extract the more relevant information, and the ability to evaluate results and correct strategies of performance. The ability of mental actions to transform elements of knowledge and the relationship among them provides, where appropriate, an opportunity to overcome the associations formed through previous learning or experience. This makes it possible to solve novel problems. In this respect all theories of learning that are based on activity principles differ from all kinds of associational psychology. This includes behaviorism, which explains mental functions and behavior exclusively in terms of associations formed in past experiences, or through behavioral acts acquired in the course of blind trial and error. The theories of learning presented in chapters 7 and 8 do not deny the role of association, however, they do not reduce the process of learning to such mechanisms. Associations are considered the result of an active process that cannot be reduced merely to pairing in time or space of stimulus and responses. Underlying association is a complex process of self-regulation. If associations are formed unconsciously, their development may be described in terms of unconscious self-regulation described in section 1.3. If associations are formed consciously as a result of automatization of conscious actions then another level of self-regulation is implicated (cf. section 2.5).

Theories of learning based on the principle of activity are intimately related to the cognitive psychology of information processing. This branch of cognitive psychology seeks to understand how people acquire new information, how they store and recall information, and how prior learning determines subsequent learning. Human beings are viewed as symbol manipulation systems. This theory treats knowledge acquisition in terms of a linear sequence of steps or stages of information processing. Special attention is devoted to the memory functions. However, the theory of activity focuses not merely on stages of processing and its reorganization in memory, but also on goal formation and motivation, and determining the content of actions and strategies of performance. Learning is treated as a nonlinear, recursive process. Special attention is paid to the process of thinking. The learner is not simply an informational system, but rather a subject with his or her own goals and relationships to reality. As can be seen, in cognitive psychology psychological processes are divided into separate stages independent of the goal of learning. By contrast to theories based on principles of activity, the learner's activity is divided into actions (mental and motor) that may be successfully mastered by the learner. This problem assumes particular importance during the period of acquisition of mental skills and knowledge. It was most successfully studied in the algo-heuristic theory of learning. Fundamental to this theory are the notions of elementary mental actions or operations. The notion of elementary mental actions depends on the stage of learning and individual idiosyncrasies. For example, specific mental actions may be elemental for one learner, but not for another. Thus, the analysis of mental processes into their components (mental actions or operations) depends on instructional objectives and the stage of development of the learner's mental processes. The problem of determining the degree and level of decomposition of cognitive processes and behavior into elementary actions is not typical for cognitive psychology or behavior approach.

Chapters 7 and 8 presented the common fundamentals of theories of learning based on activity principles as well as features that differentiate the theories. In the theory of stage-by-stage formation of mental actions and skills, the acquisition of new knowledge and skills is treated as a process of internalization. A guideline for a student's activity is considered a process that provides transformation of material or materialized actions into its mental form. From this it follows that successful constitution of mental actions requires performance of actions appropriate to a particular stage (an act based on material or materialized level, an act based on audible speech without direct support from objects, an act involving internal speech to oneself, an act using internal speech). No doubt teaching to these levels often provides a solid basis for the effective formation of corresponding mental actions. However, the approach based on these levels to the formation of mental and motor acts often became pro forma rather than substantive. These stages are not always required or useful. The teacher and student are often completely unaware of the structure of mental processes involved in solving problems. The absence of a precise method of analyzing the logical structure of knowledge and method performance is the weakness of this theory. The algo-heuristic theory of learning developed an efficient method of analyzing and describing the logical structure of mental knowledge. Further, the AHT has developed methods of algorithmic or heuristic prescriptions based on its analyses and descriptions. The theory offers a methodology of the formation of mental skills. This methodology includes: (a) inculcating learners' consciousness of the systematic nature of the mental operations involved in the thought process, (b) formation of the capacity in learners to mobilize mental operations initially through outward instructions, then through self-instructions, and (c) combining individual operations into a system that are again initially mobilized through instructions then ultimately become responsive directly to problem conditions. The initially conscious mental processes gradually become unconscious. According to AHT these newly formed mechanisms underlie a number of abilities and qualities of mind including intuition.

In both of the theories discussed previously, acquisition of knowledge and skills is provided through a conscious, verbalized process. Learning is construed as a process of internalization. However, it is well known that not all aspects of skill acquisition can be verbalized. These theories do not pay enough attention to developing perceptual and motor skills that do not admit of verbal formulation. Goal formation and motivation processes, as well as dynamic aspects of learning are not captured by these theories.

According to the self-regulation theory learning may be considered a recursive process that is best interpreted as a sequence of changes in strategies in activity performance. Thus, the structure of an activity is gradually changed. Activity is progressively developed during learning, but is not internalized. The orienting basis of an action and its executive and control parts have a complex functional structure that may be described by the self-regulation model of activity. In the learning process one can observe complicated changes in the relationship among the cognitive executive and motivational aspects of activity. Based on the influence of feedback and comparison of obtained results with the goal, the structure of long-term memory is continuously reorganized. A formation of internal mental processes occurs. Gal'perin's (1969) method of verbalization is viewed as a process of comparison among verbal, motor, and other actions. This process provides for mutual correction, as well as for awareness.

The more complicated the acquired skills and knowledge are, the more intermittent strategies should be employed by the student for their acquisition. Actions are not simply transformed into an internal plane, but are actively formed. The transition from one strategy to another is not considered internalization, but rather a reconstruction of activity, and as the transition from the unconscious level of self-regulation into conscious and vice-versa. Methods of self-control continuously change. This process requires a dynamic method of prescriptions and instructions. Unlike in Gal'perin's (1966) theory where errors are treated only as undesirable consequences, in self-regulation theory errors that occur during the training perform an indispensable information process. Teachers should restrict these errors within limits, using them to develop methods of instruction. Some of the student's strategies may not be used in the final stage of learning, but they have a positive impact because they provide a watershed to a more adequate structure of activity. These strategies are less errors than iterations toward adequate structure.

An analysis of laws governing the formation of a structure of activity at various stages of learning, as well as the final stage enables methods of teaching derived from such laws. The theories of learning presented in chapters 7 and 8 are based on the theory of activity, supplement each other, and may be used efficiently in practice.

CHAPTER NINE

Characteristics of the
Work Process

9.1 OVERVIEW AND INTRODUCTION

This chapter deals with work space organization, fatigue, vigilance, and safety, all topics that have been dealt with extensively by many ergonomics textbooks. However, the activity and self-regulation theories look at these topics from a somewhat different viewpoint.

In Western ergonomics, work space organization (i.e., control panel layout and work station arrangement) is largely a matter of anthropometry and adjustment to the overt requirements of the task (step 1, step 2, etc.). Control panel layout, for example, is a means of most efficiently implementing the perceptual (information-seeking) and response requirements of the task.

Activity theory feels that work space organization should depend not only on anthropometry, but also on psychological criteria, for example, the accuracy and performance time requirements imposed by the task. This requires translating behavioral performance indices like time and error into their physical and spatial equivalents. For example, to achieve a projected error rate in manipulating controls, how should those controls be arranged? Making this translation has also been a critical problem for Western ergonomics.

Activity theory considers that control panel and work station arrangement should consider not only routine task performance, but task performance under conditions of high workload and stressful, emergency conditions.

For Russian ergonomists, work capacity is a more global concept than what the term *fatigue* implies, although the two may be inversely related.

374

Because task performance involves many function blocks, it is necessary to select appropriate criteria in the measurement of fatigue. For example, motivation to perform the task is a critical function in task performance; it seems logical, therefore, that one measure fatigue in terms of its effects on motivation. Western ergonomists are more likely to concentrate in fatigue measurement on the overt products of task performance. The individual's strategies to deal with fatigue must also be examined. In line with this, activity theorists would examine mental as well as physiological fatigue; this requires special tests of such factors as attention and memory. Because of their interest in physiological mechanisms, Russian ergonomists have placed heavy emphasis on physiological measurement and the classification of the ease and difficulty of work in terms of these mechanisms.

Russian theorists view monotony and vigilance as highly related. In their view, stereotypy or the similarity of task performance conditions leads to a loss of vigilance that is anchored in the general arousal conditions of the individual.

Monotony, therefore, is viewed as an informational process. Activity theorists distinguish between physiologically and psychologically induced monotony. The latter rests on more variable functional aspects of behavior, such as motivation, acceptance of task goals, and evaluation of task significance, all readily recognizable from activity theory and the model of self-regulation. Physiological monotony is more unitary and homogeneous, depending on the neural state of the individual.

The key to the problem of preventing monotony and loss of vigilance is motivation. Because vigilance is an information process, providing information to the worker in the form of feedback relative to goal accomplishment and time is a key preventive mechanism.

Because motivation is a central factor in activity theory, Russian scientists view it as critical in the effort to achieve a safe environment. The motivation to achieve a goal may lead the worker to ignore potentially threatening dangers. Linked to this is peer pressure, the need to achieve group consensus with regard to task accomplishment.

Error analysis is considered a critical analytic methodology in determining danger "points" in the work situation. Error may occur at various stages in the work process and is linked to the dynamics of self-regulation, for example, incorrect formulation of the goal, incorrect interpretation of task conditions, and decision making. Inadequate strategies in the use of information may also pose dangers to the worker.

Although safety is a significant topic in its own right, it cannot be divorced from activity theory and self-regulation processes as described in earlier chapters. Despite the psychological emphasis in activity theory, it rejects the concept of the accident-prone worker and looks for causal mechanisms in situational factors.

The systemic-structural analytic method described in chapter 6 can be used to determine design features that should be modified to prevent potential errors and dangerous situations.

9.2 WORK SPACE ORGANIZATION

It is obvious that in ergonomics the correct organization of work space is extremely important. The principles used by Western ergonomists to lay out control panels and arrange multiple work stations in the work area are considered by Russian ergonomists as not sufficient, because they fail to include factors other than anthropometry and overt task requirements. Among the factors they ignore are those that will be familiar to the reader from earlier chapters of this book, for example, motivation, goal formation and acceptance, task significance, task difficulty, and so on.

Depending on the nature of the task, its spatial characteristics have an influence on the difficulty and its performance time. By changing the spatial characteristics of the task, one changes both its difficulty and its performance time. This suggests that if one can determine which work space areas require a relatively similar performance time for a particular task, one can use this similarity as the basis for locating controls, displays, and work stations.

Work space organization therefore depends not only on the operator's anthropometric characteristics but on the nature of the activity as well. Western ergonomists would agree with this premise but would ask how the ergonomist can transform motivational factors, in particular, into spatial equivalents. Russian psychologists believe that by changing the cognitive and motivational characteristics of the task, one can utilize its performance time to develop the most appropriate spatial organization of the equipment that implements the task.

The criteria used for analyzing and evaluation work space organization must therefore include not only anticipated errors but also task duration. This suggest two things. For the analysis that precedes the development of a preliminary work arrangement one needs to have two types of database, one predicting error rate for different types of tasks, and another for response time for the same tasks. Once the proposed spatiotemporal arrangement has been mocked up, empirical tests of the arrangement can be made to modify the preliminary layout. One can use any of a number of predetermined time systems for the temporal requirements.

The importance of time to Russian theorists led Zefeld, Munipov, and Tchernisheva (1980) to use the duration of moving the arm to controls and searching for indicators on a control panel as the criterion for evaluating work space arrangement. In their study, subjects had to find an illuminated indicator in a bank of indicators and turn it off. The indicators were lit randomly; individual control panel areas could be selected according to identical performance times. Because the spatial organization of the various control panel areas depended on performance time, Zefeld et al. called this the spatiotemporal organization of space.

These authors suggested that the selection of control panel work areas depends not only on biomechanical but on psychological criteria as well. For example, the total control panel work area is determined by operator anthropometry, but this approach does not suggest how to allocate the controls and displays within this area. They concluded that the criteria for organizing the control panel should be the required precision (accuracy) of the action and its speed of performance.

According to the activity approach, there are two types of operator activity: a perceptual, informational search of the control panel with immediate operator response to stimuli, and an informational search with a delayed response. In the first case the operator automatically performs actions to perceived information (in Western terms well-learned response patterns). These actions were part of the operator's inventory of responses before the signals were presented. In the second situation, there is more complex information processing because the stimuli cannot be immediately interpreted and translated into a required response; the new information must be compared with that which exists in the long-term memory; actions are then selected based on logical rules and knowledge (in Russmussen's [1986] terms). Obviously, the duration of first-case responses is shorter than that of the second case. Zefeld et al. (1980) performed the first study dealing with immediate response to stimuli. The subject's work station included a control panel on which there were 10 indicators and corresponding controls, 5 on the left-hand side and 5 on the right-hand side. When an indicator was illuminated the subject had to press a corresponding button. This arrangement could be changed to satisfy varying task requirements. Four versions of the spatial arrangement were tested (see Figs. 9.1 and 9.2). Any point in the sensory field shown in the figures designates an indicator, and any point in the motor field designates the corresponding controls. The controls and indicators are located within the bold line.

Subjects worked on all four versions of the space arrangement in the 30-minute experiment, performing 200 responses to signals presented randomly; the time interval between signals varied from 5 seconds to 8 minutes.

In a following series of studies, various complications were introduced that required remembering the sequence of signals and the names of signal positions, and delaying the response according to specified rules. This imposed a more rigorous cognitive workload. The experimenter measured the performance time of different tasks and the error rate, and interviewed the subjects.

The results of the experiments suggested that the arrangement presented in Fig. 9.2 was the best, although according to anthropometric and

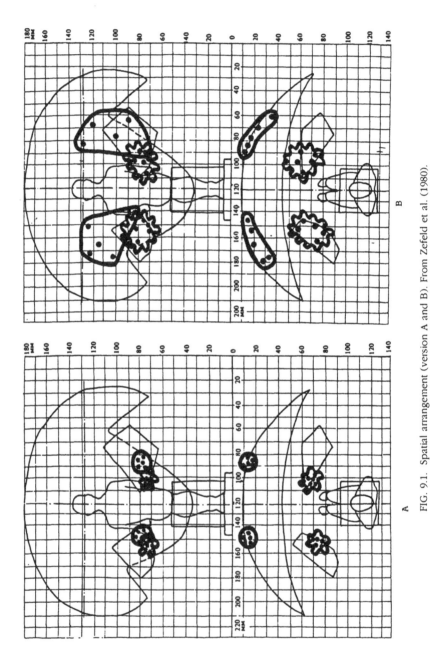

FIG. 9.1. Spatial arrangement (version A and B). From Zefeld et al. (1980).

378

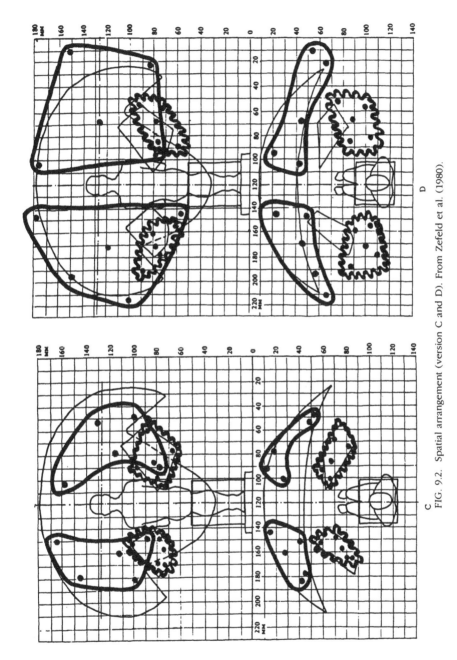

FIG. 9.2. Spatial arrangement (version C and D). From Zefeld et al. (1980).

379

biomechanical criteria the best arrangement was the version in which controls and displays were colocated. However, the experiment indicated that the subjects had to spend more time finding the required indicators and controls in version A. This was because the first version was unsatisfactory from a psychological point of view. If the cognitive workload was increased during task performance, the advantages of the second version also increased. When the workload of the working memory increased, version D became the best because it had the most dispersion of controls and indicators.

Differences in the cognitive workload appeared to influence the desirability of control panel. Taking only the normal and maximum work areas into consideration during the design of work space is not sufficient. Here, the nature of the operator's work and his or her orientation and position during the work process are also important. This requires introducing the vector-coordinate method of analysis of work space (orientation and position of the person in space) and finding the spatial surfaces that determine the person's position in relation to elements of the environment (system of coordinates in which the activity is performed). This means that a person assumes different body positions during work performance. In anthropometry the major positions are sitting and standing. Any deviation of the body from the ideal preferred position is considered undesirable. One of the major criteria for evaluating body position is the duration of a particular posture that deviates from the preferred one.

According to the activity approach, body posture that usually deviates from the ideal is determined by the nature of the task. Because of this, in addition to "ideal posture," it was recommended that the notion of "real posture" be introduced (Zefeld et al., 1980). Hence, during the organization of the work space, it is very important to discover the various postures the individual will assume during task performance. Any posture has its own system of coordinates for a particular activity and goal. The organization of the work space must be performed according to this system of coordinates.

9.3 WORK CAPACITY AND FATIGUE

In this section some ways of evaluating work capacity and fatigue, ways that may be less familiar to American specialists, are described. Work capacity and fatigue are dynamic components of activity, which may change during a work shift. Work capacity is the potential ability of a person to perform an activity under a given set of conditions. This can be evaluated as a function of his or her productivity and functional state during his or her shift. The functional state, which may change during the work process, describes the characteristics of those psychological and physiological functions on which the ability to process information and perform physical work depends. Work capacity also changes. Changes in the person's functional state during performance can be discovered based on psychological and physiological indices, for example, heart rate, blood pressure, electrical changes in skin, finger tremors, brain activity, and reaction time. These indices may also be used for workload assessment.

The notion of work capacity links the physiological functions of an organism with its behavioral activity. According to Zarakovsky and Pavlov (1987), there is a behavioral system, which is directed at achieving a conscious goal, and another functional system, which provides the energy supply for much activity. According to Anokhin (1962), the latter allows for the selective involvement and the integration of different physiological mechanisms in order to perform a particular behavioral act or function.

Fatigue is only one indication of the organism's functional state. An energetic functional system provides the selective activation of different physiological and biochemical processes and their integration into the system, which is directed toward achieving the particular goal of a task; self-regulation insures that activation should not be lower or higher than a required level. Physiological self-regulation involves stabilizing and compensatory factors. For example, sustaining the operator's body temperature when the temperature in the surrounding environment is raised can be considered a stabilizing factor. This is accomplished because the blood vessels widen and the flow of blood is increased. The latter reaction is considered a compensatory factor.

Reallocation of the organism's energetic resources depends on the nature of task activity and environmental conditions, as well as conscious or unconscious level of self-regulation. For example, when a person notices significant fatigue, he or she can consciously reduce his or her pace of work or sustain it by increasing his or her effort.

Vinogradov (1966) called work capacity and fatigue two sides of the same coin; the greater the fatigue, the less the work capacity. At the same time, other factors affecting work capacity are involved, such as the duration of warm-up and monotony. "Work capacity" for Russian ergonomics is a more general notion than "fatigue." Because work capacity and fatigue have such important consequences for productivity, it is necessary to measure them using a variety of psychological and physiological methods. The general paradigm is that measurements are performed at several work periods; the resultant data are compared and then correlated with productivity indices (Rozenblat, 1961).

Of course, there must be some initial measurement to be used as a standard of comparison. This may be the operator's functional state before the shift begins, or, as some specialists (Munipov, 1983) suggest, the worker's state after the first 2 or 3 hours. Depending, however on the measure selected, this may supply ambiguous data.

Suppose, for example, reaction time on different visual signals is used as a measure of capability (see Fig. 9.3). Reaction time is measure before the shift starts (point A), after 2 hours (point B), after 4 hours (point C), and at the 8th hour (point D). Obviously, reaction time can change during the shift. Because of the warm-up effect, reaction time initially decreased (compare points A and B). Between points B and C, reaction time was relatively stable, and until the end of the shift, reaction time significantly increased. The same reaction time can indicate a markedly different functional state. We can observe an increase in reaction time due to fatigue from point C to point D. A horizontal line can be drawn from point A to point "D." At these points, reaction time is similar. However, the operator's functional state is different in both cases. After point A, because of the warm-up, the functional state improves and reaction time decreases. At point "D," where we have the same reaction time, if the operator continues to work, reaction time increases.

It is thus important to take the dynamics of measurement indices into consideration, not only their absolute values.

Psychological and physiological functions may change at different rates. The most significant changes emerge in those systems that are most directly involved in work performance. For example, if the subject performs physical work, deterioration can be most readily observed in motor responses. The same can be seen in visual functions if the work depends on the visual sense. Changes in functions not directly involved with work performance sometimes cannot even be observed. This emphasizes the importance of correct selection of criteria specific to the work being performed. A preliminary experimental evaluation of different methods is recommended.

Because of changes in the functional state, the quality and productivity of performance may also change. Hence productivity is also used to evaluate work capacity, by comparing the person's functional state with his or her

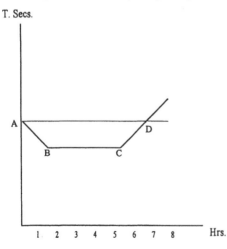

FIG. 9.3. Hypothetical changes of reaction time during shift.

productivity curve. The comparison is made in relation to varying phases, for example, in blue-collar work, the warm-up phase, the phase of stable work capacity, the phase of decreasing work capacity, and the final burst of activity. In some cases a high level of productivity can be achieved in spite of increased fatigue because, when the individual has high motivation, he or she can mobilize his or her resources. The individual may also change strategies of goal attainment. For example, Zarakovsky and Pavlov (1987) discovered that the requirement for helicopter pilots to sustain altitude within a margin of ±5 meters for 50 minutes could be achieved. However, the pilot's functional state and his or her strategies of performance had to change. These changes included an increase of heart rate with increasing flight duration, as well as with the number of control manipulations.

Energy expenditure increased when strategies of task performance changed. The two types of fatigue, mental and physical, are interconnected and influence each other. The measurement of mental fatigue is more complex than that of physical fatigue, requiring special tests of attention, memory, and response time. One method uses reaction time to different signals under varied conditions. For example, in a study by Troshikhina, Kol'chenko, and Moldavsky (1970), the subject had to react to three different stimuli in a 5-minute period with maximum speed. The operator had to distinguish among a triangle, a square, and a circle, by pressing the appropriate control. For example, if a circle, square, and triangle are presented on the screen, the subject should press the left button for a circle, the right button for a square, and give no reaction to the triangle. Under fatigue conditions, the reaction time and number of errors increased.

Zinchenko et al. (1977) studied mental fatigue from the standpoint of information-processing theory. As is customary in activity theory, function blocks were used as units of analysis. These reflect information processing at different stages of task performance. The functional model of memory used in the Zinchenko et al. study was presented in Fig. 3.6. Their idea was to discover how fatigue influences various function blocks. This was done by measuring fatigue after task performance was measured with special methods involving stimulus recognition, total reproduction, and determining an absent number in a series of numbers. The first method reflects short-term memory workload. The second method (reproduction) reflects short-term memory and the function blocks "search for memory" and "making a decision." The third method reflects long-term memory and the function blocks "search for memory" and "making a decision."

These methods can be performed in the following manner:

1. *Recognition:* The subject should be presented with a sequential list of numbers on the screen, after which postinstruction is given (instruction given after the presentation of the stimulus). The subject must press the "yes" or

"no" button depending on the number given during postinstruction and sequence of stimuli presented previously on the screen.

2. *Total reproduction:* The subject was then sequentially presented with numbers, after which she or he had to reproduce on the screen the entire array of numbers in any order. In this case, information not only was transferred into short-term memory but also had to be rehearsed.

3. *Determining absent number:* Before the presentation of stimuli, the subject was told the length of the alphabet of the numbers presented. Then, the subject was presented with a list of numbers in random order, the length of which had been shortened by one. The subject had to determine which number was absent. This task required not only the memorization of the material but also putting the numbers into order.

From Zinchenko et al.'s (1977) study, it follows that fatigue selectively influences some mental operations, such as those that require mobilizing attention. All three methods were recommended for evaluating mental fatigue.

Obviously, where particular sensori-perceptual components dominate in task performance, they require a special test, such as a visual test for tasks utilizing visual mechanisms. Electrophysiological indices, such as electroencephalogram, muscle tension, electromyogram, galvanic skin responses, and electrooculogram, may also be used (Myasnikov, 1963), but the intrusiveness of the instrumentation makes this method very difficult to use under field conditions.

When evaluating physical fatigue, such indices as heart rate, blood pressure, respiration rate, oxygen consumption, hand strength, endurance, and arm tremors (static and dynamic) are used. While comparing different functional criteria, it is important to extract the more critical ones. This is because the overloading of one functional system cannot be compensated by the underloading of another. For example, excessive workload of the cardiovascular system cannot be compensated by a lower workload of the respiratory system. Of course, the measurement of fatigue does not mean much without determining the effective cause, of which there may be many, for example, the complexity of processing information, a high pace of work, a high level of physical effort, or inefficient rest schedule. One important method of reducing fatigue and increasing work capacity is to determine a sustainable work pace and organize the work–rest schedule correctly.

An important method of reducing fatigue is to determine the length of rest or recovery time. In the former Soviet Union, Rozenblat (1975) developed an efficient and simple method of evaluating physical workload and determining the duration of the rest periods. He created quantitative criteria for estimating "heaviness" of physical work. This can be more or less fully described by pulse rates. Pulse rates have been shown to be strongly associated with muscle and neuro-emotional stress, as well as with variations in the body's thermo-regulatory systems (Rozenblat, 1975; Scherrer, Ed., 1967). Of course pulse measurement does not preclude the use of other measures. The frequency of heart rate is a linear function of oxygen consumption, which allows one to determine, with considerable precision, how a worker spends energy in different forms of work. On the other hand, the pulse rate is not an informative measure when studying static physical work because there is no linear relationship between static work and pulse rate (Lehmann, 1962; Monod & Pottier, 1967). Very often the cardiovascular system is the more overloaded system during physical work. In some cases, the pulse rate and expenditure of calories can be used as the criteria of strenuousness of workload resulting from warm environmental conditions (Klenovich, 1971; Rozenblat, 1975).

Pulse rate can be used not only to evaluate heavy physical work but also to solve other problems in ergonomics. Pulse rate can be used for evaluating the layout of work setting and designing for fitness for use (Monod & Pottier, 1967). For example, in determining the optimal height of a seat required for a driver, researchers had experimental subjects alternately sit and stand up to the beat of a metronome. They then varied seat heights, and measured pulse rates at the various seat heights as the dependent variable (Monod & Pottier, 1967).

Pulse rate also can be used for evaluating the emotional state of the operator, as with astronauts. When astronauts leave satellites or shuttles their pulse rates increase significantly. For example, the Soviet Cosmonaut A. Leonov's pulse rate was measured during "space walks" at over 162 per minute (Khrunov, Khachatur'yanz, Popov, & Ivanov, 1974).

How can one classify the strenuousness of work into different levels in terms of pulse rate? At present there are a number of approaches. Christensen (1953) suggested the classification presented in Table 9.1. Wells, Balke, and Van Fosan (1967) proposed the system shown in Table 9.2. Other systems also exist.

TABLE 9.1
Classification of Strenuousness of Work

Very Easy	Easy	Average	Heavy	Very Heavy	Unusually Heavy
Up to 75 beats/min	75-100 beats/min	100-125 beats/min	125-150 beats/min	150-175 beats/min	above 175 beats/min
2.5 kcal.min	2.5-5 kcal/min	5-7.5 kcal/min	7.5-10 kcal/min	10-12.5 kcal/min	above 12.5 kcal/min

Note. From Christensen (1953).

TABLE 9.2
Classification of Strenuousness of Work

Easier Work	Moderate	Optimal	Heavy	Very Heavy	Exhausting
Below 100 beats/min	100 - 120 beats/min	120 - 140 beats/min	140 -160 beats/min	160 - 180 beats/min	above 180 beats/min
below 4 kcal/min	4 - 7 kcal/min	7.5 - 10 kcal/min	10 - 12.5 kcal/min	12.5 - 15 kcal/min	above 15 kcal/min

Note. From Wells et al. (1957).

It is important to establish boundary levels between acceptable and unacceptable work load. According to Lehmann (1962) the amount of calories that can be expended by one person during an 8-hour work day cannot be more than 2,500 kcal, which equals 5 kcal/min. In Christensen's (1953) terms, Lehmann is calling for an easy level. Lehmann suggested that a full day's expenditure of kcal can be 4,300 kcal. From this, personnel need to allocate 2,000 kcal for 8 hours of work or 4 kcal/min. Lehmann emphasized that under these workload conditions the worker's useful work life is an extended one. Lehmann cited the German Work Research Association's recommendation that the amount of energy expended cannot be more than 4.17 kcal/min, which equals a pulse rate of 100/min. The frequency of the pulse rate of 100 beats/min and the energy expenditure of 4.17 kcal/min borders between the easy and heavy work criteria used in the former Soviet Union (Hygiene Standards, 1963). Rozenblat (1975) compared the workload in a sport situation with a production situation and concluded that the levels bordering between acceptable and unacceptable are 100 beats/min. In all cases, when the pulse rate increases beyond the standard, it is necessary to introduce an additional break time, calculated according to Lehmann's (1962) data as:

$$L_p = (\text{Actual expense}/4.17) - 1 \qquad (9.1)$$

Actual expense in this formula is presented in kcal/min: 4.17 is the level of energy expenditure in kcal/min, which corresponds to a pulse rate level of 100 bits per minute. Although evaluation of energy expenditure in a work situation is very difficult, Rozenblat suggested a simple method for acceptable break time using pulse rates that are sampled during the work shift. If the workload is stable over the shift, the measurement is conducted at the end of the first hour of work, at the end of the third hour of work, 1 hour following mealtime, and during the last hour of the work shift. The general duration between measurements is about 2 hours.

Rozenblat's (1975) method suggests calculating the following measures: (a) maximum pulse rate (MPR) during different operations, (b) average pulse rate during work (PR$_w$), (c) average pulse rate during break time (PA$_{br}$), and (d) average pulse rate during shift (PA$_{sh}$). Item (d) can be calculated by the following formula:

$$PA_{sh} = (PR_w \times T + PA_{br} \times T_{br}) / (T + T_{br}) \qquad (9.2)$$

where T is the overall duration of the actual work performed and T$_{br}$ is the duration of break time during shift.

Rozenblat (1975) also suggested the classification in Table 9.3 of strenuousness of work according to the average pulse rate during shift.

An important criterion of intervention is PA$_{sh}$. If the value for PA$_{sh}$ is less than 100 beats/min, one does not need additional break time to rest. If PA$_{sh}$ is more than 100 beats/min, additional break time can be calculated.

As a first step, the method calls for calculating a theoretical break time, designated as calculated break time BT$_{cal}$, in percentage of the shift time. Using these data and real break time (BT$_{rl}$), one determines at the next step their required break time (BT$_{rq}$) in percentage of the shift. The criterion BT$_{cal}$ assures that the average pulse rate during the shift will be less than 100 beats/min.

BT$_{cal}$ for an 8-hour shift is determined by the following formula:

$$BT_{cal} = 100 (PR_w - 100) / (PR_w - PA_{br}) \% \qquad (9.3)$$

The next stage entails calculating BT$_{rqu}$ in percentage of the shift:

$$BT_{rqu} = \frac{(BT_{rl} + BT_{cal})}{2} \qquad (9.4)$$

Formulas 9.3 and 9.4 are used for shifts that have durations between 7 and 8 hours and a general duration of 41 hr/week. In cases where the

TABLE 9.3
Classification of Strenuousness of Work

Easier Work	Average Physical Intensity of Work	Heavy Physical Intensity of Work	Very Heavy Physical Intensity of Work
90 beats/min	90 - 99 beats/min	100 -119 beats/min	over 120 beats/min

Note. From Rosenblat (1975).

duration of the shift is 6 hours and the general duration of work is 36 hr/week, after calculating BT_{cal} one should deduct 17% of the time from the shift (1 hour of the shift).

The pulse rate method can be used to estimate the cost effectiveness of environmental improvements, and to reduce the strain of physical work (Bedny, 1979). This approach can be used in situations where the average pulse rate is higher than 100 during the shift. This method is as follows: At a given level of productivity, one determines the amount of break time for those conditions. Following that, one can calculate break time under the new anticipated conditions (e.g., mechanization, improvement of climate, etc.). The difference in the break before and after implementing the changes measures the savings in work time. Once one knows the cost of work time, the effectiveness of the intervention program can be determined.

This method was used to estimate the effectiveness of improving the microclimate in a production process (Bedny, 1979). The task was to determine the cost effectiveness of introducing air conditioning in the cabin of a large earth remover excavator. The experimental procedure involved placing the excavator's cabin in the laboratory. Ten male subjects worked with this simulator, which produced a psychic and physical workload close to reality. The ambient temperature was set to approximate real working conditions. The experimental condition involved the presence or absence of air conditioning in the cabin. Testing took 2 hours. Before the task began, the subjects spent 25 minutes in the cabin, adapting to the climate. During the experimental simulation, the pulse rate (PR_w), breathing rate (B), and blood pressure were measured. In one condition (no air conditioning) air temperature was 40–41°C, wind speed lm/sec, and relative humidity 55%. In the air-conditioning stage, subjects performed the same task with air temperature at 24°C, wind speed at lm/sec, and relative humidity at 45%. Without air conditioning, the pulse rate (PR_w) was 105.1 beats higher during work time and 45 additional minutes were required for rest during the shift. With air conditioning, break time was not required (according to pulse rate criterion).

Another method of evaluating physical fatigue, which can be used in certain cases, is to measure the hand's muscle force and static endurance. Static endurance is defined as the length of time muscle force can be sustained. According to Rozenblat (1975), the arm's muscle strength at the end of a shift cannot be reduced sometimes whereas endurance can be (moderate level of fatigue). If both indices are reduced, there is strong fatigue. Maximum muscle strength should be measured first. Then specialists calculate .75 value from maximum muscle strength. In test procedures, the subjects should sustain this strength for as long as they are able. The amount of time they are able to sustain this effort is the criterion of static endurance. The subjects are timed, but they are not allowed to know how long they have been sustaining their efforts. To evaluate fatigue, the scale presented in Table 9.4

TABLE 9.4
Scale to Evaluate Fatigue According to Sttic Endurance Indices

Evaluation of Fatigue		Small	Average	Large	Very Large
Score		I	II	III	IV
Interval of Reducing	Physical Work	up to 20%	20.1% - 30%	30.1% - 40%	higher than 40%
$B_{0.75}$ (%)	Mental Work	up to 15%	15.1% - 22.5%	22.6% - 30%	higher than 30%

Note. From Rosenblat (1975).

is used. The major distinction between Russian and Western approaches to fatigue is that Russians emphasize the dynamic aspects of fatigue during the work shift and consider fatigue in the context of the self-regulation process.

9.4 MONOTONY AND VIGILANCE

Monotony and vigilance are also characteristic of work. In this section, some aspects of this problem are considered from the activity point of view.

Monotony, from an activity point of view, is considered a functional state of an organism and is conveyed by a negative emotional-motivational state as boredom, reducing motivation to achieve the goal, and increasing motivation to stop working. Its effects resemble those of fatigue, in the sense that it reduces performance. In contrast to fatigue, however, a feeling of monotony can arise when performing light physical and mental labor. Thus, fatigue and monotony result from different causes. These differences are discovered by introducing variety and novelty in work. Whereas productivity is reduced because of monotony, variety and novelty will increase it. If however, productivity is reduced because of fatigue, variety and novelty will have no positive effects.

Monotony is an idiosyncratic factor; different people feel monotony differently. Certain objective factors tend to produce monotony—primarily repetitiveness of task stimuli—although other factors may interact, such as task simplicity.

Aseev (1974) hypothesized that three groups of factors cause monotony: energetic, informational, and environmental. The energetic factors depend on the nature of a person's motor activity, characterized by repetition of similar actions, their short duration, and low physical effort. The informational factors depend on the nature of the operator's cognitive workload. Monotony is produced by repetitiveness of mental actions, their simplicity, lower significance, and a long waiting period for relevant stimuli to action. Environmental factors include such influences as unvarying environmental stimuli, such as noise or temperature. These factors reduce the stability and intensity of one's attention.

Although these factors are interconnected, their separate consideration is convenient and can help in solving practical problems. For example, in some cases monotony can be reduced by increasing physical or mental effort, and in other cases by introducing music or changing the temperature.

Another distinction is between psychologically and physiologically induced monotony. Psychological factors include how a person accepts task goals, evaluates the significance and complexity of the task, the subjective criteria of success she or he uses, and the requirement to sustain a particular level of attention. From a physiological point of view, repetitive stimulation of the same neural cells can result in a pathological human state. For example, Vinogradov (1966) discovered that if a repetitive, flashing light with a frequency of 10–25 Hz is presented, brain activity will be lowered. An encephalogram demonstrates changes similar to those that can be observed in epileptics.

Hebb (1955), while studying monotony, concluded that sensory environmental changes are required for a normal state. Moruzzi and Magoun's work (1949) is important to understand the physiological mechanisms of monotony. They proved the importance of reticular formation in these processes. From the physiological point of view, the factor of novelty is also important. Sokolov (1960) and others discovered that reticular formation and other parts of the brain have "detectors of novelty." Repetitive influences of the stimulus can change the functional state of reticular formation and thus influence a different part of the brain.

According to Aseev (1974), the feeling of monotony may create the appearance of fatigue. This can be explained by objective and subjective factors. Objective factors include the involvement of only a few muscular components and neural centers in work performance, which produces fatigue of a local character. A subjective factor would be a "feeling of tiredness," which can cause real fatigue. One kind of monotony is called "psychic saturation" (Karsten, 1928), which refers to a period when a person has a very strong desire to stop working. This psychic state has increasing effects: A person demonstrates variation in task performance, the quality of work is reduced, and there is complete inability to perform the required task.

The speed of emergence of psychic saturation depends on the task's significance and how much the person is motivated to perform it. An effect of monotony is sometimes called "boredom." One is always aware of boredom, but not necessarily of its causes.

Another factor that may produce monotony is the need to remain vigilant in the absence of external stimulation, for example, monitoring a sonar or radar screen. One who drives at night for a long time, for instance, can drive with his or her eyes open but not see anything. Sometimes this is conveyed by a hypnotic state: The driver can automatically drive the car, and correctly react to a situation, but very slowly (Aseev, 1974). The same phenomenon was observed in a study of pilots who fly at night (Grimak & Ponomarenko, 1971). Indeed, some catastrophes in aviation can be explained by this phenomenon. Khruselevsky (1971) described three types of monotony effects: (a) psychological—feelings of tiredness, saturation, boredom, and drowsiness, (b) physiological—reduced pulse rate, breathing rate, oxygen consumption, and muscle tension, and (c) productive—changed duration of task performance and reduced productivity. The author demonstrated that in production operations, the worker's physical workload cannot be too low to maintain the worker's active state (oxygen consumption cannot be lower than 500 ml/min and the expenditure of kcals cannot be lower than 2500 kcal/min). Zolina (1967) and Marchenko, Kandror, and Ryazanov (1972) developed technological criteria to evaluate the causes of monotony, including the repetitiveness of an operation, duration of a task, and the number of elements (actions) in production operations (see Table 9.5). Increasing the pace of work performance can result in decreasing monotony (Aseev, 1974).

Changes in productivity may reflect monotony, although it may be difficult to distinguish these changes from normal work variability. The first to propose using a production curve to evaluate monotony was Wyatt (1927) who suggested that there is a typical productivity curve for monotony. However,

TABLE 9.5
Criteria for the Evaluation of the Degree of Monotony

Task Characteristics	Degree of Monotony			
	Low - 1	2	3	4 - High
Repetitions (per hour)	180	181 - 300	301 - 600	over 600
Duration (seconds)	100	100 - 41	40 - 21	20 - 2
Number of elements	10	10 - 7	6 - 4	3 - 2

Note. From Asseev (1974).

Rothe (1946) later showed that there may be several such curves. Despite this, the productivity curve, in combination with other criteria, can be useful in ergonomics. Of course, productivity curves cannot be used as a criterion of monotony when decreasing productivity results technological factors. Among psychological factors that relate to monotony (but do not unequivocally determine it) is the worker's feeling of dissatisfaction. For this purpose, an interview or questionnaire can be used.

Self-regulation mechanisms suggest how to solve the problem of ensuring against monotony. The key is motivation, which is inherent in the work process and is a consequence of achieving the goal. Two aspects of motivation follow from the self-regulation of activity. One is connected with the procedural components of motivation, and the other with the consequential aspects of motivation. In the first case, the major source of motivation is the work process itself. In the second case, the motivation derives from the consequences of activity. For example, if a small child plays with a toy, the major motivation is the process of play. The major motivation for the adult performing a job that lacks inherent interest for her or him is the salary paid and the rewards that salary can provide. In some cases, as in monotonous work, the work process may produce negative motivation. In this situation, the motivational components connected with achieving a particular goal become very important.

Bedny (1979) and Bedny and Zelenin (1988) studied poor motivation in a monotonous production operation in both a production plant and a vocational school. Three series of experiments were conducted. The first series was performed in a laboratory environment, the second and third under production conditions.

In laboratory conditions, five subjects performed special tasks for 2 hours (1,800 tasks) per day for 2 days. The subject's emotional behavior was observed, and the number of operations and pauses for every 30 minutes was calculated. After the task subjects were interviewed. They worked without watches and meters for the first day, and with them on the second day. Because of this, the subjects had no information about what they accomplished on the first day. On the second, they knew what time it was and how many operations had been performed in a particular time. This permitted them to compare how much they had done with how much had yet to be accomplished.

On the first day monotony appeared after the first 30 minutes. Subjects varied the pace of work; they inquired of the experimenter how many production operations they had performed and or how long they had been working; they wanted to stop working and only continued under duress. After 2 hours, the average productivity was 1,420 operations.

On the second (information) day, monotony appeared after only 1 hour of work, but not as strongly as on the first day. Subjects varied the pace of work less, and none attempted to stop working. They achieved the required result after 1 hour and 28 minutes.

The study demonstrated that when personnel have a precise goal of what they should do and information about task performance, the feeling of monotony is sharply reduced. Information about task performance motivates to sustain a particular work pace. Interviews with subjects indicated that, on the first day, all tasks were considered the same. When information about the number of completed tasks was provided, each successive task was considered new. This permitted the subjects to reduce the contradictory motivational tendencies: to stop work and to continue work.

After the laboratory study, a field study was conducted. Ten women used a punch press to produce small parts from scrap metal, a process that could not be automatized. The quota was 15,000–18,000 pieces during one shift. The finished pieces were dropped into a collection barrel. Task simplicity and its high level of repetitiveness caused a strong feeling of monotony. All workers calculated the number of pieces they produced in the following manner: For every 100 pieces they dropped into the collection barrel, they laid one piece aside. Counting up to 18,000 pieces per shift resulted in a high level of exhaustion and mental fatigue. Nevertheless, in discussions with workers it was discovered that counting had become involuntary; they even counted in their sleep. Counting during the shift had become a way of overcoming monotony. Because of this, special counters to display the number of finished pieces were recommended. Under these conditions, workers stopped their counting. As a result, mental fatigue was sharply reduced. Apparently, information about goal achievement performed important motivational functions.

The third group of experiments was conducted with 12 students (6 experimental and 6 control) at a vocational school. Each group worked on a lathe for 3 days, a highly repetitive task that resulted in monotony. In the control group, only one requirement was given: to achieve the required quality of performance. The experimental group had to achieve a certain quantity of work as well. They had to finish 60 pieces during their shift.

The results demonstrated that, in the control group, productivity decreased from day to day (36.1, 33.8, 26.8 pieces per shift). In the experimental group, productivity increased from day to day (76.5, 83.1, 87.1). Introducing a precise quantitative goal increased productivity and reduced the feeling of monotony. The experimental subjects even introduced their own new goals each day—to produce more than the day before.

Integrating these studies, one can come to the following conclusion: In performing a monotonous job, the motivational aspect of self-regulation has a special meaning. Any monotonous work can be characterized by two contradicting tendencies. Achieving the required goal is assisted by positive motivation, whereas the process of achieving this goal is hampered by negative motivation. Repetition of the same operations results in distortion of goal-directed activity and increases the negative components of the motivational process. To overcome negative motivation and increase positive motivation, one must form a precise goal, which includes quantitative and qualitative requirements, determine the objective criteria of success, and provide information about approaching the final goal. In this situation feedback about achieving the goal has not only a cognitive but also a motivational role. Because of all these factors (precise goal, success criteria, and feedback) subjects can regulate the pace of activity performance more effectively.

Another method of overcoming monotony from the position of self-regulation of activity is to introduce intermediate goals. This method requires that a task with a long performance time be divided into smaller (part) tasks with their own (sub) goals. Achieving an intermediate goal gives the worker information about approaching the final goal, which bolsters motivation. Aseev (1974) studied blue-collar workers working on a conveyor belt. Because of the high level of repetitiveness and short duration of production operations, this type of work created a feeling of monotony. Under ordinary conditions, raw material was delivered to the worker once per shift. It was suggested that the workers be given raw material for only 1 hour of work. After they finished their work, the next portion of raw material was delivered. The amount of raw material delivered in each hour therefore depended on the workers and, as a result, productivity increased. For example, during the first hour the raw material given equaled 94% of the worker's average productivity per hour. In the second hour, they were given 104%, 103% in the third and fourth hours, 97% in the fifth and sixth hours, 101% in the seventh hour, and 96% in the eighth hour. As a result, work performance became more rhythmical, monotony was reduced (Asseev, 1974), and productivity increased approximately 4% to 5%.

Another method of reducing monotony is to preserve the meaning and significance of task performance. Dividing production into small operations may destroy the meaning and sense of task performance, which in turn increases monotony. It is thus important to provide logical completeness to the task. Aseev (1974) had subjects place small pegs into holes in any order for 2 hours. After 30 minutes of performance, they claimed that they felt monotony and wanted to stop working. In a second condition subjects had to put the same pegs into holes in a designated order, for example, shapes. Consequently, feelings of monotony were sharply decreased and the task was completed within 2 hours. It was hypothesized that in this case, increasing the complexity of the task made it more meaningful and significant for subjects.

Another method of reducing monotony involves changing the strategy of task performance. Where workers possesses a high skill level, they can change their method of performing the same task by using their reserve of attention to think about something other than the task. Aseev (1974) conducted a survey of excavator operators whose job was highly monotonous because of the repeated motor (levers and pedals) actions required. The survey discovered that experienced operators learn to think about something unrelated to the work without reducing productivity or reliability of performance. As a result, feelings of monotony decreased and productivity was sustained at the required level. The process of automatizing skills enables the operator to change his or her strategy of goal attainment by switching attention and, as a result, reduce monotony.

In automatic systems, monotony can emerge as a problem associated with vigilance. The operator serves initially and primarily as a monitor, which requires vigilance (Montmollin, 1967). In Russian sources, more attention is paid to those tasks involving expectation of emergencies in task performance. The operator monitors, a task which has a low workload; she or he must anticipate for periods of time, and when an emergency arises, become immediately involved in system operations.

The requirement to sustain vigilance continually is important for drivers who make long trips, particularly at night, railroad operators, and ship navigators. A study demonstrates that loss of vigilance has a short duration; it can be regained quickly. In spite of this, it can be very dangerous for the operator because accidents can occur during the nonvigilant period. One important condition of sustaining vigilance is anticipation—when an operator can predict ongoing events. This means that, in some cases, special displays that permit the operator to predict future events will be useful.

Pushkin and Nersesyan (1972) developed a method of measuring vigilance where vigilance is defined as the readiness to act in an unexpected situation. They demonstrated that people can have similar reactions times, but they react differently to unpredictable signals. Because of this, reaction time for unpredictable stimuli is not an acceptable means to measure vigilance. They suggested measuring vigilance (V) as the relationship of average reaction time (R_{av}) for the individual to his or her reaction time at a particular moment of the shift (R_t):

$$V = R_{av}/Rt \qquad (9.5)$$

For example, the average reaction time for a particular person, R_{av}, may equal 200 ms. If she or he reacts to a signal with the same speed, readiness or vigilance equals 100%. If reaction time increases to 400 ms, V is 50%.

In one experiment conducted by Pushkin and Nersesyan (1972), a paper drum with special symbols on it (squares, diamonds, triangles, circles, etc.) was presented to the operator through a screen. The drum's speed was variable. The subjects had to count the symbols. A light illuminated the screen, but was turned off irregularly, which meant that subjects were unable to see the symbols. They had to then press a button as quickly as possible to turn the light on again. Reaction time for these unpredictable events was measured. The experiment lasted for several hours, sometimes as long as a work shift. In this experiment, monotonous work and accidental situations were combined. Before conducting the experiment, the reaction time of all subjects was measured during 30–50 trials, permitting the researchers to calculate average reaction time (R_{av}). The coefficient V was then calculated for the various unpredictable events. Figure 9.4 presents vigilance or readiness for these unpredictable events. Note that vigilance never reached 100% and there were periods when vigilance sharply decreased.

The vigilance curve depends not only on the nature of the task, but on the operator's idiosyncratic features. Nevertheless, it is possible to assess vigilance for groups of workers. For example, this procedure has been used to evaluate a railroad operator's vigilance. Occasionally, a special device known as a "vigilance apparatus," which stimulates the worker by presenting false signals, is introduced into the workplace. However, this apparatus may impose an additional workload on the operator because it requires special attention. The notion is that random signals will diffuse the operator's monotony and stimulate his or her vigilance. These signals are artificial in the sense that they are not the ones the operator should respond to in case of an actual emergency. One would not wish the operator to act upon them, because this would cause a "false positive" reaction, making the operator respond as if the emergency actually existed. The U.S. Navy experimented in the past with such vigilance devices by inserting an artificial signal on a sonar scope, but rejected the notion because of the possibility it would set off an actual naval operation.

9.5 ERGONOMICS ASPECTS OF SAFETY

Danger is an undesirable component of work activity. According to Schultz and Schultz (1990), more than 14,000 people are killed annually and 2.2 million people are disabled each year as a result of industrial accidents. Billions of dollars are lost to worker compensation and medical benefits.

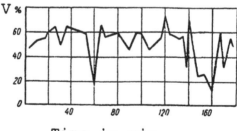

Time in min.

FIG. 9.4. Vigilance curve. From Pushkin and Nersesyan (1972).

One aspect of accident prevention is equipment design. Poor design results in accidents. Proper attention to equipment design, method of task performance, and the work environment can help reduce the frequency and severity of accidents.

Some aspects of safety were discussed in earlier chapters. However, preventing accident and injury requires special consideration. For example, adapting the work space to a person's body size plays an important role in providing safety in the workplace. Anthropometric data can be used to prevent workers from reaching dangerous areas in the workplace. This problem becomes especially important when the workplace is confining.

An important aspect of safety is to provide the motivation for safe behavior. Many motives influence the worker, including rewards (salary, bonuses), social (prestige, self-worth), conveniences (ease), satisfaction, and unity (desire to act like others in the group). All of these may impact the potential dangerous situation, but safety is the most important. However, the relationship between the various motives also influences safe behavior. For example, in a work group where dangerous methods of performance exist, the desire to work as others do (unity) can result in an accident.

When one studies safety from a motivational viewpoint, it is helpful to analyze the relationship between the desire to both achieve goal and avoid a failure. Some workers neglect safety requirements because the desire to achieve a goal dominates and they do not notice their failures. Under the influence of the strong need to maintain habitual performance, people can inadvertently deceive themselves (Kotik, 1987). The contradiction between productivity and safety is one motivational reason for violating safety requirements.

In some cases, one encounters risk-prone behavior because the worker seeks to project brave characteristics. The worker's behavior in a risky situation depends on both subjective and objective factors. Perception of the risk involved can change, but the process is largely unconscious, based on perceived goal, motivation to perform, and idiosyncratic factors.

Another important aspect of safety is the ability to adequately reflect the dynamic components of a situation. This is especially the case during a difficult, stressful situation with time limits. The dynamic aspect of the situation is connected, first of all, with the function block "relevant task conditions." In a dangerous situation, normal operation of this function block becomes very difficult. Very often, the operator involuntarily focuses his or her attention on the secondary features of the situation, and thus misses the more important ones. Errors are usually discovered after action is taken.

Kotik (1987) studied the cause of accidents in the construction industry and discovered that they are connected with the worker's inability to correctly orient her or himself and select the situation's essential features. If the worker develops anticipatory skills and the ability to quickly discover the more essential indicative features of a situation, erroneous actions in a dangerous situation can be reduced.

Another important aspect of safety from the self-regulation point of view is the factor of significance. During skill acquisition the worker selects certain aspects of a task as being especially significant, and cognitive and emotional reactions are associated with these. The recognition of danger mobilizes these associations and the worker responds differentially to these aspects. The more significant and difficult a particular task is and the more dangerous the situation, the higher the worker's motivational level becomes. Underestimating the significance level of the task can result in an inadequate use of the worker's resources, which will cause erroneous actions. In the same way, an excessive mobilization of the worker's energetic resources is also undesirable, as it can result in an inadequate formation of the program of performance. Because of this, the worker does not correctly allocate his or her attention and performs uncoordinated, tense actions.

Kotik (1987) observed the significance factor when performing a study of safety. The purpose of his study was to discover the relationship between three factors: the actual danger level of particular actions, the workers' evaluation of their danger, and the number of accidents during actions involving differing danger levels. The study involved electricians who serviced high-voltage power lines. They performed tasks involving benchwork repair, which had a low danger level, and tasks on high-voltage wires that could result in death. The workers evaluated the danger of the following kinds of work: service and repair of high-voltage wires (35–330 kv), service and repair of low-voltage wires (6–10 kv), service and repair of lower voltage wires (.4 kv), and the usual benchwork tasks.

The electricians evaluated work on low-voltage wires to be the most dangerous, work on high-voltage wires ranking next in danger level, followed by work on the lowest voltage wires. Benchwork tasks were considered the least dangerous.

The percentage of accidents for each type of work and the percentage of accidents resulting in death was then determined; based on these data, they discovered that when evaluating the danger factor, workers did not pay attention to the frequency of accidents, but rather their consequences. For example, work on lines of 6–10 kv only resulted in 4.6% of total accidents, whereas accidents resulting in death amounted to 40% of accidents when working on these types of lines. When working on lines of 35–330 kv, accidents accounted for 13.9% of the total number with death resulting in 20% of them. In benchwork, the number of accidents were 60% of the total, but none resulted in death.

It appears then that the most dangerous work was also the most significant to the workers. When they worked on lines of 6–10 kv, they worked more cautiously. This is why the number of accidents was less here than when they worked on 35–330 kv lines. At the same time, benchwork, which was perceived as less dangerous and less significant, was characterized by a greater frequency of accidents. From this it follows that the factor of significance essentially influences the frequency and character of accidents.

This is why after completing a very dangerous task without accident, the worker may experience trauma during a relatively safe part of the work. After completing dangerous work, the worker relaxes and thus may commit relatively obvious errors. An awareness of the significance to the worker of different work components is important in preventing accidents.

An analysis of actual and potential errors is fundamental to safety analysis. From the self-regulation point, errors should be analyzed in the context of strategies used by the worker while achieving specific goals. The error analysis must also consider the actual and potential consequences of the error. Some errors merely affect productivity, others may cause injury or system failure. Safety specialists normally pay attention only to the second category. Error analysis follows the error taxonomy in chapter 4. The analyst will consider the actual cause of the error (if that information can be retrieved from work records) or, in the case of potential errors, the most important factors that could lead to error (ignoring idiosyncratic factors like rashness, because these are specific to the individual). Error consequences (actual or potential) are determined next. The final stage in the analysis is the consideration of steps to be taken that will reduce error likelihood, for example, equipment redesign, warning signs, instruction. From the self-regulation point of view, errors can occur at different stages of that process. They can be produced by an inadequate orientation (incorrect formulation of the goal or task, incorrect extraction of relevant task conditions, etc.), incorrect decision making and execution, and incorrect evaluation and correction of the result of performance. In most cases, erroneous actions are multidetermined. Erroneous actions can emerge as a result of insufficient or irrelevant information or failure to use or incorrect use of this information (Hacker, 1980). Failure to use information is, from the position of self-regulation, connected with an incorrect formulation of the goal of observation, an inadequate extraction of subjectively relevant task conditions, and incorrect evaluation of the significance of information. For example, in overload conditions, an operator cannot deal with all the information presented to him or her and selects that information that seems the most important. This can result in inefficient strategies of gathering information. An operator can ignore objectively important information as a result of underestimating its significance, an attentional failure. All of these errors are involuntary and should thus be distinguished from the premeditated ignoring of information. For example, a quality control inspector begins to neglect information when time is limited. The function blocks "subjectively relevant task conditions" and "forming a program of task performance" relate to the errors connected with incorrect anticipation of forthcoming system events, hypotheses formation, and fore-casting. Errors connected with incorrect anticipation are of two types: not using available information and using that information incorrectly. The first type results from an inadequate orientation in a situation, and the second from an incorrect anticipation of the program of performance and making wrong decisions. For example, incorrect orientation in a situation can be explained by perceiving information incorrectly, false interpretation of in-formation, or incorrect expectations (Hacker, 1980). Sometimes tasks are very similar, but require somewhat different performance programs; if the operator cannot distinguish between the task or between the programs, she or he will select the wrong program and thus perform inadequately.

To prevent errors requires determining the goal of the task and the actions required to achieve the goal, evaluating the significance and complexity of the task, and specifying relevant task conditions. The motivational compo-nents of an activity and their role in the occurrence of errors must be inferred. After this, one determines the preferred strategies of performance. In essence, the same kind of analysis described previously in relation to activity theory and self-regulation generally is utilized in performing an error analysis.

The goal of this analysis is to determine the critical (dangerous) points of the work process as indicated by a high probability of accident occurrence. Such a probability may arise because of incorrect equipment design, an inadequate workplace arrangement, and inefficient methods of task per-formance. Such critical points may be indicated by a higher than normal frequency of accidents.

As far as the idiosyncratic causes of error are concerned, there may an inherent tendency to error (the so-called "accident-prone" individual) or there may not (Hacker, 1980; Kotik, 1987). If there is no inherent tendency, emphasis must be put on the determination of the situational factors that lead to error. If it were possible to determine in advance who is error prone, this would be useful in selecting people for a type of work (Klimov, 1969; Merlin, 1973). Unfortunately, there seems as yet to be no way of making such predictions. Fatigue and monotony also enter into erroneous actions and error is certainly more likely under these circumstances.

Safety analysis may involve a variety of methods, not much different from those in general ergonomics practice, but error analysis as it has been de-scribed previously is most important.

In general form, an analysis of errors is preformed using the following steps: selecting erroneous actions, or those that result in errors; separating the errors resulting in accident or injury from others; analyzing the frequency of errors connected to safety with regard to general number of errors; performing a qualitative analysis of errors; selecting errors stipulated by the specifity of the person's individual features, strategies of performance, and the equipment's physical configuration; and developing a method of preventing errors.

One important means of increasing safety is to design equipment from the standpoint of safety requirements. The first question that arises is connected with a method of presenting information in an accidental situation. The work station should be examined in terms of the need for warning signals, and whether they should alert individuals or a group. Acoustical or visual signals are commonly used. Such signals should perform the following functions: attract attention to the emergency situation, supply information about what has occurred, allow sufficient time for decision making, and avoid error when the operator transfers from normal activity to that of the emergency. Although signals should attract attention, they should not be excessive, lest they cause panic.

Attracting attention quickly does not always provide understanding and correct decision making. Because warning signals should prompt a person to act, this means that their main purpose is to provide understanding, meaning, and (if possible) cause of what has happened. Which warning signal is selected depends on the nature of the situation. Acoustical signals provide less information than do visual ones. For example, it may be desirable to indicate the location of a problem; in that case visual signals should be used.

Equipment should be designed to satisfy safety requirements, which means determining what those requirements are. For example, a press operator in the standing position works with two arm controls and one pedal control. A special table on the left holds uncut pieces whereas one on the right holds finished pieces. She or he can switch from arm control to leg control by manipulating a two-position switch with the right hand. A study was performed by Bedny (1987) to analyze the transition from arm to leg control.

According to the systemic-structural analysis procedure described in chapter 6, one begins by verbally describing what the worker has to do. The worker turns the switch to the required position with his or her right hand, takes one piece from the left with both hands, moves it to the work surface of the press, pushes it to the stopper, then turns the press on by simultaneously pressing buttons with his or her right and left hands. When the cutting process is completed, the operator releases the buttons, takes the finished piece with both hands, and moves it to the table on the right. When the worker takes an unfinished piece or puts a finished one on the table, he turns his body at a 45° angle. In Table 9.6, an algorithmical description of this process is shown. (Recall that 1 refers to the logical condition, O^{ε} to the efferent operator, $O^{\alpha w}$ to the afferent operator connected with perceiving the slider's movement and waiting until it finishes.)

After the algorithmical analysis, the specialist analyzes the time structure of the task, after which there is a quantitative analysis of task complexity. In the following, only those measures of complexity that permit one to evaluate the safety of task performance are presented: (a) time for algorithm execution, T

TABLE 9.6
Algorithm of Working on a Press

Members of Algorithm	Description
1_1	Make a decision about switching the press from pedal to hand control
0^ε_1	Move the switch to the required position with the right hand
0^ε_2	Take an uncut piece from the left side with both hands and put it on the work surface of the press
0^ε_3	Push the piece to the stopper
0^ε_4	Turn on the press with the two-handed control
0^ε_5	Wait based on visual control
0^ε_6	After finishing the work cycle, release the two buttons, and move the finished piece to the right-hand table

= 6.11 seconds, (b) time for performance of logical conditions, $L_g = .26$, (c) time for performance of efferent operator, $T_{ex} = 5.85$ seconds, (d) proportion of time for logical conditions to time for executive activity, $N_1 = .04$, (e) measure of changeability of logical processing of information, $L_{ch} - 1$, and (f) portion of time for logical components of work activity depending on information from long-term memory, rather than exteroceptive information, $L_{ltm} = 1$.

The logical components of work activity during the transition from pedal to hand control have only a short duration. However, this period of time is more important from the safety point of view. It has the following specifity: The measure of changeability of logical processing information conditions is 1. This means that this logical condition is flexible and occurs randomly during the various components of the work process, which destroys the habitual sequence of actions. The last measure, L_{ltm}, indicates that the worker must rely on long-term memory, which means that an external stimulus that tells the worker what he or she should do is lacking, complicating the decision-making process and making it less reliable. This means that the logical component of activity connected with the decision-making process is the task's danger point, which can result in injury. The situation can worsen because the worker may ignore the safety requirements when she or he transfers from pedal to hand control. From this it follows that, when the worker removes the safety guard, the press automatically reverses to hand control. In this situation, the logical component of the algorithm connected with the decision-making process of transferring from pedal to hand control is eliminated (performed automatically by technical components of a system).

Another way to increase safety is by training, but this is obviously not as efficient as eliminating a danger situation by more efficient design. To train for safe performance, it is important to analyze potential errors that can result in injury. Errors are either psychomotor or decision making (Powell, 1972). The first type destroy the regulative functions of motor actions; such actions may be linked to the incorrect formation of the orientational basis of an action. Decision-making errors either restrict choice or leave the choice open-ended. In the first case, the operator is required to choose from a number of existing actions with which she or he was familiar beforehand. The second case is connected with the situation when, after a decision is made, a new decision-making process emerges. After performing one task, the worker must perform a new task. In training for this type of situation, which requires various levels of knowledge—such as the ability to recognize and identify various phenomena, use knowledge in a particular application, and evaluate phenomena critically—the instructor should transfer the student from one level of knowledge to another sequentially, for example, from knowledge to application, and from there to evaluation. One weakness of safety training is that students do not usually proceed to the application level. As a result, they cannot use their safety knowledge in practice and cannot forecast and evaluate the consequences of their actions.

For more adequate safety training, it is important to create a model of the emergency situation and use special equipment that simulates the situation, for example, engine failure in an aircraft. Such simulators are necessary because emergencies in real life occur only at rare intervals. Because time is often a critical variable in responding to the emergency, the training process often involves performing a particular task under restricted time conditions.

An important aspect of safety training is the automatization of actions. As actions become simpler for the operator, his or her attention to the task is reduced and the method of performance becomes stereotyped. At the same time, though, his or her thinking about the task also becomes stereotyped. If an unstereotyped typical situation occurs under dangerous conditions, the operator may have an inadequate strategy for coping with it. This is most likely to occur when time is limited and conditions are stressful. To prevent this, it is important in training to emphasize possible unusual conditions.

9.6 CONCLUDING COMMENTS

The topics addressed in this chapter, as indeed in all the previous chapters, are ones traditionally addressed in ergonomics texts. Activity theory in itself unearths no new topics but suggests a somewhat different way of looking at them. In particular, greater attention is paid by activity theory and self-regulation to the cognitive and motivational aspects of work performance, more so than is ordinarily found in Western ergonomics.

That there are advantages and disadvantages to the Russian methodology has been pointed out; the advantages lie in a more comprehensive and more detailed explanation of work phenomena, particularly as they relate to the psychic structure of the individual operator.

The comprehensive nature of activity theory results in a nontraditional (from a Western standpoint) methodology. At one extreme of the methodological continuum one finds a concentration on molecular behaviors, particularly motor actions; and at the other end of the continuum is the effort to include in explanations of work behavior complex psychic processes such as motivation and work strategies. The Russian need to achieve a high level of precision in behavioral measurement leads to emphasis on time as a measure and as an explanatory mechanism, although, as seen in this and previous chapters, error is not ignored.

As an illustration, in work space organization there is the customary emphasis on performance time as a criterion for layout of the work station. This is linked with a distinction between immediate recognition and interpretation of familiar stimuli presented by the equipment, and a delayed response based on the operator's need to interpret unfamiliar stimuli. Obviously, the first takes much less time than the second. Of course, the use of such response times as a guide to work station layout presupposes that one knows which stimuli will evoke which responses and what times are associated with them.

The reliance on a broad, abstract theory and the necessity to incorporate every function block in that theory in explanations of work performance makes it difficult for Russian theorists to develop pragmatic guidelines like those used by American ergonomists. Western design guidelines are almost always qualitative; Russian specialists prefer more often use of quantitative procedures.

As one finds on first encountering any new conceptual structure, it is not necessary for the reader to accept or reject the totality of that structure. It is possible, for example, for the Westerner to admire the comprehensiveness of activity theory without necessarily adopting all elements of its behavioral design analysis. The reader will find it most useful to use activity concepts as a counterpoise to Western concepts; to examine Western ergonomics in terms of Russian concepts to see how well the former "stacks up" against the latter, and in this way to cast additional light on Western methods; to use activity theory and self-regulation as a means of evaluating the strengths and weaknesses of Western ergonomics. Both conceptual and methodological structures have strengths and weaknesses, and a comparison of the two approaches will illuminate both.

References

Ackoff, R. (1980). Towards a system of systems concepts. In H. R. Smith, A. B. Carroll, A. G. Kefalas, & H. J. Watson (Eds.), *Management: Making organizations perform* (pp. 43–87). New York: Macmillan.

Adams, J. A. (1968). Response feedback and learning. *Psychological Bulletin, 70,* 486–504.

Adams, J. A. (1987). Historical review and appraisal of research on learning, retention and transfer of human motor skills. *Psychological Bulletin, 101*(1), 47–74.

Adams, M. J., Tenney, Y. J., & Pew, R. W. (1995). Situation awareness and the cognitive management of complex systems. *Human Factors, 37*(1), 85–104.

Akhutin, V. M. (1977). Stage modeling and synthesis of adaptive biotechnical and ergotical systems. In B. F. Lomov, V. F. Rubakhin, & V. F. Venda (Eds.), *Engineering psychology: Theory, methodology, practical applications* (pp. 149–181). Moscow: Manufacturing Publishers.

Aladjanova, N. A., Slotintseva, T. V., & Khomskaya, E. D. (1979). Relationship between voluntary attention and evoked potentials of brain. In E. D. Khomskaya (Ed.), *Neuropsychological mechanisms of attention* (pp. 168–173). Moscow: Science Publishers.

Anan'ev, B. G. (1960). *The psychology of sense recognition.* Moscow: Academy of Pedagogical Science RSFSR.

Anan'ev, B. G., Vekker, L. M., Lomov, B. F., & Yarmolenko, A. V. (1959). *Sensing by touch in cognition and work.* Moscow: Academy of Pedagogical Science RSFSR.

Anderson, B. F. (1975). *Cognitive psychology: The study of knowing, learning and thinking.* New York: Academic Press.

Anderson, J. R. (1985). *Cognitive psychology and its application* (2nd ed.). New York: Freeman.

Anokhin, P. K. (1935). *The problem of center and periphery in the physiology of higher nervous activity.* Gorky, Russia: Gorky Publishers.

Anokhin, P. K. (1955). Features of the afferent apparatus of the conditioned reflex and their importance in psychology. *Problems of Psychology, 6,* 16–38.

Anokhin, P. K. (1962). *The theory of functional systems as a prerequisite for the construction of physiological cybernetics.* Moscow: Academy Science of the USSR.

405

Aseev, V. G. (1974). *Overcoming monotony in industry*. Moscow: Economika Publishers.

Asratyan, E. A. (1966). Functional architecture of instrumental conditions reflexes. In Symposium 4, Classical and instrumental conditions, Eighteenth International Congress. Moscow, pp. 35–53.

Atkinson, J. (1957). Motivational determinants of risk-taking behavior. *Psychological Review, 64*, 359–371.

Atkinson, R. C., & Shiffrin, R. M. (1971). The control of short-term memory. *Scientific American, 225*(2), 82–90.

Ausubel, D. P. (1968). Educational psychology: A cognitive view. New York: Holt, Rinehart & Winston.

Baddely, A. D., & Patterson, K. A. (1971). The relationship between long-term and short-term memory. *British Medical Bulletin, 27*, 237–242.

Bandura, A. (1977). *Social learning theory*. Englewood Cliffs, NJ: Prentice-Hall.

Bandura, A. (1982). Self-efficacy mechanism in human agency. *American Psychologist, 37*(2), 122–147.

Bardin, K. V. (1969). Using reaction time to study sensomotor processes. In M. A. Kotik (Ed.), *On the problem of experimental study of reaction times* (pp. 10–12). Tartu, Estonia: Valgus Publishers.

Bardin, K. V. (1982). The observer's performance in a threshold area. *Psychological Journal, 1*, 52–59.

Bardin, K. V., Sadov, V. A., & Szen N. K. (1984). New data about the threshold phenomenon. In K. V. Bardin & Y. M. Zabrodin (Eds.), *The psychophysics of sensory and sensory-motor processes* (pp. 40–70). Moscow: Science Publishers.

Bardin, K. V., & Voytenko, T. P. (1985). The phenomenon of simple discrimination. In Y. M. Zabrodin & A. P. Pakhomov (Eds.), *The psychophysics of discrete and continual tasks* (pp. 73–95). Moscow: Science Publishers.

Bardin, K. V., & Zabrodina, T. A. (1988). Changing sensory sensitivity when solving problems. *Problems of Psychology, 1*, 149–154.

Barnard, P., Wilson, M., & McLean, A. (1988). Approximate modeling of cognitive activity with an expert system: A theory-based strategy for developing an interactive design tool. *The Computer Journal, 31*, 448–456.

Bartlett, F. C. (1932). *Remembering: A study in experimental and social psychology*. Cambridge, England: Cambridge University Press.

Bassin, F. V. (1973). The problem of meaning and sense. *Problems of Psychology, 6*, 14–23.

Batishev, S. Y. (1977). Teaching workers in industry. In S. Y. Batishev & S. A. Shaporinsky (Eds.), *The fundamentals of vocational pedagogy* (pp. 412–503). Moscow: Higher Education Publishers.

Bayton, J. A., & Conley, H. W. (1957). The duration of success background and the effect of failure on performance. *Journal of General Psychology, 56*(2), 29–36.

Bedny, G. Z. (1975). *Engineering psychology in industry*. Kiev, Ukraine: Kiev University Publishers.

Bedny, G. Z. (1976). Individual style of activity and features of personality. In V. G. Aseev (Ed.), *Psychological problems of personality and student activity* (pp. 41–50). Irkutsk, Russia: Irkutsk University Press.

Bedny, G. Z. (1979). *Psychophysiological aspects of a time study*. Moscow: Economika Publishers.

Bedny, G. Z. (1981). *The psychological aspects of a timed study during vocational training*. Moscow: Higher Education Publishers.

Bedny, G. Z. (1985). Some mechanisms of self-regulation in the formation of a professional pace of activity. *Problems of Psychology, 1*, 74–80.

Bedny, G. Z. (1987). *The psychological foundations of analyzing and designing work processes*. Kiev, Ukraine: Higher Education Publishers.

Bedny, G. Z., & Podnevich, A. A. (1975). On the psychological microanalysis of production operations. *Problems of Psychology, 5,* 133–137.

Bedny, G. Z., & Voskoboynikov, F. A. (1975). Problems of how a person adapts to the objective requirements of an activity. In V. G. Aseev (Ed.), *Psychological problems of personality* (Vol. 2, pp. 18–30). Irkutsk, Russia: Irkutsk University Press.

Bedny, G. Z., & Zelenin, M. P. (1980). *The ergonomics of merchant marine transportation.* Moscow: Transportation Publishers.

Bedny, G. Z., & Zelenin, M. P. (1986). *Formation of professional pace during vocational training.* Moscow: All Union Research Center of Vocational Training.

Bedny, G. Z., & Zelenin, M. P. (1988). *Ergonomical analysis of work activity and the problem of safety in merchant marine transportation.* Moscow: Merchant Marine Publishers.

Bedny, G. Z., & Zelenin, M. P. (1989). *An analysis of how an operator processes information* (Lectures for the Institute of Continuing Education for the Merchant Marine Ministry). Odessa, Ukraine: Institute of Continuing Education for the Merchant Marine Ministry.

Bedny, G. Z., Zgidelev, M. A., Figanov, I. S. (1972). Selection of task and time study in vocational training. In M. I. Zgidelev (Ed.), *Acquisition of actions in vocational training* (pp. 172–200). Moscow: Higher Education Publishers.

Beregovoy, G. T. (1981). Orientation in the atmosphere and outer space. *Aviation and Cosmonaut,* pp. 34–35.

Beregovoy, G. T., Zavalova, N. D., Lomov, B. F., & Ponomarenko, V. A. (1978). *Experimental-psychology in aviation and aeronautics.* Moscow: Science Publishers.

Beringer, D. B., Williges, R. S., & Roscoe, S. N. (1975). The transition of experienced pilots to a frequency-separated aircraft attitude display. *Human Factors, 7*(4), 401–414.

Bernshtein, N. A. (1935). The problem of the relationship between coordination and localization. *Archives of Biological Science, 38*(1), 1–34.

Bernshtein, N. A. (1947). *On the structure of movement.* Moscow: Medical Publishers.

Bernshtein, N. A. (1966). *The physiology of movement and activity.* Moscow: Medical Publishers.

Bertalanffy, L. (1969). General system theory: A critical review. In I. V. Blauberg & V. N. Sadovsky (Eds.), *Study of general system theory* (pp. 23–82). Moscow: Mir Publishing House.

Bespal'ko, V. P. (1977). *Basics of the theory of pedagogical systems.* Voronezh, Russia: Voronezh University Press.

Bezdyuk, S. V., & Tsiganov, O. S. (1985). Ideomotor evaluation of sensory motor skills. In U. M. Zabrodin & A. M. Pakhomov (Eds.), *The psychophysics of discrete and continual tasks* (pp. 42–46). Moscow: Science Publishers.

Blackwell, H. R. (1952). Studies of psychophysical methods for measuring visual thresholds. *Journal of the Optical Society of America, 42*(9), 606–616.

Bloch, V. (1966). Level of wakefulness and attention. In P. Fraisse & J. Piaget (Eds.), *Experimental psychology* (Vol. 3, pp. 97–146). Paris: University Press of France.

Bodrov, V. A., & Orlov, V. Y. (1985). Classification of human operator errors. *Technical Esthetics, 7,* 22–24.

Borger, R. (1963). The refractory period and serial choice-reactions. *Quarterly Journal of Experimental Psychology, 15*(1), 1–12.

Broadbent, D. E. (1958). *Perception and communication.* London: Pergamon.

Bruner, J. S. (1960). *The process of education.* New York: Vantage Books.

Brushlinsky, A. V. (1979). *Thinking and forecasting.* Moscow: Science Publishers.

Brusnichkina, R. I. (1975). The specificity of orientational reactions from the character of the information presented to the pilot. *Biology in Aeronautical and Aviation Medicine, 5,* 49–54.

Chebisheva, V. V. (1969). *The psychology of vocational training.* Moscow: Education Publishers.

Checkland, P. B. (1981). *Systems thinking, systems practice.* New York: Wiley.

Christensen, E. H. (1953). Psychophysiological valuation of work in the Nykroppa iron works. In W. F. Floyd & A. T. Welford (Eds.), *Ergonomics Society Symposium on Fatigue* (pp. 93–108). London: Lewis.

Cooper, G. E., & Harper, R. P. (1969). The use of pilot ratings in the evaluation of aircraft handling qualities (AGARD Rep. No. 567). London: Technical Editing and Reproduction, Ltd.

Davidov, V. V. (1972). *Types of generalization in teaching (logical-psychological problems of developing different subjects)*. Moscow: Pedagogical Publishers.

Deitrich, Y. (1981). *Engineering and design*. Moscow: World Publishers.

Demersky, V. Y. (1955). On using imaginative actions. *Problems of Psychology, 6,* 49–61.

Dguravlev, G. E. (1981). Classification of actions performed by an operator. In B. F. Lomov & V. F. Venda (Eds.), *Methodology of engineering psychology and the psychology of work* (pp. 218–230). Moscow: Science Publishers.

Dhillon, B. S., & Singh, C. (1981). *Engineering reliability: New techniques and applications*. New York: Wiley.

Dickerson, R. (1965). *The fundamentals of legal drafting*. Boston/Toronto: Little, Brown.

Dmitrieva, M. A. (1964a). Psychological analysis of aviation dispatcher. In B. F. Lomov (Ed.), *Problems of general and engineering psychology* (pp. 100–108). Leningrad: Association of Psychology Publishers.

Dmitrieva, M. A. (1964b). Speed and accuracy of information processing and their dependence on signal discrimination. In B. F. Lomov (Ed.), *Problems of engineering psychology* (pp. 121–126). Leningrad: Association of Psychology Publishers.

Dobrinin, N. F. (1958). Voluntary and involuntary attention. In *Scientific works of the Moscow State Pedagogical Institute* (Vol. 8, pp. 34–52). Moscow: Pedagogical Publishers.

Dobrolensky, U. P., Zavalova, N. D., Ponomarenko, V. A., & Tuvaev, V. A. (1975). *Methods of engineering psychological study in aviation*. Moscow: Manufacturing Publishers.

Donders, F. S. (1862). Die schnelligkeit psychischer processe. *Arch. Anat. Physiol, 657*–681.

Dushkov, B. A. (1981). *Industrial educational psychology*. Moscow: Higher Education Publishers.

Eastman Kodak Company Ergonomics Group. (1986). *Ergonomic design for people at work* (Vol. 2). New York: Van Nostrand Reinhold.

Edwards, W. (1987). Decision-making. In G. Salvendy (Ed.), *Handbook of human factors* (pp. 1061–1104). New York: Wiley.

El'kin, D. G. (1962). *Perception of time*. Moscow: Pedagogical Publishers.

Endsley, M. R. (1995). Toward a theory of situation awareness in dynamic systems. *Human Factors, 37*(1), 32–64.

Epps, B. W. (1986). Comparison of six control devices based on Fitt's law of models. In *Proceedings of the Human Factors Society's 30th annual meeting* (pp. 327–331). Santa Monica, CA: The Human Factors Society.

Ericson, K. A., & Simon, H. A. (1984). *Protocol analysis: Verbal reports as data*. Cambridge, MA: MIT Press.

Fitts, P. M. (1951). Engineering psychology and equipment design. In S. S. Stevens (Ed.), *Handbook of experimental psychology* (pp. 1287–1340). New York: Wiley.

Fitts, P. M. (1954). The information capacity of the human motor system in controlling the amplitude of movement. *Journal of Experimental Psychology, 47,* 381–391.

Fitts, P. M., & Petterson, J. R. (1964). Information capacity of discrete motor responses. *Journal of Experimental Psychology, 67*(2), 103–112.

Fleishman, E. A. (1975). *Toward a taxonomy of human performance. American Psychologist, 30,* 1127–1149.

Fogel, I. L. (1961). Levels of intelligence in decision-making. *Annals of the New York Academy of Sciences, 89*(5), 27–40.

Freud, S. (1960). *A general introduction to psychoanalysis* (2nd ed.). New York: Washington Square Press. (Original work published 1920)

Gagné, R. M. (1985). *The cognitive psychology of school learning*. Boston: Little, Brown.

Galaktionov, A. I. (1978). *The fundamentals of engineering in the psychological design of ASGTP*. Moscow: Energy Publishers.

Gal'perin, P. Y. (1955). The problem of internal speech. In *Materials of the Conference on Psychology* (pp. 22–30). Moscow: Academy of Pedagogical Science RSFSR.

Gal'perin, P. Y. (1957). The mental act as the basis for the formation of ideas and images. *Problems of Psychology, 6,* 12–26.

Gal'perin, P. Y. (1959). *The development of research on the formation of cognitive actions* (Vol. 1). Moscow: Academy of Pedagogical Science RSFSR.

Gal'perin, P. Y. (1966). The psychology of thinking and study of the stages in the formation of cognitive actions. In A. N. Leont'ev (Ed.), *Research on thinking in Soviet psychology* (pp. 236–237). Moscow: Academy of Pedagogical Science RSFSR.

Gal'perin, P. Y. (1969). Stages in the development of mental acts. In M. Cole & I. Maltzman (Eds.), *A handbook of contemporary Soviet psychology* (pp. 249–273). New York: Basic Books.

Gal'perin, P. Y., & Sachko, N. N. (1968). Formation of motor skills. In P. Y. Gal'perin (Ed.), *Stage by stage formation of knowledge and skills* (pp. 3–54). Moscow: Moscow University Publishers.

Gal'sev, A. D. (1973). *Time study and scientific management of work in manufacturing.* Moscow: Manufacturing Publishers.

Gaverdorksy, Y. (1987). Horizontal bar exercises. In Y. Gaverdosksy (Ed.), *All-around gymnastics* (pp. 336–478). Moscow: Physical Culture and Sport.

Gaverdorsky, Y. (1985). Motor imagination of gymnasts. *Gymnastics* (pp. 22–30). Moscow: Physical Culture and Sport Publishers.

Geer, E. G. (1981). *Human engineering procedures guide* (Rep. No. AFAMRL-TR-81-35). Wright-Patterson AFB, OH: Aerospace Medical Division.

Gellershtein, S. G. (1958). *Feeling of time and speed of motor reactions.* Moscow: Medical Publishers.

Gellershtein, S. G. (1966). Anticipation and the problem of unconsciousness. In F. V. Bassin (Ed.), *The problem of unconsciousness* (pp. 305–360). Moscow: Science Publishers.

Gibson, J. T., & Chandler, L. A. (1988). *Educational psychology.* Boston: Allyn & Bacon.

Gilbreth, F. B. (1911). *Motion study.* Princeton, NJ: Van Nostrand.

Gil'bukh, U. Z. (1979). *Simulator devices in vocational training.* Kiev, Ukraine: Higher Education Publishers.

Glezer, V. D., & Nevskaya, A. A. (1964). On the simultaneous and sequential processing of information in a visual system. In V. D. Glezer (Ed.), *Collected papers of the Academy of Science of the USSR* (Vol. 155, pp. 97–105). Moscow: The Academy of Science of the USSR.

Glezer, V. D., & Tsukkerman, I. I. (1961). *Information and vision.* Moscow: Academia of Science USSR.

Goldstein, I. I. (1974). *Training program development and evaluation.* Monterey, CA: Brooks/Cole.

Golliwitzer, P. M. (1996). The volitional benefits of planning. In P. M. Golliwitzer & J. A. Bargh (Eds.), *The psychology of action* (pp. 188–192). New York: Guilford Press.

Golubev, G. G. (1953). *Issues in the methods of flight instruction.* Moscow: Defense Publishers.

Gordeeva, N. D., Devishvily, V. M., & Zinchenko, V. P. (1975). *The microstructural analysis of executive activity.* Moscow: All Union Research Institute of Technical Esthetics.

Gordeeva, N. D., & Zinchenko, V. P. (1982). *Functional structure of action.* Moscow: Moscow University Publishers.

Green, D. M., & Swets, J. A. (1966). *Signal detection theory and psychophysics.* New York: Wiley.

Green, E. J. (1962). *The learning process and programming instruction.* New York: Appleton-Century.

Grimak, L. P., & Ponomarenko, V. A. (1971). Psychic state of pilots and its influence on flight. *Military Medical Journal, 1,* 43–57.

Gubinsky, A. I., & Efgrafov, V. G. (1977). *Ergonomic design of ship systems.* Leningrad: Ship Construction Publishers.

Hacker, W. (1980). *Engineering psychology and work psychology.* Berlin: V. E. Deutscher Verlag der Wissenschafter.

Hacker, W. (1994). Moving from cognition to action? Control theory and beyond. In M. Frese (Ed.), *Applied psychology: An international review* (Vol. 43, pp. 379–381). Hillsdale, NJ: Lawrence Erlbaum Associates.

Hall, A. D. (1962). A methodology for systems engineering. Princeton, NJ: Van Nostrand.

Hasbrock, A. H., & Rasmussen, P. G. (1973). *Inflight performance of civilian pilots using moving-aircraft and moving-horizon attitude indicators* (Rep. FAA-AM-739). Oklahoma City, Federal Administration Aviation Office.

Hassan, M., & Block, S. M. (1968). Study of simultaneous motions. *L'Etude du Travail, 196,* 29–36.

Hebb, D. O. (1955). Drives and the central nervous system (CNS). *Psychological Review, 62,* 243–254.

Heckhausen, H. (1986). *Motivation and activity.* Moscow: Pedagogical Publishers.

Hick, W. E. (1952). On the role of gain of information. *Quarterly Journal of Experimental Psychology, 4,* 11–26.

Hilgard, R., Atkinson R. L., & Atkinson, R. C. (1979). *Introduction to psychology.* New York: Harcourt Brace.

Howard, R. A. (1968). The foundations of decision analysis. *IEEE Transactions on Systems Science and Cybernetics, 3,* 211–219.

Hoyos, C. G. (1987). Motivation. In G. Salvendy (Ed.), *Handbook of human factors* (pp. 108–123). New York: Wiley.

Hrunov, L. S., Hrachaturynz, L. S., & Popov, V. A. (1974). *Human-operator in cosmos flight.* Moscow: Manufacturing Publishers.

Hygiene Standards (1963). (Rep. HS0245-63). Moscow, USSR.

Hyman, R. (1953). Stimulus information as determined of reaction time. *Journal of Experimental Psychology, 45*(3), 188–196.

Il'in, E. P. (1986). Skills and habits: Unsolved problems. *Problems of Psychology, 2,* 138–148.

Isakov, P. K. (1974). Actual problems in aviation medicine. *Aeronautical Biology and Medicine, 6,* 35–41.

Itelson, L. B. (1972). *Lectures about temporal problems of learning.* Vladimir, Russia: Vladimir Educational University.

Kahneman, D. (1973). *Attention and effort.* Englewood Cliffs, NJ: Prentice-Hall.

Kalikinsky, O. A. (1968). *Developing sensory skills during vocational training.* Moscow: Higher Education Publishers.

Kalmikova, Z. I. (1979). *Psychological principles of development teaching.* Moscow: Knowledge Publishers.

Kanfer, R. (1996). Self-regulation and other non-ability determinants of skill acquistion. In P. M. Golliwitzer & J. A. Bargh (Eds.), *The psychology of action* (pp. 404–423). New York: Guilford Press.

Karger, D. W., & Bayha, F. H. (1977). *Engineering work measurement* (3rd ed.). New York: Industrial Press.

Karsten, A. (1928). Psychological saturation. *Psychology Research, 10,* 142–245.

Khachatur'yanz, L. S., Grimak, L. P., & Khrunov, E. V. (1976). *Experimental psychophysiology in aeronautical study.* Moscow: Science Publishers.

Khodikina, S. L., & Portnoy, M. P. (1985). *Applying rhythmical stimulation while training on a keyboard.* Moscow: Higher Education Publishers.

Khrunov, E. V., Khachatur'yanz, L. S., Popov, V. A., & Ivanov, E. A. (1974). *Man-operator in astronautics flights.* Moscow: Manufacturing Publishers.

Khruselevsky, B. V. (1971). Electromiographical study of muscle activity during the performance of monotonous work. In V. G. Aseev (Ed.), *Problems of the psychophysiology of work* (pp. 43–52). Irkutsk, Russia: State Pedagogical Institute.

Kirwan, B., & Ainsworth, L. K. (Eds.). (1992). *A guide to task analysis*. London: Taylor & Francis.

Kitov, A. I. (1981). Experience of developing a psychological theory of management. *Psychological Journal, 2*(4), 21–32.

Klatsky, R. L. (1975). *Human memory: Structures and processes*. San Francisco: Freeman.

Kleinback, U., & Schmidt, K. H. (1990). The translation of work motivation into performance. In V. Kleinback, H.-H. Quast, H. Thierry, & H. Hacker (Eds.), *Work motivation* (pp. 27–40). Hillsdale, NJ: Lawrence Erlbaum Associates.

Klenovich, S. (1971). Selective problems in work psychology. In V. F. Venda (Ed.), *Ergonomics* (pp. 122–181). Moscow: Mir Publishing House.

Klimov, E. A. (1969). *Individual style of activity*. Kazan, Russia: Kazahnsky State University Press.

Klinger, E. (1996). Emotional influences on cognitive processing with implications for theories of both. In P. M. Golliwitzer & J. A. Bargh (Eds.), *The psychology of action* (pp. 404–423). New York: Guilford Press.

Kojin, A. M., Tsukkerman, B. G., & Pakhomov, A. F. (1961). On the formation of skills of reading different displays. *Issues in the Academy of Pedagogical Science RSFSR, 3*, 91–96.

Kondrat'eva, L. L., & Titova, I. P. (1965). *Written instructions in vocational training*. Moscow: Higher Education Publishers.

Konopkin, O. A. (1962). Temporal variability and pace of performance of production operations. In D. A. Oshanin (Ed.), *Industrial psychology* (pp. 37–50). Moscow: Academia of Educational Science of RSFSR.

Konopkin, O. A. (1980). *Psychological mechanisms of regulation of activity*. Moscow: Science Publishers.

Konopkin, O. A., Engels, I. L., & Stephansky, V. T. (1983). On shaping of subjective standards of success. *Problems of Psychology, 6*, 109–114.

Konopkin, O. A., & Luchkov, V. V. (1973). Influence of subjective evaluation of indeterminism about duration of work on specificity of activity regulation. In O. A. Konopkin (Ed.), *Psychological questions of activity regulation* (pp. 129–153). Moscow: Pedagogical Publishers.

Kossov, V. V., & Rotanova, T. A. (1965). Subjective redundancy of information perceived by man. In *Collective works of engineering psychology* (Vol. 2, pp. 65–71). Leningrad: Psychology Association.

Kostuk, I. P. (Ed.). (1962). *General psychology*. Kiev, Ukraine: Higher Education.

Kotik, M. A. (1974). *Self-regulation and reliability of operator*. Tallin, Estonia: Valgus.

Kotik, M. A. (1978). *Textbook of engineering psychology*. Tallin, Estonia: Valgus.

Kotik, M. A. (1987). *Psychology and safety*. Tallin, Estonia: Valgus.

Kovalev, F. L. (1950). *Experience of studying efficient methods of performance*. Moscow: Educational Publishers.

Kozeleski, J. (1979). *Psychological theory of decision making*. Moscow: Progress.

Krilov, A. A. (1972). *Man in automated systems of control*. Leningrad: Leningrad University.

Krilov, A. A., & Pakhomov, A. F. (1965). Influence of interval magnitude between two stimuli on duration of simple sensory motor reaction. In V. F. Lomov (Ed.), *The problem of engineering psychology* (Vol. 2, pp. 86–97). Leningrad: Psychology Association.

Krinchik, E. P., & Risakova, S. L. (1965). Influence of significance factor of signals on specificity of processing information. In V. P. Zinchenko, A. N. Leont'ev, & D. Y. Panov (Eds.), *Engineering psychology* (pp. 155–159). Moscow: Moscow University Publishers.

Krukov, N., & Kremen, M. (1983a). Method of supporting points. *Aviation and Cosmonaut, 6*, 26–27.

Krukov, N., & Kremen, M. (1983b). Method of supporting points. *Aviation and Cosmonaut, 7*, 27–28.

Kudryavsev, T. V. (1975). *Psychology of technical thinking*. Moscow: Pedagogical Publishers.

Kuhl, J. (1992). A theory of self-regulation: Action versus state orientation, self-discrimination and some applications. *Applied Psychology: An International Review, 41,* 97–129.

Kulyutkin, Y. N. (1970). *Heuristic method in the structure of decision.* Moscow: Science Publishers.

Landa, L. M. (1974). *Algorithmization in learning and instruction.* Englewood Cliffs, NJ: Educational Technology Publications.

Landa, L. M. (1976). *Instructional regulation and control: Cybernetics, algorithmization and heuristic in education.* Englewood Cliffs, NJ: Educational Technology Publications.

Landa, L. N. (1978). Some problems in algo-heuristic of thinking, learning and instruction. In J. Scandura & C. Brainerd (Eds.), *Structural process/models of complex human behavior* (pp. 115–121). Slijthoff & Noordhoff, The Netherlands: Alphen aan den Rijn.

Landa, L. N. (1983). Descriptive and prescriptive theory of learning and instruction. In C. M. Reigeluth (Ed.), *Instructional design, theories and models: An overview of their current state* (pp. 55–74). Hillsdale, NJ: Lawrence Erlbaum Associates.

Landa, L. N. (1984). Algo-heuristic theory of performance, learning and instruction: Subject, problems, principles. *Contemporary Educational Psychology, 9,* 235–245.

Landy, F. J. (1989). *Psychology of work behavior.* Pacific Grove, CA: Brooks/Cole.

Lazareva, V. V., Svederskaya, N. E., & Khomskaya, E. D. (1979). Electrical activity of brain during mental workload. In E. D. Khomskaya (Ed.), *Neuropsychological mechanisms of attention* (pp. 151–168). Moscow: Science Publishers.

Lazootkin, V. I., & Magazannik, V. (1984). Engineering psychological design of poly-functional informational models. In *6th Soviet Union Conference of Engineering Psychologists* (pp. 25–27). Moscow: Science Publishers.

Lee, T. W., Locke, E. A., & Lathman, G. P. (1989). Goal setting, theory and job performance. In A. Pervin (Ed.), *Goal concepts in personality and social psychology* (pp. 291–326). Hillsdale, NJ: Lawrence Erlbaum Associates.

Leeper, R. W. (1943). *Lewin's topological and vector psychology.* Eugene, OR.

Lehmann, G. (1962). *Practical work physiology.* Stuttgart, Germany: George Theme Verlag.

Leont'ev, A. N. (1957). Teaching as problem of psychology. *Problems of Pychology, 1,* 17–22.

Leont'ev, A. N. (1971). *Needs, motive, emotions.* Moscow: Moscow University Publishers.

Leont'ev, A. N. (1977). *Activity, consciousness and personality.* Moscow: Political Publishers.

Leont'ev, A. N., & Krinchik, E. P. (1964). Processing information by man in decision making. In A. N. Leont'ev (Ed.), *Engineering psychology* (pp. 195–324). Moscow: Moscow University Publishers.

Leplat, J. (1963). Sensorimotor connections. In P. Fraisse & J. Piaget (Eds.), *Experimental psychology* (pp. 375–427). Paris: University Press of France.

Liapunov, A. A., & Shestopal, G. A. (1957). The algorithmic description of control processes. *Mathematical Education, 2,* 14–27.

Lindsay, P. H., & Norman, D. A. (1992). *Human information processing* (2nd ed.). San Diego: Harcourt Brace.

Linhart, J. (1970). *Process and structure of human learning.* Moscow: Progress.

Litvak, I. I., Lomov, B. F., & Soloveichik, I. E. (1975). In A. Y. Breitbart (Ed.), *Basics of design apparatus for presenting information in automatic systems.* Moscow: Soviet Radio.

Lobanov, V. L. (1983). *Handbook for technical exploration of the bottom of the ocean.* Leningrad: Ship Construction Publishers.

Locke, E. A., & Lathman, G. P. (1984). *Goal setting: A motivational technique that works.* Englewood Cliffs, NJ: Prentice-Hall.

Locke, E. A., & Lathman, G. P. (1990). Work motivation: The high performance cycle. In V. Kleinbeck et al. (Eds.), *Work motivation* (pp. 3–26). Hillsdale, NJ: Lawrence Erlbaum Associates.

Lomov, B. F. (1966). *Man and machine.* Moscow: Soviet Radio.

Lomov, B. F. (1977). Directions on developing theories of engineering psychology based on a systemic approach. In B. F. Lomov, V. F. Rubakhin, & V. F. Venda (Eds.), *Engineering psychology* (pp. 31–55). Moscow: Science Publishers.

Lomov, B. F. (Ed.). (1982). *Handbook of engineering psychology*. Moscow: Manufacturing Publishers.

Lomov, B. F. (1984). *Methodological and theoretical problems of psychology*. Moscow: Science Publishers.

Lomov, B. F., & Surkov, E. N. (1980). *Anticipation in structure of activity*. Moscow: Science Publishers.

Lomov, B. F., Vasil'ev, A. A., Oficerov, V. V., & Rubakhin, V. F. (Eds). (1970). *Military engineering psychology*. Moscow: Defense Publishers.

Lomov, B. F., Zavalova, N. D., & Ponomarenko, V. A. (1977). Principle of the active operator. In B. F. Lomov, V. F. Rubakhin, & V. F. Venda (Eds.), *Engineering psychology* (pp. 119–134). Moscow: Science Publishers.

Long, J. (1976). Visual feedback and skilled keying: Differential effects of masking the printed copy and keyboard. *Ergonomics, 19*(1), 39–110.

Loos, V. G. (1974). *Industrial psychology*. Kiev: Technical Publishers.

Lord, R. G., & Levy, P. E. (1994). Moving from cognition to action: A control theory perspective. In M. Frese (Ed.), *Applied psychology: An international review* (Vol. 43, pp. 335–336). Hillsdale, NJ: Lawrence Erlbaum Associates.

Luria, A. R. (1975). *Cognitive development: Its cultural and social foundations*. Cambridge, MA: Harvard University Press.

Magnusson, D., & Endler, N. S. (1977). *Personality and the crossroads: Current issues in interactional psychology*. Hillsdale, NJ: Lawrence Erlbaum Associates.

Manwaring, T., & Wood, S. (1985). The ghost in the labor process. In O. Knight, H. W. Willmott, & D. Collinson (Eds.), *Job redesign: Organization and control of the labor process* (pp. 101–196). Aldershot, England: Gower.

Marchenko, E. N., Kandror, I. S., & Ryazanov, S. (1972). Principle of classification of physical work according to physiological criteria. *Work Hygiene and Professional Sickness, 3*, 30–36.

Maslow, A. H. (1954). *Motivation and personality*. New York: Harper & Row.

Matyushkin, A. M. (1972). *Problem situations in thought and instructions*. Moscow: Pedagogical Publishers.

Maynard, H. B., Stegemerten, G. J., & Schwab, J. L. (1948). *Methods-time measurement*. New York: McGraw-Hill.

McCormick, E. J., & Ilgen, D. R. (1985). *Industrial and organizational psychology*. Englewood Cliffs, NJ: Prentice-Hall.

McCormick, E. J., & Sanders, M. S. (1982). *Human factors in engineering and design*. New York: McGraw-Hill.

McLeod, R. W., & Sherwoods-Jones, B. M. (1992). Stimulation to predict operator work load in a command system. In B. Kirwan & L. K. Ainsworth (Eds.), *A guide to task analysis* (pp. 301–310). London: Taylor & Francis.

Medvedev, G. (1991). *The truth about Chernobyl*. New York: HarperCollins.

Meister, D. (1971). *Human factors: Theory and practice*. New York: Wiley.

Meister, D. (1976). *Behavioral foundations of system development*. New York: Wiley.

Meister, D. (1985). *Behavioral analysis and measurement methods*. New York: Wiley.

Meister, D. (1989). *Conceptual aspects of human factors*. Baltimore, MD: Johns Hopkins University Press.

Meister, D. (1992). Special problems of human factors test and evaluation in advanced nuclear power design. In *Proceedings, International Conferences on Design and Safety of Advanced Nuclear Power Plants, Anp '92* (pp. 34.3-1–34.3-5).

Meister, D. (1993). Human reliability database and future systems. In *Proceedings, Annual Reliability and Maintainability Symposium* (pp. 276–280).

Meister, D. (1995). Cognitive behavior of nuclear reactor operators. *International Journal of Industrial Ergonomics, 16,* 109.

Meister, D. (in press). Human factors test and evaluation in the twenty-first century. In T. G. O'Brien & S. G. Charlton (Eds.), *Handbook of human factors and testing and evaluation.* Mahwah, NJ: Lawrence Erlbaum Associates.

Meister, D., & Rabideau, G. F. (1965). *Human factor evaluation in system development.* New York: Wiley.

Mel'nik, I. M. (1976). Influence of objective characteristics of presented information on short-term memory. In B. A. Smirnov (Ed.), *Problems of general and engineering psychology* (Vol. 132, pp. 36–42). Kharkov, Ukraine: Kharkov University Publishers.

Merlin, V. S. (1973). *Outlines of a theory of temperament.* Perm, Russia: Perm Pedagogical Institute.

Mileryan, E. A. (1974). Psychological evaluation of equipment. In E. A. Mileryan (Ed.), *Work psychology* (pp. 187–206). Moscow: Science Publishers.

Miller, G. A. (1956). The magical number seven, plus or minus two: Some limits on our capacity for processing information. *Psychological Review, 63,* 81–97.

Miller, G. A., Galanter, E., & Pribram, K. M. (1960). *Plans and the structure of behavior.* New York: Holt.

Miller, R. B. (1953). Method for man-machine task analysis (Rep. No. WADC-TR-53-137). Wright-Patterson AFB, OH: Wright Air Development Center.

Mirabella, A., & Wheaton G. R. (1974). *Effects of task index variations on transfer of training criteria* (Rep. No. NAVTRAEQUIPEN 72-C-0126-1). Orlando, FL: Naval Training Equipment Center.

Monod, H., & Pottier, M. (1967). Circulatory and respiratory adjustments to physical work. In J. Sherrez (Ed.), *Physiologie du Travial* (Vol. II, pp. 171–253). Paris: Nasson.

Montmollin, M. de (1967). *Man and machine system.* Paris: University Press of France.

Morgan, C. T., Chapanis, A., Cook, J. S., & Lund, M. W. (Eds). (1963). *Human engineering guide to equipment design.* New York: McGraw-Hill.

Morosanova, V. I. (1974). Operative characteristics of perception of dynamic objects. *Problems of Psychology, 6,* 128–131.

Morosanova, V. I., & Stepansky, V. I. (1982). Relationship between subjective and objective criteria of success and its influence on result of performance. In *Collective Works: New Study in Psychology* (pp. 14–17). Moscow: Pedagogical Publishers.

Morosanova, V. I., Stepansky, V. I., & Sablin, V. N. (1980). Calculation of objective indices in quality of gun shooting proficiency. In *Collection of scientific work of VNIIFK* (pp. 52–62). Moscow: Higher Education Publishers.

Moruzzi, G., & Magoun, H. W. (1949). Brain stem reticular formation and activation of the EEG. *Clinical Neurophysiology, 1,* 455–473.

Munipov, V. M. (Ed.). (1983). *Ergonomics: Principles and recommendations.* Moscow: All Union National Research Institute of Technical Aesthetics.

Myasnikov, V. I. (1963). Multi-factor method of evaluating the functional state of the organism. In *International Conference in Psychology* (Vol. 3, pp. 120–123). Moscow: Psychology Association.

Navon, D., & Gopher, D. (1979). On the economy of the human processing system. *Psychological Review, 86,* 254–255.

Nayanko, N. I. (1976). *Psychic tension.* Moscow: Moscow University Publishers.

Nebylitsin, V. D. (1965). *Basic features of man's neural system.* Moscow: Education Publishers.

Neumann, J., & Timpe, K. P. (1975). Organization of work. *Psychophysiological problems.* Moscow: Economics.

Neumin, Y. G. (1984). *Models in science and technic.* Leningrad: Science Publishers.

Nevel'skii, P. V. (1969). The capacity of memory and the quantity of information. In V. P. Zinchenko, P. I. Zinchenko, & B. F. Lomov (Eds.), *Problems of engineering psychology* (Vol. 3, pp. 71–80). Leningrad.

Newell, A., & Simon, H. A. (1972). *Human problem solving*. Englewood Cliffs, NJ: Prentice-Hall.

Newtson, D. (1976). Foundations of attribution, the perception of ongoing behavior. In J. H. Harvey, W. V. Ickes, & R. F. Kidd (Eds.), *New directions in attribution research* (pp. 172–187). Hillsdale, NJ: Lawrence Erlbaum Associates.

Nikiforov, G. S. (1977). *Self-control as a method of human reliability*. Leningrad: Leningrad University.

Nikolaev, V. I. (1973). *Information theory of control and operation*. Leningrad: Merchant Marine.

Nojivin, U. S. (1974). On psychological self-regulation of sensory motor actions. In V. D. Shadrikov (Ed.), *Engineering and psychology* (Vol. 1, pp. 206–210). Yaroslav, Russia: Yaroslav University.

Norman, D. A. (1976). *Memory and attention*. New York: Wiley.

Norman, D. A. (1981). Catagorization of action slips. *Psychology Review, 88*, 1–15.

Novikov, A. I., Sidorova, I. N., & Fedorov, I. V. (1980). Study of operator activity in conditions of destroying feedback connections. *Ergonomics, 19*, 32–39.

Novikov, A. M. (1986). *Process and method of formation of vocational skills*. Moscow: Higher Education Publishers.

Ol'shannikova, A. E. (1962). Demonstration of law of force in different operator working conditions. *Problems of Psychology, 5*, 31–44.

Ormrod, J. E. (1990). *Human learning. Theories, principles and educational applications*. New York: Macmillan.

Oshanin, D. A. (1977). Concept of operative image in engineering and general psychology. In B. F. Lomov, V. F. Rubakhin, & V. F. Venda (Eds.), *Engineering psychology* (pp. 134–148). Moscow: Science Publishers.

Patrick, J. (1992). *Training: Research and practice*. London: Academic Press.

Pavlov, I. P. (1927). *Conditioned reflex*. London: Oxford University Press.

Pervin, L. A. (1989). Goal concepts, themes, issues and questions. In L. A. Pervin (Ed.), *Goal concepts in personality and social psychology* (pp. 173–180). Hillsdale, NJ: Lawrence Erlbaum Associates.

Petrovsky, A. V. (Ed.). (1986). *General psychology*. Moscow: Education Publishers.

Piaget, J. (1952). *The origins of intelligence in children*. New York: International University Press.

Platonov, K. K. (1970). *Problems of work psychology*. Moscow: Magazine Publishers.

Poltorak, M. I. (1974). Temporal characteristics of processing information presented by multi-modal signals. In P. I. Kornilov (Ed.), *Problem of engineering psychology and ergonomics* (pp. 218–221). Yaroslav, Russia: Yaroslav University.

Ponomarenko, V. A., & Rudney, N. M. (1973). Readiness to act in emergency situation. *Aviation and Aeronautics, 6*, 10–12.

Ponomarenko, V. A., & Zavalova, N. D. (1966). The pilot and problems of engineering psychology. *Issues of Aeronautical Defense, 1*, 48–53.

Ponomarenko, V. A., & Zavalova, N. D. (1970). Readiness to danger. *Aviation and aeronautics, 3*, 11–13.

Ponomarenko, V. A., & Zavalova, N. D. (1981). Study of psychic image as regulator of operator actions. In B. F. Lomov & V. F. Venda (Eds.), *Methodology of engineering psychology and psychology of work of management* (pp. 30–41). Moscow: Science Publishers.

Ponomarev, Y. A. (1976). *Psychology of creativity and education*. Moscow: Pedagogical Publishers.

Popov, G. P. (1971). *Engineering psychology in radio-location*. Moscow: Soviet Radio.

Powell, P. J. (1972). *Accident prevention: A practical framework*. Royal Society for the Prevention of Accidents (pp. 14–19). Surrey, England.

Proctor, R. W., & Van Zandt, T. (1994). *Human factors in simple and complex systems*. Boston: Allyn & Bacon.

Pushkin, V. N. (1965). *Operative thinking in large systems*. Moscow: Science Publishers.

Pushkin, V. N. (1978). Construction of situational concepts in activity structure. In A. A. Smirnov (Ed.), *Problem of general and educational psychology* (pp. 106–120). Moscow: Pedagogical Publishers.

Pushkin, V. N., & Nersesyan, L. S. (1972). *Psychology of the railroad.* Moscow: Transportation.

Radwin, R. G., Vanderheiden, G. C., & Lin, M. L. (1990). A method for evaluating head-controlled computer input devices using Fitt's Law. *Human Factors, 32*(4), 423–438.

Rasmussen, J. (1986). *Information processing and human-machine interaction: An approach to cognitive engineering.* New York: North-Holland.

Reason, J. (1990). *Human error.* Cambridge, England: Cambridge University Press.

Repkina, G. V. (1967). On capacity of working memory. In A. N. Leont'ev & V. P. Zinchenko (Eds.), *Engineering psychology* (pp. 133–140). Moscow: Science Publishers.

Reykovski, J. (1979). *Experimental psychology of emotion.* Moscow: Progress.

Rogovin, M. S., & Gromov, N. M. (1979). Comparative analysis of the factor of significance in analyzing x-rays. In M. C. Rogovin & L. P. Urvanzev (Eds.), *Psychological problems of rationalization of activity* (pp. 118–126). Yaroslav, Russia: Yaroslav University.

Rothe, H. F. (1946). Output rates among butter wrappers: Work curves and their stability. *Journal of Applied Psychology, 3*(3), 67–75.

Rouse, W. B., & Rouse, S. H. (1983). Analysis and classification of human error. *IEEE Transactions on Systems, Man, and Cybernetics, 14,* 539–549.

Roze, N. A. (1963). Experimental investigation of finger micromovements in manipulating control apparatus. In B. G. Anan'ev & B. F. Lomov (Eds.), *Problems of general and industrial psychology* (pp. 35–43). Leningrad: Leningrad University.

Rozenblat, V. V. (1961). *The problem of fatigue.* Moscow: Medicine Publishers.

Rozenblat, V. V. (1975). Principle of physiological evaluation of hard labor based on pulse measurement procedures. In V. V. Rozenblat (Ed.), *Function of organism in work process* (pp. 112–126). Moscow: Medicine Publishers.

Rubakhin, V. F. (1974). *Psychological foundations of processing input.* Moscow: Science Publishers.

Rubinshtein, S. L. (1946). *Foundations of general psychology.* Moscow: Academic Pedogogical Science.

Rubinshtein, S. L. (1958). *About thinking and methods of its development.* Moscow: Academic Science.

Rubinshtein, S. L. (1959). *Principles and directions of developing psychology.* Moscow: Academic Science.

Rubinshtein, S. L. (1973). *Problems of general psychology.* Moscow: Academic Science.

Rutman, E. M. (1974). About mechanisms of trade-off between precision or speed during choice reactions. In V. D. Shadikov (Ed.), *Problem of engineering psychology and ergonomics* (Vol. 1, pp. 43–51). Yaroslav, Russia: Yaroslav University.

Salvendy, G. (Ed.). (1987). *Handbook of human factors.* New York: Wiley.

Sanders, M. S., & McCorme, E. J. (1987). *Human factors in engineering and design* (6th ed.). New York: McGraw-Hill.

Scherrer, J. (Ed.). (1967). *Physiology of work* (Vols. I & II). Paris: Nasson et Cie.

Schmidt, R. A. (1975). A schema theory of discrete motor skill learning. *Psychological Review, 82*(4), 225–260.

Schmidt, R. A., & Russell, D. G. (1972). Movement velocity and movement time as determinant of degree of preprogramming in simple movements. *Journal of Experimental Psychology, 96,* 315–320.

Schroder, H. M., Driver, M. J., & Streufert, S. (1967). *Human information processing: Individuals and groups functioning in complex social situations.* New York: Holt, Rinehart & Winston.

Schultz, D. P., & Schultz, S. E. (1990). *Psychology and industry today.* New York: Macmillan.

Seashore, H., & Bavelas, A. (1941). The functioning of knowledge of results in Thorndike's experiment. *Psychological Review, 84,* 155–164.

Sergeev, G. A., & Romanenko, A. F. (1965). Statistical method of evaluating efficiency of transition functions of the operator. *Problems of Psychology, 4,* 140–150.

Shadrikov, V. D. (1979). *Psychology of vocational training.* Yaroslav, Russia: Yaroslav University.

Shadrikov, V. D. (1982). *Development of systemic vocational training.* Moscow: Science Publishers.

Shaporinsky, S. A. (1981). *Theory of vocational training.* Moscow: Pedagogical Publishers.

Shebicheva, V. V. (1969). *Psychology of vocational training.* Moscow: Higher Education Publishers.

Shekhter, M. S. (1967). *Psychological problems of recognition.* Moscow: Pedagogical Publishers.

Shibanov, G. P. (1983). *Quantitative evaluation of operator's work activity in man-machine systems.* Moscow: Manufacturing Press.

Sibulevsky, I. E. (1962). Operator delays during processing of visual information. *Automation and Telemechanics, 11,* 1513–1526.

Siegal, A. I., & Wolf, J. J. (1969). *Man-machine simulation models.* New York: Wiley.

Simonov, P. V. (1982). Need-informational theory of emotions. *Problems of Psychology, 6,* 44–56.

Sipachev, N. O. (1976). Strategy of sensory motor reactions on stimuli sequence with probability. *New Study in Psychology, 2,* 29–32.

Skinner, B. F. (1974). *About behaviorism.* New York: Knopf.

Smirnov, B. A., & Samoshkina, N. M. (1976). *Man, labor and techniques.* Kharkov, Ukraine: Flag Publisher.

Smirnov, S. D. (1985). *Psychology of image: The problem of active psychic reflections.* Moscow: Moscow University Publishers.

Smith, K. U. (1962). *Delay, sensory, feedback, and behavior.* Philadelphia: Saunders.

Smith, K. V., & Hancock, P. A. (1995). Situational awareness is adaptive, externally directed consciousness. *Human Factors, 37*(1), 137–148.

Sokolov, E. N. (1960). Neural models and the orienting reflex. In A. B. Brazierc (Ed.), *The central nervous system and behavior* (pp. 187–276). New York: Josiah Macy, Jr. Foundation.

Sokolov, E. N. (1963). *Perception and conditioned reflex.* New York: Macmillan.

Sperling, G. (1960). The information available in brief visual presentations. *Psychological Monographs, 74*(Whole No. 11).

Stein, J. (Ed.). (1980). *Random House College Dictionary* (rev. ed.). New York: Random House.

Stepansky, V. I. (1974). Influence of the significance factor on time reaction. *New Studies in Psychology, 3,* 9–10.

Stepansky, V. I. (1976). Functional role of feedback and criteria of success in self-regulation of activity. In E. A. Klimov (Ed.), *Theoretical and applicational study of psychological self-regulation* (pp. 24–30). Kazan, Russia: Kazan University.

Sternberg, S. (1969). The discovery of processing stages: Extension of Donder's method. *Acta Psychologica, 30,* 276–315.

Sternberg, S. (1969). Memory-scanning, mental processes revealed by reaction-time experiments. *American Scientist, 57,* 421–457.

Strelkov, U. K. (1972). Microstructure analysis of processing information. *Ergonomics, 3,* 112–121.

Strukov, G. A. (1972). Mechanism of anticipation in activity of the operator. In E. P. Zinchenko (Ed.), *Problems of engineering psychology* (Vol. 3, pp. 47–51). Moscow: National Research Institute of Technical Aesthetics.

Strukov, G. A., Gritsevsky, M. A., & Konopkin, O. A. (1971). Evaluation of reliability of expectation of signal in operator performance. In *The Problem of Engineering Psychology; Third Soviet Union Conference* (Vol. 2, pp. 190–194). Moscow: APS USSR.

Swets, J. A. (1964). *Signal detection and recognition by human observers.* New York: Wiley.

Talizina, N. F. (1975). *The management of processing skills and knowledge.* Moscow: Moscow University Publishers.

Tanner, W. P., Jr., & Birdsall, T. G. (1955). *The evidence for a decision making theory of visual detection* (Tech. Rep. No. 40). Electronic Difference Group of the University of Michigan.

Teplov, B. M. (1961). The theory of types of higher nervous activity and psychology. *Problem of Psychology, 1,* 3–16.

Thorndike, E. L. (1932). *The fundamentals of learning.* New York: Teachers College Press.

Tikhomirov, O. K. (1984). *Psychology of thinking.* Moscow: Moscow University Publishers.

Tolman, E. C. (1932). *Purposive behavior in animals and men.* New York: Century.

Troshihkina, B. A., Kol'chenko, S. I., & Moldavsky, S. I. (1970). New methods for studying topological specificity of man's higher neural system. *Pavlov's Journal of Higher Neural Systems, 4,* 37–42.

Tversky, A., & Kahneman, D. (1973). Availability: A heuristic for judging frequency and probability. *Cognitive Psychology, 5,* 207–232.

Uznadze, D. N. (1961). *Experimental basis of the psychology of set.* Tbilisi: Georgia Academy of Science.

Van Santen, J. H., & Philips, N. Y. (1970). Method and time study of mental work. *Work study and management services, 14*(1), 21–25.

Vekker, L. M. (1964). Restructuring sensory and motor functions in the remote control of railroad transport. In B. G. Anan'ev & B. F. Lomov, (Eds.), *Problems of general and industrial psychology* (pp. 87–100). Leningrad: Leningrad University.

Vekker, L. M., & Paley, I. M. (1971). Information and energy in psychological reflection. In B. G. Ana'nev (Ed.), *Experimental psychology* (Vol. 3, pp. 61–66). Leningrad: Leningrad University.

Velichkovsky, B. M. (1982). *Contemporary cognitive psychology.* Moscow: Moscow University Publishers.

Venda, V. F. (1975). *Engineering psychology and synthesis of information.* Moscow: Moscow Manufacturing.

Venda, V. F., & Ribal'chenko, M. V. (1983). Transformational theory of learning and design of engineering—Psychological experiment. In V. F. Venda & V. A. Vavilov (Eds.), *Theory and experiment in analysis of operator works* (pp. 147–157). Moscow: Science Publishers.

Vertut, J., & Coiffet, P. (1985). Remote manipulation of robots. *Teleoperations.* Paris: Hermes Publishing.

Vinogradov, M. I. (1966). *Work physiology* (2nd ed.). Moscow: Medicine Publishers.

Vishepan, L. I., Vishepan, L. N. (1975). Operators coefficient of workload. In A. P. Chernishev (Ed.), *Applied problems of engineering psychology* (pp. 24–37). Russia: Taganrog.

Volkova, I. M., Goncharov, I. E., Gordhkov, C. I., Zelenkin, A. N., Kokhanova, N. A., & Shardakova, E. F. (1979). *Ergonomics in production.* Moscow: Medicine Publishers.

Vorona, A. A., Aleshin, S. V., & Safronov, A. M. (1984). On the physiological nature of "feeling of aircraft." *Cosmobiology and Aviation Medicine, 5,* 14–18.

Vygotsky, L. S. (1956). *Selected psychological research.* Moscow: Academy of Pedagogical Science RSFSL.

Vygotsky, L. S. (1960). *Developing higher order psychic functions.* Moscow: Academy of Pedagogical Science RSFSR.

Vygotsky, L. S. (1962). *Thought and language.* Cambridge, MA: MIT Press.

Wade, C., & Tavris, C. (1990). *Psychology* (2nd ed.). New York: HarperCollins. (Includes task bank)

Walker, N., Meyer, D. E., & Smelcer, J. B. (1993). Spatial and temporal characteristics of rapid cursor-positioning movements with electromechanical mice in human-computer interaction. *Human Factors, 35*(3), 431–458.

Welford, A. T. (1960). The measurement of sensory-motor performance: Survey and reappraisal of twelve years' progress. *Ergonomics, 3,* 189–230.

Welford, A. T. (1961). On the human demands of automation, mental work, conceptual model, satisfaction and training. In *Industrial and business psychology* (Vol. 5, pp. 182–194). Copenhagen: Munksgaard.

Welford, A. T. (1974). On the sequencing of action. *Brown Research, 71,* 381–392.

Wells, J. G., Balke, B., & Van Fosan, D. D. (1957). Lactic acid accumulation during work. A suggested standardization of work classification. *Journal of Applied Physiology, 10,* 51–55.

Wickens, C. D. (1992). *Engineering psychology and human performance.* New York: HarperCollins.

Wickens, C. D., Sandry, D., & Vidulich, M. (1983). Compatibility and resource competition between modalities of input, central processing and output. *Testing and Model of Complex Task Performance, 25,* 227–248.

Wiener, N. (1958). *Cybernetics.* Moscow: Soviet Radio.

Wiener, N., & Rosenblueth, A. (1950). Purposeful and non-purposeful behavior. *Philosophy of Science, 17,* 20–36.

Wyatts, S. (1927). An experimental study of repetitive process. *The British Journal of Psychology. Gen. Sec., V. XVII,* 267–273.

Yaroshevsky, M. G. (1985). *History of psychology* (3rd. ed.). Moscow: Thinking.

Yastrebov, S. Y. (Ed.). (1977). *Underwater works.* Leningrad: Ship Construction.

Yntema, D. (1963). Keeping track of several things at once. *Human Factors, 6,* 7–17.

Yur'nev, A. P. (1975). *Unmanned underwater apparatus.* Moscow: Unitary Publishers.

Zabrodin, Y. M. (1985). Methodological and theoretical problems of psychophysics. In B. F. Lomov & Y. M. Zabrodin (Eds.), *Psychophysics of discrete and continual tasks* (pp. 3–26). Moscow: Science Publishers.

Zabrodin, Y. M., & Chernishov, A. P. (1981). On losing information during description of operators' transfer functions. In B. F. Lomov & V. F. Venda (Eds.), *Methodology of engineering psychology and psychology of work and management* (pp. 244–249). Moscow: Science Publishers.

Zabrodin, Y. M., & Lebedev, A. M. (1977). *Psychophysiology and psychophysics.* Moscow: Science Publishers.

Zaporozhets, A. V., & Zinchenko, V. P. (1982). *Perception, motions and actions.* Moscow: Pedagogical Publishers.

Zarakovsky, G. M. (1966). *Psychophysiological analysis of work activity; logical-probability approach.* Moscow: Science Publishers.

Zarakovsky, G. M., Korolev, B. A., Medvedev, V. I., & Shlaen, P. Y. (1974). *Introduction to ergonomics.* Moscow: Soviet Radio.

Zarakovsky, G. M., & Magazannik, V. D. (1981). Psychological criteria of complex decision making processes. In B. F. Lomov & V. F. Venda (Eds.), *Methodology of engineering psychology and psychology of work management* (pp. 63–79). Moscow: Science Publishers.

Zarakovsky, G. M., & Medvedev, V. I. (1971). *Psychological evaluation of efficiency of man-machine system.* Leningrad: Psychology Society.

Zarakovsky, G. M., & Medvedev, V. I. (1979). Classification of operator's errors. *Technical Aesthetics, 10,* 5–6.

Zarakovsky, G. M., & Pavlov, V. V. (1987). *Laws of functioning man-machine systems.* Moscow: Soviet Radio.

Zavalishina, D. N. (1977). Activity of operator in tie deficit conditions. In B. F. Lomov, V. F. Rubakhin, & V. F. Venda (Eds.), *Engineering psychology* (pp. 190–218). Moscow: Science Publishers.

Zavalishina, D. N. (1985). *Psychological analysis of operator's thinking.* Moscow: Science Publishers.

Zavalishina, D. N., & Pushkin, V. N. (1964). On operative thinking. *Problems of Psychology, 3,* 87–100.

Zavalova, N. D., Lomov, B. F., & Ponomarenko, V. A. (1971). Principle of active operator and function allocation. *Problems of Psychology, 3,* 3–12.

Zavalova, N. D., Lomov, B. F., & Ponomarenko, V. A. (1986). *Image in regulation of activity.* Moscow: Science Publishers.

Zavalova, N. D., & Ponomarenko, V. A. (1968). Some questions of operator reliability in automated control systems during breakdown. *Problems of Psychology, 4,* 49–56.

Zavalova, N. D., & Ponomarenko, V. A. (1970). Characteristics of pilots' behavior in emergency flight conditions *Problems of Psychology, 5,* 111–121.

Zavalova, N. D., & Ponomarenko, V. A. (1980). Structure and content of psychic image as mechanisms of regulation of actions. *Psychological Journal, 1*(2), 5–18.

Zavalova, N. D., & Ponomarenko, V. A. (1984). Specificity of psychic images in disturbed feedback conditions. *Problem of Psychology, 2,* 26–35.

Zefeld, V. V., Munipov, V. M., & Tchernyshova, O. N. (1980). *Project ergonomic modeling.* Moscow: VNIITE.

Zimkin, N. V. (1934). Method of studying precision of reactions and visual acuity. *Soviet Psychotechnique, 3,* 13–21.

Zinchenko, P. I. (1961). *Involuntary memorization.* Moscow: Pedagogical Publishers.

Zinchenko, T. P. (1981). *Identification and coding.* Leningrad: Leningrad University.

Zinchenko, V. P. (1978). Functional structure of executive perceptual-motor actions. In V. P Zinchenko & V. M. Munipov, (Eds.), *Study of cognitive and executive actions* (Vol. 16, pp. 3–40).

Zinchenko, V. P., Leonova, A. B., & Strelkov, U. K. (1977). *Psychometry of fatigue.* Moscow: Moscow University Publishers.

Zinchenko, V. P., & Munipov, V. M. (1979). *Fundamentals of ergonomics.* Moscow: Moscow University Publishers.

Zinchenko, V. P., & Vergiles, N. Y. (1969). *Creation of visual image.* Moscow: Moscow University Publishers.

Zolina, Z. M. (1967). Physiological basis of work performance on conveyers. Moscow: Medicine Publishers.

Author Index

A

Ackoff, R., 52
Adams, J. A., 234
Adams, M. J., 54, 55
Ainsworth, L. K., 246, 294
Akhutin, V. M., 161
Aladjanova, N. A., 266, 267
Aleshin, S. V., 358
Anan'ev, B. G., 61, 131
Anderson, B. F., 106
Anderson, J. R., 131, 307, 308
Anokhin, P. K., 7, 11, 44, 66, 342, 343, 351, 381
Aseev, V. G., 389, 390, 391, 394
Asratyan, E. A., 302
Atkinson R. L., 34
Atkinson, J., 46, 120
Atkinson, R. C., 34, 106
Ausubel D.P., 81

B

Baddely, A. D., 106
Balke, B., 385
Bandura, A., 51, 67, 83, 87, 342, 366
Bardin, K. V., 100, 101, 102, 175
Barnard, P., 60
Bartlett, F. C., 21, 55
Bassin, F. V., 38
Batishev, S. Y., 301
Bavelas, A., 67
Bayha, F. H., 27, 162, 247, 250
Bayton, J. A., 34
Bedny, G. Z., 15, 25, 27, 29, 42, 71, 74, 79, 125, 126, 127, 146, 155, 162, 233, 237, 238, 246, 247, 252, 259, 272, 275, 282, 305, 307, 315, 317, 317, 318,
320, 327, 341, 343, 349, 352, 353, 354, 355, 363, 366, 388, 392, 401
Beregovoy, G. T., 6, 89, 90, 136, 192, 197, 221, 227, 229
Beringer, D. B., 226
Bernshtein, N. A., 7, 11, 13, 52, 66, 316, 342, 342
Bertalanffy, L.,143
Bespal'ko, V. P, 309
Bezdyuk, S. V., 132
Blackwell, H. R., 45
Bloch, V., 265, 267
Block, S.M., 269, 270
Bodrov, V. A., 164
Borger, R., 70
Bridsall, T. G. ,98
Broadbent, D. E., 250
Bruner, J. S., 305
Brushlinsky, A. V., 18,116
Brusnichkina, R. I. , 191

C

Chandler, L. A., 51
Chapanis, A.,143
Chebisheva, V. V., 175, 363
Chernishov, A. P., 68
Christensen, E. H., 385, 386
Coiffet, P., 57
Conley, H. W., 34
Cook, J. S.,143
Cooper, G. E.,192

D

Davidov, V. V., 304
Deitrich, Y.,146
Demersky, V. Y., 136

421

Subject Index